COMMUNICATION SYSTEMS

By the Same Author

SIGNALS, SYSTEMS AND COMMUNICATION (1965)

COMMUNICATION SYSTEMS

B. P. LATHI

Associate Professor of Electrical Engineering
Bradley University

John Wiley & Sons, Inc.
New York · London · Sydney

164890

Library of Congress Catalog Card Number: 68-11093
Printed in the United States of America

Preface

The purpose of this book is to introduce the student to communication systems and the broad principles of modern communication theory at an early stage in the undergraduate curriculum. It begins with the study of specific communication systems and gradually develops the underlying role of the signal-to-noise ratio and the bandwidth in limiting the rate of information transmission.

Since the book is intended for an introductory course, it was necessary to ignore many of the finer points regarding the power density spectra of random processes. The student is introduced to the concept of the power density spectrum of nonrandom signals. This concept is then extended to random signals without any formal development. A rigorous treatment of random processes is deemed unnecessarily distracting in such an introductory course, for it would defeat its very purpose. After completing this course, a student can then fruitfully undertake a rigorous course in communication theory using statistical concepts.

Throughout the book, the stress is on a physical appreciation of the concepts rather than mathematical manipulation. In this respect the book closely follows the philosophy of my earlier book, *Signals, Systems and Communication*. Wherever possible, the concepts and results are interpreted intuitively. The basic concepts of information theory are not introduced as axioms but are developed heuristically.

v

Communication Systems can be used for a semester or a quarter by judiciously choosing the topics. Any of the following four combinations of chapters will form a well balanced first course in communication systems.

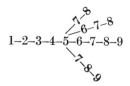

Other combinations will no doubt prove suitable in some cases. Chapter 1 (Signal Analysis) is essentially a review. The Fourier series is introduced as a representation of a signal in orthogonal signal space. This is done because of the growing importance of geometrical representation of signals in communication theory. This aspect, however, is not essential for the material covered in this book. Thus the student may skip the first 30 pages (Sections 1.1 through 1.3).

The book is self-contained and there are no prerequisites whatsoever. No knowledge of probability theory is assumed on the part of students. The modicum of probability theory that is required in Chapter 9 (on digital communication) is developed in that chapter.

I would like to thank Mr. Ivar Larson for assisting me in proof-reading, Professors J. L. Jones and R. B. Marxheimer for helpful suggestions, and Professor Philip Weinberg, the department head, for making available to me the time to complete this book. I am also pleased to acknowledge the assistance of Mrs. Evelyn Kahrs for typing the manuscript.

B. P. LATHI

Peoria, Illinois
January, 1968

Contents

COMMUNICATION SYSTEMS

chapter 1

Signal Analysis

There are numerous ways of communicating. Two people may communicate with each other through speech, gestures, or graphical symbols. In the past, communication over a long distance was accomplished by such means as drumbeats, smoke signals, carrier pigeons, and light beams. More recently, these modes of long distance communication have been virtually superceded by communication by electrical signals. This is because electrical signals can be transmitted over a much longer distance (theoretically, any distance in the universe) and with a very high speed (about 3×10^8 meters per second). In this book, we are concerned strictly with the latter mode, that is, communication by electrical signals.

The engineer is chiefly concerned with efficient communication. This involves the problem of transmitting messages as fast as possible with the least error. We shall treat these aspects quantitatively throughout this book. It is, however, illuminating to discuss qualitatively the factors that limit the rate of communication. For convenience, we shall consider the transmission of symbols (such as alpha-numerical symbols of English language) by certain electrical waveforms. In the process of transmission, these waveforms are contaminated by omnipresent noise signals which are generated by numerous natural and man-made events. Man-made events such as faulty contact switches, turning on and off of electrical equipment, ignition radiation, and fluorescent lighting continuously radiate random noise signals. Natural phenomena such as lightning, electrical storms, the sun's radiation, and intergalactic radiation are the sources of noise signals. Fluctuation noise such as thermal noise in resistors and shot noise in active devices is also an important source of noise in all electrical systems. When the message-bearing signals are transmitted over a channel, they are corrupted with random noise signals and may consequently become unidentifiable at the receiver. To avoid this difficulty, it is necessary to increase the power of the message-bearing waveforms. A certain ratio of signal power to noise power must be maintained. This ratio, S/N, is an important parameter in evaluating the performance of a system.

We shall now consider increasing the speed of transmission by compressing the waveforms in time scale so that we can transmit more messages during a given period. When the signals are compressed, their variations are rapid, that is, they wiggle faster. This naturally increases their frequencies. Hence compressing a signal gives rise to the problem of transmitting signals of higher frequencies. This necessitates the increased bandwidth of the channel over which the messages are transmitted. Thus the rate of communication can be increased by increasing the channel bandwidth. In general, therefore, for faster and more accurate communication, it is desirable to increase S/N, the signal-to-noise power ratio, and the channel bandwidth.

These conclusions are arrived at by qualitative reasoning and are hardly surprising. What is surprising, however, is that the bandwidth and the signal-to-noise ratio can be exchanged. We shall show later that to maintain a given rate of communication with a given accuracy, we can exchange the S/N ratio for the bandwidth, and vice versa. One may reduce the bandwidth if he is willing to increase the S/N ratio. On the other hand, a small S/N ratio may be adequate if the bandwidth of the channel is increased correspondingly. This is expressed by the Shannon-Hartley law,

$$C = B \log \left(1 + \frac{S}{N}\right)$$

where C is the channel capacity or the rate of message transmission (to be discussed later), and B is the bandwidth of the channel (in Hz). For a given C, we may increase B and reduce S/N, and vice versa.

In order to study communication systems we must be familiar with various ways of representing signals. We shall devote this chapter to signal analysis.

1.1 ANALOGY BETWEEN VECTORS AND SIGNALS

A problem is better understood or better remembered if it can be associated with some familiar phenomenon. Therefore we always search for analogies when studying a new problem. In the study of abstract problems, similarities are very helpful, particularly if the problem can be shown to be analogous to some concrete phenomenon. It is then easy to gain some insight into the new problem from the knowledge of the corresponding phenomenon. Fortunately, there is a perfect analogy between vectors and signals which leads to a better understanding of signal analysis. We shall now briefly review the properties of vectors.

Vectors

A vector is specified by magnitude and direction. We shall denote all vectors by boldface type and their magnitudes by lightface type; for example, $\mathbf{A}$ is a certain vector with magnitude A. Consider two vectors $\mathbf{V}_1$ and $\mathbf{V}_2$ as shown in Fig. 1.1. Let the component of $\mathbf{V}_1$ along $\mathbf{V}_2$ be given by $C_{12}V_2$. How do we interpret physically the component of one vector along the other vector? Geometrically the component of a vector $\mathbf{V}_1$ along the

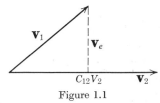

Figure 1.1

vector $\mathbf{V}_2$ is obtained by drawing a perpendicular from the end of $\mathbf{V}_1$ on the vector $\mathbf{V}_2$, as shown in Fig. 1.1. The vector $\mathbf{V}_1$ can now be expressed in terms of vector $\mathbf{V}_2$.

$$\mathbf{V}_1 = C_{12}\mathbf{V}_2 + \mathbf{V}_e \qquad (1.1a)$$

However, this is not the only way of expressing vector $\mathbf{V}_1$ in terms of vector $\mathbf{V}_2$. Figure 1.2 illustrates two of the infinite alternate possibilities. Thus, in Fig. 1.2a,

$$\mathbf{V}_1 = C_1\mathbf{V}_2 + \mathbf{V}_{e_1} \qquad (1.1b)$$

and in Fig. 1.2b,

$$\mathbf{V}_1 = C_2\mathbf{V}_2 + \mathbf{V}_{e_2} \qquad (1.1c)$$

In each representation, $\mathbf{V}_1$ is represented in terms of $\mathbf{V}_2$ plus another vector, which will be called the error vector. If we are asked to approximate the vector $\mathbf{V}_1$ by a vector in the direction of $\mathbf{V}_2$, then $\mathbf{V}_e$ represents the error in this approximation. For example, in Fig. 1.1 if we approximate $\mathbf{V}_1$ by $C_{12}\mathbf{V}_2$, then the error in the approximation is $\mathbf{V}_e$. If $\mathbf{V}_1$ is

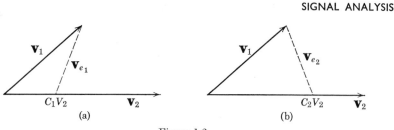

Figure 1.2

approximated by C_1V_2 as in Fig. 1.2a, then the error is given by $\mathbf{V}_{e_1}$, and so on. What is so unique about the representation in Fig. 1.1? It is immediately evident from the geometry of these figures that the error vector is smallest in Fig. 1.1. We can now formulate a quantitative definition of a component of a vector along another vector. The component of a vector $\mathbf{V}_1$ along the vector $\mathbf{V}_2$ is given by $C_{12}\mathbf{V}_2$, where C_{12} is chosen such that the error vector is minimum.

Let us now interpret physically the component of one vector along another. It is clear that the larger the component of a vector along the other vector, the more closely do the two vectors resemble each other in their directions, and the smaller is the error vector. If the component of a vector $\mathbf{V}_1$ along $\mathbf{V}_2$ is $C_{12}\mathbf{V}_2$, then the magnitude of C_{12} is an indication of the similarity of the two vectors. If C_{12} is zero, then the vector has no component along the other vector, and hence the two vectors are mutually perpendicular. Such vectors are known as *orthogonal vectors*. Orthogonal vectors are thus independent vectors. If the vectors are orthogonal, then the parameter C_{12} is zero.

For convenience, we define the dot product of two vectors $\mathbf{A}$ and $\mathbf{B}$ as

$$\mathbf{A} \cdot \mathbf{B} = AB \cos \theta$$

where θ is the angle between vectors $\mathbf{A}$ and $\mathbf{B}$. It follows from the definition that

$$\mathbf{A} \cdot \mathbf{B} = \mathbf{B} \cdot \mathbf{A}$$

According to this notation,

$$\text{the component of } \mathbf{A} \text{ along } \mathbf{B} = A \cos \theta = \frac{\mathbf{A} \cdot \mathbf{B}}{B}$$

and

$$\text{the component of } \mathbf{B} \text{ along } \mathbf{A} = B \cos \theta = \frac{\mathbf{A} \cdot \mathbf{B}}{A}$$

Similarly,

$$\text{the component of } \mathbf{V}_1 \text{ along } \mathbf{V}_2 = \frac{\mathbf{V}_1 \cdot \mathbf{V}_2}{V_2}$$

$$= C_{12} V_2$$

Therefore

$$C_{12} = \frac{\mathbf{V}_1 \cdot \mathbf{V}_2}{V_2{}^2} = \frac{\mathbf{V}_1 \cdot \mathbf{V}_2}{\mathbf{V}_2 \cdot \mathbf{V}_2} \tag{1.2}$$

Note that if $\mathbf{V}_1$ and $\mathbf{V}_2$ are orthogonal, then

$$\mathbf{V}_1 \cdot \mathbf{V}_2 = 0 \tag{1.3}$$

and

$$C_{12} = 0$$

Signals

The concept of vector comparison and orthogonality can be extended to signals.* Let us consider two signals, $f_1(t)$ and $f_2(t)$. Suppose we want to approximate $f_1(t)$ in terms of $f_2(t)$ over a certain interval $(t_1 < t < t_2)$ as follows:

$$f_1(t) \simeq C_{12} f_2(t) \qquad \text{for} \quad (t_1 < t < t_2) \tag{1.4}$$

How shall we choose C_{12} in order to achieve the best approximation? Obviously, we must find C_{12} such that the error between the actual function and the approximated function is minimum over the interval $(t_1 < t < t_2)$. Let us define an *error function* $f_e(t)$ as

$$f_e(t) = f_1(t) - C_{12} f_2(t) \tag{1.5}$$

One possible criterion for minimizing the error $f_e(t)$ over the interval t_1 to t_2 is to minimize the average value of $f_e(t)$ over this interval; that is, to minimize

$$\frac{1}{(t_2 - t_1)} \int_{t_2}^{t_2} [f_1(t) - C_{12} f_2(t)] \, dt$$

However, this criterion is inadequate because there can be large positive and negative errors present that may cancel one another in the process of averaging and give the false indication that the error is zero. For

---

* We shall often use the terms signals and functions interchangeably. A signal is a function of time. However, there is one difference between signals and functions. A function $f(t)$ can be a multivalued function of variable t. But the physical signal is always a single-valued function of t. Hence, whenever we use a term function, it will be understood that it is a single-valued function of the independent variable.

example, if we approximate a function $\sin t$ with a null function $f(t) = 0$ over an interval 0 to 2π, the average error will be zero, indicating wrongly that $\sin t$ can be approximated to zero over the interval 0 to 2π without any error. This situation can be corrected if we choose to minimize the average (or the mean) of the square of the error instead of the error itself. Let us designate the average of $f_e^2(t)$ by ε.

$$\varepsilon = \frac{1}{(t_2 - t_1)} \int_{t_1}^{t_2} f_e^2(t)\, dt = \frac{1}{(t_2 - t_1)} \int_{t_1}^{t_2} [f_1(t) - C_{12} f_2(t)]^2\, dt \quad (1.6)$$

To find the value of C_{12} which will minimize ε, we must have

$$\frac{d\varepsilon}{dC_{12}} = 0 \quad (1.7)$$

that is,

$$\frac{d}{dC_{12}} \left\{ \frac{1}{(t_2 - t_1)} \int_{t_1}^{t_2} [f_1(t) - C_{12} f_2(t)]^2\, dt \right\} = 0 \quad (1.8)$$

Changing the order of integration and differentiation, we get

$$\frac{1}{(t_2 - t_1)} \left[\int_{t_1}^{t_2} \frac{d}{dC_{12}} f_1^2(t)\, dt - 2 \int_{t_1}^{t_2} f_1(t) f_2(t)\, dt + 2C_{12} \int_{t_1}^{t_2} f_2^2(t)\, dt \right] = 0 \quad (1.9)$$

The first integral is obviously zero, and hence Eq. 1.9 yields

$$C_{12} = \frac{\displaystyle\int_{t_1}^{t_2} f_1(t) f_2(t)\, dt}{\displaystyle\int_{t_1}^{t_2} f_2^2(t)\, dt} \quad (1.10)$$

Observe the similarity between Eqs. 1.10 and 1.2, which expresses C_{12} for vectors.

By analogy with vectors, we say that $f_1(t)$ has a component of waveform $f_2(t)$, and this component has a magnitude C_{12}. If C_{12} vanishes, then the signal $f_1(t)$ contains no component of signal $f_2(t)$, and we say that the two functions are orthogonal over the interval (t_1, t_2). It therefore follows that the two functions $f_1(t)$ and $f_2(t)$ are *orthogonal* over an interval (t_1, t_2) if

$$\int_{t_1}^{t_2} f_1(t) f_2(t)\, dt = 0 \quad (1.11)$$

Observe the similarity between Eq. 1.11 derived for orthogonal functions and Eq. 1.3 derived for orthogonal vectors.

We can easily show that the functions $\sin n\omega_0 t$ and $\sin m\omega_0 t$ are orthogonal over any interval $(t_0, t_0 + 2\pi/\omega_0)$ for integral values of n

and m. Consider the integral I:

$$I = \int_{t_0}^{t_0+2\pi/\omega_0} \sin n\omega_0 t \sin m\omega_0 t \, dt$$

$$= \int_{t_0}^{t_0+2\pi/\omega_0} \tfrac{1}{2}[\cos (n-m)\omega_0 t - \cos (n+m)\omega_0 t] \, dt$$

$$= \frac{1}{2\omega_0}\left[\frac{1}{(n-m)} \sin (n-m)\omega_0 t - \frac{1}{(n+m)} \sin (n+m)\omega_0 t\right]_{t_0}^{t_0+2\pi/\omega_0}$$

Since n and m are integers, $(n-m)$ and $(n+m)$ are also integers. In such a case the integral I is zero. Hence the two functions are orthogonal. Similarly, it can be shown that $\sin n\omega_0 t$ and $\cos m\omega_0 t$ are orthogonal functions and $\cos n\omega_0 t$, $\cos m\omega_0 t$ are also mutually orthogonal.

Example 1.1

A rectangular function $f(t)$ is defined by (Fig. 1.1):

$$f(t) = \begin{cases} 1 & (0 < t < \pi) \\ -1 & (\pi < t < 2\pi) \end{cases}$$

Approximate this function by a waveform $\sin t$ over the interval $(0, 2\pi)$ such that the mean square error is minimum.

Solution. The function $f(t)$ will be approximated over the interval $(0, 2\pi)$, as

$$f(t) \simeq C_{12} \sin t$$

We shall find the optimum value of C_{12} which will minimize the mean square error in this approximation. According to Eq. 1.10, to minimize the mean square error:

$$C_{12} = \frac{\displaystyle\int_0^{2\pi} f(t) \sin t \, dt}{\displaystyle\int_0^{2\pi} \sin^2 t \, dt}$$

$$= \frac{1}{\pi}\left[\int_0^{\pi} \sin t \, dt + \int_{\pi}^{2\pi} -\sin t \, dt\right]$$

$$= \frac{4}{\pi}$$

Thus

$$f(t) \simeq \frac{4}{\pi} \sin t$$

represents the best approximation of $f(t)$ by a function $\sin t$ which will minimize the mean square error.

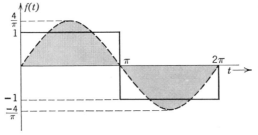

Figure 1.3

By analogy with vectors, we may say that the rectangular function $f(t)$ shown in Fig. 1.3 has a component of function $\sin t$ and the magnitude of this component is $4/\pi$.

What is the significance of orthogonality of two functions? In the case of vectors, orthogonality implies that one vector has no component along the other. Similarly, a function does not contain any component of the form of the function which is orthogonal to it. If we try to approximate a function by its orthogonal function, the error will be larger than the original function itself, and it is better to approximate a function with a null function $f(t) = 0$ rather than with a function orthogonal to it. Hence the optimum value of $C_{12} = 0$ in such a case.

Graphical Evaluation of a Component of One Function in the Other

It is possible to evaluate the component of a function in the other function by graphical means, using Eq. 1.10. Suppose two functions $f_1(t)$ and $f_2(t)$ are known graphically, and it is desired to evaluate the component of waveform $f_2(t)$ contained in signal $f_1(t)$ over a period $(0, T)$. We know that this component is given by $C_{12}f_2(t)$; that is, $f_1(t)$ contains the component of function $f_2(t)$ of magnitude C_{12}, given by

$$C_{12} = \frac{\displaystyle\int_0^T f_1(t)f_2(t)\,dt}{\displaystyle\int_0^T f_2{}^2(t)\,dt}$$

The integral in the numerator in this equation can be found by multiplying the two functions and evaluating the area under the product curve as shown in Fig. 1.4. The denominator integral can be evaluated by finding the area under the function $[f_2(t)]^2$ in a similar way.

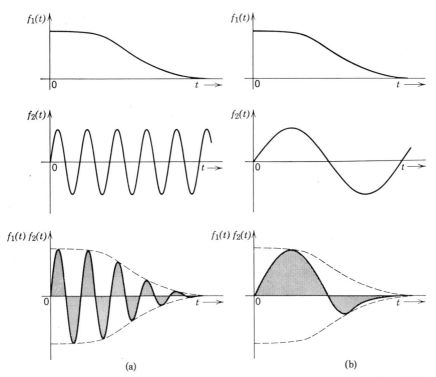

Figure 1.4 Graphical evaluation of the component of waveform $f_2(t)$ in a signal $f_1(t)$.

It is evident that if $f_1(t)$ varies much more slowly than $f_2(t)$, the area under the curve $f_1(t)f_2(t)$ will be very small since the positive and negative areas will be approximately equal and will tend to cancel each other as shown in Fig. 1.4a. Hence $f_1(t)$ contains a small component of $f_2(t)$. If, however, $f_1(t)$ varies at about the same rate as $f_2(t)$, then the area under the product curve $f_1(t)f_2(t)$ will be much larger, as shown in Fig. 1.4b, and hence $f_1(t)$ will contain a large component of function $f_2(t)$. This result is also intuitively obvious, since if two functions vary at about the same rate, there must be a great deal of similarity between the two functions, and hence $f_1(t)$ will contain a large component of the function $f_2(t)$.

Orthogonal Vector Space

The analogy between vectors and signals may be extended further. Let us now consider a three-dimensional vector space described by rectangular coordinates, as shown in Fig. 1.5. We shall designate a vector

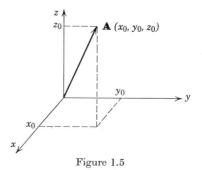

Figure 1.5

of unit length along the x axis by $\mathbf{a}_x$. Similarly, unit vectors along the y and z axes will be designated by $\mathbf{a}_y$ and $\mathbf{a}_z$, respectively. Since the magnitude of vectors $\mathbf{a}_x$, $\mathbf{a}_y$, and $\mathbf{a}_z$ is unity, it follows that for any general vector $\mathbf{A}$:

The component of $\mathbf{A}$ along the x axis $= \mathbf{A} \cdot \mathbf{a}_x$

The component of $\mathbf{A}$ along the y axis $= \mathbf{A} \cdot \mathbf{a}_y$

The component of $\mathbf{A}$ along the z axis $= \mathbf{A} \cdot \mathbf{a}_z$

A vector $\mathbf{A}$ drawn from the origin to a general point (x_0, y_0, z_0) in space has components x_0, y_0, and z_0 along the x, y, and z axes, respectively. We can express this vector $\mathbf{A}$ in terms of its components along the three mutually perpendicular axes:

$$\mathbf{A} = x_0\mathbf{a}_x + y_0\mathbf{a}_y + z_0\mathbf{a}_z$$

Any vector in this space can be expressed in terms of the three vectors $\mathbf{a}_x$, $\mathbf{a}_y$, and $\mathbf{a}_z$.

Since the three vectors $\mathbf{a}_x$, $\mathbf{a}_y$, and $\mathbf{a}_z$ are mutually perpendicular, it follows that

$$\mathbf{a}_x \cdot \mathbf{a}_y = \mathbf{a}_y \cdot \mathbf{a}_z = \mathbf{a}_z \cdot \mathbf{a}_x = 0$$

and   (1.12)

$$\mathbf{a}_x \cdot \mathbf{a}_x = \mathbf{a}_y \cdot \mathbf{a}_y = \mathbf{a}_z \cdot \mathbf{a}_z = 1$$

The properties of the three vectors, as expressed by Eq. 1.12, can be succinctly expressed by

$$\mathbf{a}_m \cdot \mathbf{a}_n = \begin{cases} 0 & m \neq n \\ 1 & m = n \end{cases}$$ (1.13)

where m and n can assume any value x, y, and z.

Now we make an important observation. If the coordinate system has only two axes, x and y, then the system is inadequate to express a general vector $\mathbf{A}$ in terms of the components along these axes. This system can only express two components of vector $\mathbf{A}$. Therefore it is necessary that to express any general vector $\mathbf{A}$ in terms of its coordinate components, the system of coordinates must be complete. In this case there must be three coordinate axes.

A single straight line represents a one-dimensional space; a single plane represents a two-dimensional space; and our universe, in general, has three dimensions of space. We may extend our concepts as developed here to a general n-dimensional space. Such a physical space, of course, does not exist in nature. Nevertheless, there are many analogous problems that may be viewed as n-dimensional problems. For example, a linear equation in n-independent variables may be viewed as a vector expressed in terms of its components along n mutually perpendicular coordinates.

If unit vectors along these n mutually perpendicular coordinates are designated as $\mathbf{x}_1, \mathbf{x}_2, \ldots, \mathbf{x}_n$ and a general vector $\mathbf{A}$ in this n-dimensional space has components $C_1, C_2, \ldots, C_n$, respectively, along these n coordinates, then

$$\mathbf{A} = C_1\mathbf{x}_1 + C_2\mathbf{x}_2 + C_3\mathbf{x}_3 + \cdots + C_n\mathbf{x}_n \qquad (1.14)$$

All the vectors $\mathbf{x}_1, \mathbf{x}_2, \ldots, \mathbf{x}_n$ are mutually orthogonal, and the set must be complete in order for any general vector $\mathbf{A}$ to be represented by Eq. 1.14. The condition of orthogonality implies that the dot product of any two vectors $\mathbf{x}_n$ and $\mathbf{x}_m$ must be zero, and the dot product of any vector with itself must be unity. This is the direct extension of Eq. 1.13 and can be expressed as

$$\mathbf{x}_m \cdot \mathbf{x}_n = \begin{cases} 0 & m \neq n \\ 1 & m = n \end{cases} \qquad (1.15)$$

The constants $C_1, C_2, C_3, \ldots, C_n$ in Eq. 1.14 represent the magnitudes of the components of $\mathbf{A}$ along the vectors $\mathbf{x}_1, \mathbf{x}_2, \mathbf{x}_3, \ldots, \mathbf{x}_n$, respectively. It follows that

$$C_r = \mathbf{A} \cdot \mathbf{x}_r \qquad (1.16)$$

This result can also be obtained by taking the dot product of both sides in Eq. 1.14 with vector $\mathbf{x}_r$. We have

$$\mathbf{A} \cdot \mathbf{x}_r = C_1\mathbf{x}_1 \cdot \mathbf{x}_r + C_2\mathbf{x}_2 \cdot \mathbf{x}_r + \cdots + C_r\mathbf{x}_r \cdot \mathbf{x}_r + \cdots + C_n\mathbf{x}_n \cdot \mathbf{x}_r \quad (1.17)$$

From Eq. 1.15 it follows that all the terms of the form $C_j \mathbf{x}_j \cdot \mathbf{x}_r (j \neq r)$ on the right-hand side of Eq. 1.17 are zero. Therefore

$$\mathbf{A} \cdot \mathbf{x}_r = C_r \mathbf{x}_r \cdot \mathbf{x}_r = C_r \tag{1.18}$$

We call the set of vectors $(\mathbf{x}_1, \mathbf{x}_2, \ldots, \mathbf{x}_n)$ an *orthogonal vector space*. In general, the product $\mathbf{x}_m \cdot \mathbf{x}_n$ can be some constant k_m instead of unity. When k_m is unity, the set is called normalized orthogonal set, or *orthonormal vector space*. Therefore, in general, for orthogonal vector space $\{\mathbf{x}_r\} \cdots (r = 1, 2, \ldots, n)$ we have

$$\mathbf{x}_m \cdot \mathbf{x}_n = \begin{cases} 0 & m \neq n \\ k_m & m = n \end{cases} \tag{1.19}$$

For an orthogonal vector space, Eq. 1.18 is modified to

$$\mathbf{A} \cdot \mathbf{x}_r = C_r \mathbf{x}_r \cdot \mathbf{x}_r = C_r k_r$$

and

$$C_r = \frac{\mathbf{A} \cdot \mathbf{x}_r}{k_r}$$

We shall now summarize the results of our discussion. For an orthogonal vector space $\{\mathbf{x}_r\} \cdots (r = 1, 2, \ldots)$,

$$\mathbf{x}_m \cdot \mathbf{x}_n = \begin{cases} 0 & m \neq n \\ k_m & m = n \end{cases} \tag{1.20}$$

If this vector space is complete, then any vector $\mathbf{F}$ can be expressed as

$$\mathbf{F} = C_1 \mathbf{x}_1 + C_2 \mathbf{x}_2 + \cdots + C_r \mathbf{x}_r + \cdots \tag{1.21}$$

where

$$C_r = \frac{\mathbf{F} \cdot \mathbf{x}_r}{k_r} = \frac{\mathbf{F} \cdot \mathbf{x}_r}{\mathbf{x}_r \cdot \mathbf{x}_r} \tag{1.22}$$

Orthogonal Signal Space

We shall now apply certain concepts of vector space to gain some intuition about signal analysis. We have seen that any vector can be expressed as a sum of its components along n mutually orthogonal vectors, provided these vectors formed a complete set of coordinate system. We therefore suspect that it may be possible to express any function $f(t)$ as a sum of its components along a set of mutually orthogonal functions if these functions form a complete set. We shall now show that this indeed is the case.

Approximation of a Function by a Set of Mutually Orthogonal Functions

Let us consider a set of n functions $g_1(t)$, $g_2(t)$, $\ldots$, $g_n(t)$ which are orthogonal to one another over an interval t_1 to t_2; that is,

$$\int_{t_1}^{t_2} g_j(t)g_k(t) \, dt = 0 \qquad j \neq k \qquad (1.23a)$$

and let

$$\int_{t}^{t_2} g_j{}^2(t) \, dt = K_j \qquad (1.23b)$$

Let an arbitrary function $f(t)$ be approximated over an interval (t_1, t_2) by a linear combination of these n mutually orthogonal functions.

$$f(t) \simeq C_1 g_1(t) + C_2 g_2(t) + \cdots + C_k g_k(t) + \cdots + C_n g_n(t)$$

$$= \sum_{r=1}^{n} C_r g_r(t)$$

For the best approximation, we must find the proper values of constants $C_1, C_2, \ldots, C_n$ such that ε, the mean square of $f_e(t)$, is minimized.

By definition,

$$f_e(t) = f(t) - \sum_{r=1}^{n} C_r g_r(t)$$

and

$$\varepsilon = \frac{1}{t_2 - t_1} \int_{t_1}^{t_2} \left[f(t) - \sum_{r=1}^{n} C_r g_r(t) \right]^2 dt \qquad (1.24)$$

It is evident from Eq. 1.24 that ε is a function of $C_1, C_2, \ldots, C_n$ and to minimize ε, we must have

$$\frac{\partial \varepsilon}{\partial C_1} = \frac{\partial \varepsilon}{\partial C_2} = \cdots = \frac{\partial \varepsilon}{\partial C_j} = \cdots = \frac{\partial \varepsilon}{\partial C_n} = 0$$

Let us consider the equation:

$$\frac{\partial \varepsilon}{\partial C_j} = 0 \qquad (1.25)$$

Since $(t_2 - t_1)$ is constant, Eq. 1.25 may be expressed as

$$\frac{\partial}{\partial C_j} \left\{ \int_{t_1}^{t_2} \left[f(t) - \sum_{r=1}^{n} C_r g_r(t) \right]^2 dt \right\} = 0 \qquad (1.26)$$

When we expand the integrand, we note that all the terms arising due to the cross product of the orthogonal functions are zero by virtue of orthogonality; that is, all the terms of the form $\int g_j(t)g_k(t)\, dt$ are zero, as expressed in Eq. 1.23. Similarly, the derivative with respect to C_j of all the terms that do not contain C_j is zero; that is,

$$\frac{\partial}{\partial C_j} \int_{t_1}^{t_2} f^2(t)\, dt = \frac{\partial}{\partial C_j} \int_{t_1}^{t_2} C_r^2 g_r^2(t)\, dt = \frac{\partial}{\partial C_j} \int_{t_1}^{t_2} C_r f(t) g_r(t)\, dt = 0$$

This leaves only two nonzero terms in Eq. 1.26 as follows:

$$\frac{\partial}{\partial C_j} \int_{t_1}^{t_2} [-2C_j f(t) g_j(t) + C_j^2 g_j^2(t)]\, dt = 0 \qquad (1.27)$$

Changing the order of differentiation and integration in Eq. 1.27, we get

$$2\int_{t_1}^{t_2} f(t) g_j(t)\, dt = 2C_j \int_{t_1}^{t_2} g_j^2(t)\, dt$$

Therefore

$$C_j = \frac{\displaystyle\int_{t_1}^{t_2} f(t) g_j(t)\, dt}{\displaystyle\int_{t_1}^{t_2} g_j^2(t)\, dt} \qquad (1.28a)$$

$$= \frac{1}{K_j} \int_{t_1}^{t_2} f(t) g_j(t)\, dt \qquad (1.28b)$$

We may summarize this result as follows. Given a set of n functions $g_1(t), g_2(t), \ldots, g_n(t)$ mutually orthogonal over the interval (t_1, t_2), it is possible to approximate an arbitrary function $f(t)$ over this interval by a linear combination of these n functions.

$$f(t) \simeq C_1 g_1(t) + C_2 g_2(t) + \cdots + C_n g_n(t)$$

$$= \sum_{r=1}^{n} C_r g_r(t) \qquad (1.29)$$

For the best approximation, that is, the one that will minimize the mean of the square error over the interval, we must choose the coefficients $C_1, C_2, \ldots, C_n$, etc., as given by Eq. 1.28.

Evaluation of Mean Square Error

Let us now find the value of ε when optimum values of coefficients $C_1, C_2, \ldots, C_n$ are chosen according to Eq. 1.28. By definition,

$$\varepsilon = \frac{1}{(t_2 - t_1)} \int_{t_1}^{t_2} \left[f(t) - \sum_{r=1}^{n} C_r g_r(t) \right]^2 dt$$

$$= \frac{1}{(t_2 - t_1)} \left[\int_{t_1}^{t_2} f^2(t)\, dt + \sum_{r=1}^{n} C_r^2 \int_{t_1}^{t_2} g_r^2(t)\, dt - 2 \sum_{r=1}^{n} C_r \int_{t_1}^{t_2} f(t) g_r(t)\, dt \right] \tag{1.30}$$

But from Eqs. 1.28a and 1.28b it follows that

$$\int_{t_1}^{t_2} f(t) g_r(t)\, dt = C_r \int_{t_1}^{t_2} g_r^2(t)\, dt = C_r K_r \tag{1.31}$$

Substituting Eq. 1.31 in Eq. 1.30, we get

$$\varepsilon = \frac{1}{(t_2 - t_1)} \left[\int_{t_1}^{t_2} f^2(t)\, dt + \sum_{r=1}^{n} C_r^2 K_r - 2 \sum_{r=1}^{n} C_r^2 K_r \right]$$

$$= \frac{1}{(t_2 - t_1)} \left[\int_{t_1}^{t_2} f^2(t)\, dt - \sum_{r=1}^{n} C_r^2 K_r \right] \tag{1.32}$$

$$= \frac{1}{(t_2 - t_1)} \left[\int_{t_1}^{t_2} f^2(t)\, dt - (C_1^2 K_1 + C_2^2 K_2 + \cdots + C_n^2 K_n) \right] \tag{1.33}$$

One can therefore evaluate the mean-square error by using Eq. 1.33.

Representation of a Function by a Closed or a Complete Set of Mutually Orthogonal Functions

It is evident from Eq. 1.33 that if we increase n, that is, if we approximate $f(t)$ by a larger number of orthogonal functions, the error will become smaller. But by its very definition, ε is a positive quantity; hence in the limit as the number of terms is made infinity, the sum $\sum_{r=1}^{\infty} C_r^2 K_r$ may converge to the integral

$$\int_{t_1}^{t_2} f^2(t)\, dt$$

and then ε vanishes. Thus

$$\int_{t_1}^{t_2} f^2(t)\,dt = \sum_{r=1}^{\infty} C_r{}^2 K_r \tag{1.34}$$

Under these conditions $f(t)$ is represented by the infinite series:

$$f(t) = C_1 g_1(t) + C_2 g_2(t) + \cdots + C_r g_r(t) + \cdots$$

The infinite series on the right-hand side of Eq. 1.34 thus converges to $f(t)$ such that the mean square of the error is zero. The series is said to *converge in the mean*. Note that the representation of $f(t)$ is now exact.

A set of functions $g_1(t)$, $g_2(t)$, ..., $g_r(t)$ mutually orthogonal over the interval (t_1, t_2) is said to be a complete or a closed set if there exists no function $x(t)$ for which it is true that

$$\int_{t_1}^{t_2} x(t) g_k(t)\,dt = 0 \qquad \text{for} \quad k = 1, 2, \ldots$$

If a function $x(t)$ could be found such that the above integral is zero, then obviously $x(t)$ is orthogonal to each member of the set $\{g_r(t)\}$ and, consequently, is itself a member of the set. Evidently the set cannot be complete without $x(t)$ being its member.

Let us now summarize the results of this discussion. For a set $\{g_r(t)\}$, $(r = 1, 2, \ldots)$ mutually orthogonal over the interval (t_1, t_2),

$$\int_{t_1}^{t_2} g_m(t) g_n(t)\,dt = \begin{cases} 0 & \text{if} \quad m \neq n \\ K_m & \text{if} \quad m = n \end{cases} \tag{1.35}$$

If this function set is complete, then any function $f(t)$ can be expressed as

$$f(t) = C_1 g_1(t) + C_2 g_2(t) + \cdots + C_r g_r(t) + \cdots \tag{1.36}$$

where

$$C_r = \frac{\int_{t_1}^{t_2} f(t) g_r(t)\,dt}{K_r} = \frac{\int_{t_1}^{t_2} f(t) g_r(t)\,dt}{\int_{t_1}^{t_2} g_r{}^2(t)\,dt} \tag{1.37}$$

Comparison of Eqs. 1.35 to 1.37 with Eqs. 1.20 to 1.22 brings out forcefully the analogy between vectors and signals. Any vector can be expressed as a sum of its components along n mutually orthogonal vectors, provided these vectors form a complete set. Similarly, any function $f(t)$ can be expressed as a sum of its components along mutually orthogonal functions, provided these functions form a closed or a complete set.

In the comparison of vectors and signals, the dot product of two vectors is analogous to the integral of the product of two signals, that is

$$\mathbf{A} \cdot \mathbf{B} \sim \int_{t_1}^{t_2} f_A(t) f_B(t) \, dt$$

It follows that the square of the length A of a vector $\mathbf{A}$ is analogous to the integral of the square of a function, that is,

$$\mathbf{A} \cdot \mathbf{A} = A^2 \sim \int_{t_1}^{t_2} f_A{}^2(t) \, dt$$

If a vector is expressed in terms of its mutually orthogonal components, the square of the length is given by the sum of the squares of the lengths of the component vectors. An analogous result holds true for signals. This is precisely expressed by Eq. 1.34 (Parseval's theorem). Since the component functions are not orthonormal, the right-hand side is $\Sigma \, C_r{}^2 K_r{}^2$ instead of $\Sigma \, C_r{}^2$. For an orthonormal set, $K_r = 1$. Equation 1.34 is thus analogous to the case where a vector is expressed in terms of its components along mutually orthogonal vectors whose length squares are $K_1, K_2, \ldots, K_r \ldots$, etc.

Equation 1.36 shows that $f(t)$ contains a component of signal $g_r(t)$, and this component has a magnitude C_r. Representation of $f(t)$ by a set of infinite mutually orthogonal functions is called *generalized Fourier series representation* of $f(t)$.

Example 1.2

As an example we shall again consider the rectangular function of Example 1.1 as shown in Fig. 1.3. This function was approximated by a single function $\sin t$. We shall now see how the approximation improves when a large number of mutually orthogonal functions are used. It was shown previously that functions $\sin n\omega_0 t$ and $\sin m\omega_0 t$ are mutually orthogonal over the interval $(t_0, t_0 + 2\pi/\omega_0)$ for all integral values of n and m. Hence it follows that a set of functions $\sin t$, $\sin 2t$, $\sin 3t$, etc., are mutually orthogonal over the interval $(0, 2\pi)$. The rectangular function in Fig. 1.3 will now be approximated by a finite series of sinusoidal functions.

$$f(t) \simeq C_1 \sin t + C_2 \sin 2t + \cdots + C_n \sin nt$$

The constants C_r can be evaluated by using Eq. 1.28

$$C_r = \frac{\displaystyle\int_0^{2\pi} f(t) \sin rt \, dt}{\displaystyle\int_0^{2\pi} \sin^2 rt \, dt}$$

$$= \frac{1}{\pi} \left(\int_0^{\pi} \sin rt \, dt - \int_{\pi}^{2\pi} \sin rt \, dt \right)$$

$$= \frac{4}{\pi r} \qquad \text{if } r \text{ is odd}$$

$$= 0 \qquad \text{if } r \text{ is even}$$

Thus $f(t)$ is approximated by

$$f(t) = \frac{4}{\pi} (\sin t + \tfrac{1}{3} \sin 3t + \tfrac{1}{5} \sin 5t + \tfrac{1}{7} \sin 7t + \cdots) \qquad (1.38)$$

Figure 1.6 shows the actual function and the approximated function when the function is approximated with one, two, three, and four terms, respectively, in Eq. 1.38. For the given number of terms of the form $\sin rt$, these are the optimum approximations which minimize the mean-square error. As we increase the number of terms, the approximation improves and the mean-square error diminishes. For infinite terms the mean-square error is zero.*

Let us evaluate the error ε in these approximations. From Eq. 1.33,

$$\varepsilon = \frac{1}{t_2 - t_1} \left[\int_{t_2}^{t_2} f^2(t) \, dt - C_1^2 K_1 - C_2^2 K_2 - \cdots \right]$$

In this case

$$(t_2 - t_1) = 2\pi$$

$$f(t) = \begin{cases} 1 & (0 < t < \pi) \\ -1, & (\pi < t < 2\pi) \end{cases}$$

Therefore

$$\int_0^{2\pi} f^2(t) \, dt = 2\pi$$

Also

$$C_r = \begin{cases} \dfrac{4}{\pi r} & \text{if } r \text{ is odd} \\ 0 & \text{if } r \text{ is even} \end{cases}$$

* The Fourier series fails to converge at the points of discontinuity, and hence even though the number of terms is increased, the approximated function shows large amounts of ripples at the points of discontinuity. This is known as the *Gibbs phenomenon*.

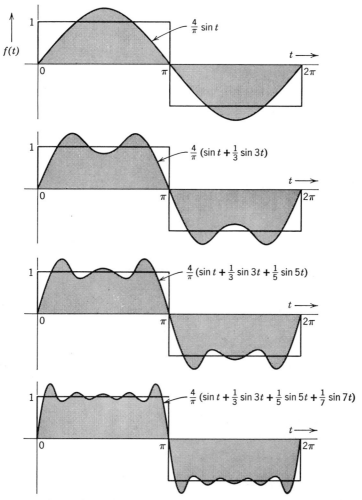

Figure 1.6 Approximation of a rectangular function by orthogonal functions.

and

$$K_r = \int_0^{2\pi} \sin^2 rt \, dt = \pi$$

Therefore, for one-term approximation,

$$\varepsilon_1 = \frac{1}{2\pi}\left[2\pi - \left(\frac{4}{\pi}\right)^2 \pi\right] = 0.19$$

For two-term approximation,

$$\varepsilon_2 = \frac{1}{2\pi}\left[2\pi - \left(\frac{4}{\pi}\right)^2 \pi - \left(\frac{4}{3\pi}\right)^2 \pi\right] = 0.1$$

For three-term approximation,

$$\varepsilon_3 = \frac{1}{2\pi}\left[2\pi - \left(\frac{4}{\pi}\right)^2 \pi - \left(\frac{4}{3\pi}\right)^2 \pi - \left(\frac{4}{5\pi}\right)^2 \pi\right] = 0.0675$$

and

$$\varepsilon_4 = \frac{1}{2\pi}\left[2\pi - \left(\frac{4}{\pi}\right)^2 \pi - \left(\frac{4}{3\pi}\right)^2 \pi - \left(\frac{4}{5\pi}\right)^2 \pi - \left(\frac{4}{7\pi}\right)^2 \pi\right] = 0.051$$

and so on.

It can be easily seen that in this case the mean-square error diminishes rapidly as the number of terms is increased.

Orthogonality in Complex Functions

In the previous discussion, we considered only real functions of real variables. If $f_1(t)$ and $f_2(t)$ are complex functions of real variable t, then it can be shown that $f_1(t)$ can be approximated by $C_{12}f_2(t)$ over an interval (t_1, t_2).

$$f_1(t) \simeq C_{12}f_2(t)$$

The optimum value of C_{12} to minimize the mean-square error magnitude is given by†

$$C_{12} = \frac{\displaystyle\int_{t_1}^{t_2} f_1(t)f_2{}^*(t)\,dt}{\displaystyle\int_{t_1}^{t_2} f_2(t)f_2{}^*(t)\,dt} \tag{1.39}$$

where $f_2{}^*(t)$ is a complex conjugate of $f_2(t)$.

It is evident from Eq. 1.39 that two complex functions $f_1(t)$ and $f_2(t)$ are orthogonal over the interval (t_1, t_2) if

$$\int_{t_1}^{t_2} f_1(t)f_2{}^*(t)\,dt = \int_{t_1}^{t_2} f_1{}^*(t)f_2(t)\,dt = 0 \tag{1.40}$$

For a set of complex functions $\{g_r(t)\}$, $(r = 1, 2, \ldots)$ mutually orthogonal over the interval (t_1, t_2):

$$\int_{t_1}^{t_2} g_m(t)g_n{}^*(t)\,dt = \begin{cases} 0 & \text{if } m \neq n \\ K_m & \text{if } m = n \end{cases} \tag{1.41}$$

If this set of functions is complete, then any function $f(t)$ can be expressed as

$$f(t) = C_1g_1(t) + C_2g_2(t) + \cdots + C_rg_r(t) + \cdots \tag{1.42}$$

† See, for instance, S. Mason and H. Zimmerman, *Electronic Circuits, Signals and Systems*, pp. 199–200, John Wiley and Sons, New York, 1960.

where

$$C_r = \frac{1}{K_r} \int_{t_1}^{t_2} f(t) g_r^*(t) \, dt \qquad (1.43)$$

If the set of functions is real, then $g_r^*(t) = g_r(t)$ and all results for complex functions reduce to those obtained for real functions in Eqs. 1.35 to 1.37.

1.2 SOME EXAMPLES OF ORTHOGONAL FUNCTIONS

Representation of a function over a certain interval by a linear combination of mutually orthogonal functions is called Fourier series representation of a function. There exist, however, a large number of sets of orthogonal functions, and hence a given function may be expressed in terms of different sets of orthogonal functions. In vector space this is analogous to the representation of a given vector in different sets of coordinate systems. Each set of orthogonal functions corresponds to a coordinate system. Some of the examples of sets of orthogonal functions are trigonometric functions, exponential functions, Legendre polynomials, and Jacobi polynomials. Bessel functions also form a special kind of orthogonal functions.†

Legendre Fourier Series

A set of Legendre polynomials $P_n(x)$, $(n = 0, 1, 2, \ldots)$ forms a complete set of mutually orthogonal functions over an interval $(-1 < t < 1)$. These polynomials can be defined by Rodrigues' formula:

$$P_n(t) = \frac{1}{2^n n!} \frac{d^n}{dt^n} (t^2 - 1)^n \qquad (n = 0, 1, 2, \ldots)$$

It follows from this equation that

$$P_0(t) = 1 \qquad\qquad P_1(t) = t$$
$$P_2(t) = (\tfrac{3}{2}t^2 - \tfrac{1}{2}) \qquad P_3(t) = (\tfrac{5}{2}t^3 - \tfrac{3}{2}t)$$

and so on.

---

† Bessel functions are orthogonal with respect to a weighting function. See, for instance, W. Kaplan, *Advanced Calculus*, Addison-Wesley, Reading, Mass., 1953.

We may verify the orthogonality of these polynomials by showing that

$$\int_{-1}^{1} P_m(t)P_n(t)\, dt = \begin{cases} 0 & m \neq n \\ \dfrac{2}{2m+1} & m = n \end{cases} \qquad (1.44)$$

We can express a function $f(t)$ in terms of Legendre polynomial over an interval $(-1 < t < 1)$ as

$$f(t) = C_0 P_0(t) + C_1 P_1(t) + \cdots + \cdots$$

where

$$C_r = \frac{\displaystyle\int_{-1}^{1} f(t)P_r(t)\, dt}{\displaystyle\int_{-1}^{1} P_r{}^2(t)\, dt}$$

$$= \frac{2r+1}{2}\int_{-1}^{1} f(t)P_r(t)\, dt \qquad (1.45)$$

Note that although the series representation is valid over the region -1 to 1, it can be extended to any region by the appropriate change in variable.

Example 1.3

Let us consider the rectangular function shown in Fig. 1.7. This function can be represented by Legendre Fourier series:

$$f(t) = C_0 P_0(t) + C_1 P_1(t) + \cdots + C_r P_r(t) + \cdots$$

The coefficients $C_0, C_1, C_2, \ldots, C_r$, etc., may be found from Eq. 1.45. We have

$$f(t) = \begin{cases} 1 & (-1 < t < 0) \\ -1 & (0 < t < 1) \end{cases}$$

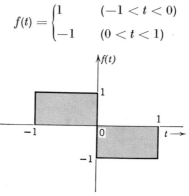

Figure 1.7

and

$$C_0 = \frac{1}{2} \int_{-1}^{1} f(t)\, dt = 0$$

$$C_1 = \frac{3}{2} \int_{-1}^{1} t f(t)\, dt$$

$$= \frac{3}{2} \left(\int_{-1}^{0} t\, dt - \int_{1}^{0} t\, dt \right)$$

$$= -\frac{3}{2}$$

$$C_2 = \frac{5}{2} \int_{-1}^{1} f(t) \left(\frac{3}{2} t^2 - \frac{1}{2} \right) dt$$

$$= \frac{5}{2} \left[\int_{-1}^{0} \frac{3}{2} t^2 - \frac{1}{2}\, dt + \int_{0}^{1} - \left(\frac{3}{2} t^2 - \frac{1}{2} \right) dt \right]$$

$$= 0$$

It can be shown, in general, that for even values of r,

$$C_r = 0$$

that is,

$$C_0 = C_2 = C_4 = C_6 = \cdots = 0$$

$$C_3 = \frac{7}{2} \int_{-1}^{1} f(t) \left(\frac{5}{2} t^3 - \frac{3}{2} t \right) dt$$

$$= \frac{7}{2} \left[\int_{-1}^{0} \frac{5}{2} t^3 - \frac{3}{2} t\, dt + \int_{0}^{1} - \left(\frac{5}{2} t^3 - \frac{3}{2} t \right) \right] dt$$

$$= \frac{7}{8}$$

Similarly, the coefficients C_5, C_7, ..., etc., may be evaluated. We now have

$$f(t) = -\tfrac{3}{2} t + \tfrac{7}{8}(\tfrac{5}{2} t^3 - \tfrac{3}{2} t) + \cdots$$

Trigonometric Fourier Series

We have already shown that functions $\sin \omega_0 t$, $\sin 2\omega_0 t$, etc., form an orthogonal set over any interval $(t_0, t_0 + 2\pi/\omega_0)$. This set, however, is not complete. This is evident from the fact a function $\cos n\omega_0 t$ is orthogonal to $\sin m\omega_0 t$ over the same interval. Hence to complete the set, we must include cosine as well as sine functions. It can be shown that the composite set of functions consisting of a set $\cos n\omega_0 t$ and $\sin n\omega_0 t$ for $(n = 0, 1, 2, \ldots)$ forms a complete orthogonal set. Note that for $n = 0$, $\sin n\omega_0 t$ is zero, but $\cos n\omega_0 t = 1$. Thus we have a completed orthogonal set represented by functions 1, $\cos \omega_0 t$, $\cos 2\omega_0 t$, $\ldots$, $\cos n\omega_0 t$, $\ldots$; $\sin \omega_0 t$, $\sin 2\omega_0 t$, $\ldots$, $\sin n\omega_0 t$, $\ldots$, etc. It therefore follows that any function $f(t)$ can be represented in terms of these functions over any interval $(t_0, t_0 + 2\pi/\omega_0)$. Thus

$$f(t) = a_0 + a_1 \cos \omega_0 t + a_2 \cos 2\omega_0 t + \cdots + a_n \cos n\omega_0 t + \cdots$$
$$+ b_1 \sin \omega_0 t + b_2 \sin 2\omega_0 t + \cdots + b_n \sin n\omega_0 t + \cdots$$
$$(t_0 < t < t_0 + 2\pi/\omega_0)$$

For convenience we shall denote $2\pi/\omega_0$ by T. The preceding equation can be expressed as

$$f(t) = a_0 + \sum_{n=1}^{\infty} (a_n \cos n\omega_0 t + b_n \sin n\omega_0 t) \qquad (t_0 < t < t_0 + T)$$

$$(1.46)$$

Equation 1.46 is the trigonometric Fourier series representation of $f(t)$ over an interval $(t_0, t_0 + T)$. The various constants a_n and b_n are given by

$$a_n = \frac{\int_{t_0}^{(t_0+T)} f(t) \cos n\omega_0 t \, dt}{\int_{t_0}^{(t_0+T)} \cos^2 n\omega_0 t \, dt} \qquad (1.47a)$$

and

$$b_n = \frac{\int_{t_0}^{(t_0+T)} f(t) \sin n\omega_0 t \, dt}{\int_{t_0}^{(t_0+T)} \sin^2 n\omega_0 t \, dt} \qquad (1.47b)$$

If we let $n = 0$ in Eq. 1.47a, we get

$$a_0 = \frac{1}{T} \int_{t_0}^{(t_0+T)} f(t) \, dt \qquad (1.48a)$$

We also have

$$\int_{t_0}^{(t_0+T)} \cos^2 n\omega_0 t \, dt = \int_{t_0}^{(t_0+T)} \sin^2 n\omega_0 t \, dt = \frac{T}{2}$$

Therefore

$$a_n = \frac{2}{T} \int_{t_0}^{(t_0+T)} f(t) \cos n\omega_0 t \, dt \qquad (1.48b)$$

$$b_n = \frac{2}{T} \int_{t_0}^{(t_0+T)} f(t) \sin n\omega_0 t \, dt \qquad (1.48c)$$

The constant term a_0 in the series is given by Eq. 1.48a. It is evident that a_0 is the average value of $f(t)$ over the interval $(t_0, t_0 + T)$. Thus a_0 is the d-c component of $f(t)$ over this interval.

The trigonometric series (Eq. 1.46) may be represented in compact form as

$$f(t) = \sum_{n=0}^{\infty} c_n \cos(n\omega_0 t + \varphi_n) \tag{1.49}$$

where

$$c_n = \sqrt{a_n^2 + b_n^2} \tag{1.50}$$

and $\varphi_n = -\tan^{-1}(b_n/a_n)$.

Example 1.4

We shall now expand a function $f(t)$ shown in Fig. 1.8a by trigonometric Fourier series over the interval $(0, 1)$. It is evident that $f(t) = At$ $(0 < t < 1)$, the interval $T = 1$, and $\omega_0 = 2\pi/T = 2\pi$. We must choose $t_0 = 0$. Thus

$$f(t) = a_0 + a_1 \cos 2\pi t + a_2 \cos 4\pi t + \cdots + a_n \cos 2\pi nt + \cdots$$
$$+ b_1 \sin 2\pi t + b_2 \sin 4\pi t + \cdots + b_n \sin 2\pi nt + \cdots \tag{1.51}$$

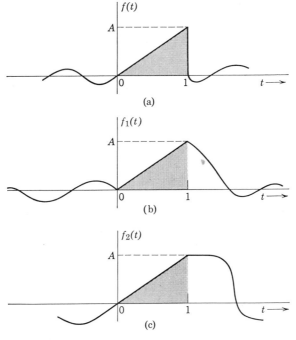

Figure 1.8

Various coefficients in the series in Eq. 1.51 can be found by using Eqs. 1.48a to 1.48c.

$$a_0 = \frac{1}{T} \int_0^T f(t) \, dt = 1 \int_0^1 At \, dt = \frac{A}{2} \qquad (1.52a)$$

Similarly,

$$a_n = \frac{2}{1} \int_0^1 At \cos 2\pi n t \, dt$$

$$= \frac{A}{2\pi^2 n^2} [\cos 2\pi n t + 2\pi n t \sin 2\pi n t]_0^1 = 0 \qquad (1.52b)$$

$$b_n = \frac{2}{1} \int_0^1 At \sin 2\pi n t \, dt$$

$$= \frac{A}{2\pi^2 n^2} [\sin 2\pi n t - 2\pi n t \cos 2\pi n t]_0^1 = \frac{-A}{\pi n} \qquad (1.52c)$$

Since $a_n = 0$ for all values of n, all the cosine terms in series in Eq. 1.51 are zero. The coefficients of the sine terms are given by Eq. 1.52c. The series in Eq. 1.51 may now be expressed as

$$f(t) = \frac{A}{2} - \frac{A}{\pi} \sin 2\pi t - \frac{A}{2\pi} \sin 4\pi t - \frac{A}{3\pi} \sin 6\pi t - \cdots$$

$$- \frac{A}{n\pi} \sin 2\pi n t - \cdots \qquad (0 < t < 1)$$

$$= \frac{A}{2} - \frac{A}{\pi} \sum_{n=1}^{\infty} \frac{\sin 2\pi n t}{n} \qquad (0 < t < 1) \qquad (1.53)$$

We have thus expressed $f(t)$ in terms of its components over the interval $(0, 1)$. From Eq. 1.53 it is evident that $f(t)$ has a d-c component $A/2$. In addition, $f(t)$ has components of sinusoidal functions $\sin 2\pi t$, $\sin 4\pi t$, etc., with magnitudes of $-A/\pi$, $-A/2\pi$, etc., respectively.

We note here that the series in Eq. 1.51 will also represent any other functions identical to $f(t)$ over the interval $(0, 1)$. Thus the functions $f_1(t)$ and $f_2(t)$ in Figs. 1.8b and 1.8c are identical to $f(t)$ over the interval $(0, 1)$, and hence both these functions can also be represented over the interval $(0, 1)$ by the series in Eq. 1.53.

Exponential Fourier Series

It can be shown easily that a set of exponential functions $\{e^{jn\omega_0 t}\}$, $(n = 0, \pm 1, \pm 2, \ldots)$ is orthogonal over an interval $(t_0, t_0 + 2\pi/\omega_0)$ for any value of t_0. Note that this is a set of complex functions. We can demonstrate the orthogonality of this set by considering the integral

$$I = \int_{t_0}^{t_0+2\pi/\omega_0} (e^{jn\omega_0 t})(e^{jm\omega_0 t})^* \, dt = \int_{t_0}^{t_0+2\pi/\omega_0} e^{jn\omega_0 t} e^{-jm\omega_0 t} \, dt$$

If $n = m$, the integral I is given by

$$I = \int_{t_0}^{t_0+2\pi/\omega_0} dt = \frac{2\pi}{\omega_0}$$

If $n \neq m$, the integral I is given by

$$I = \frac{1}{j(n-m)\omega_0} e^{j(n-m)\omega_0 t} \Big|_{t_0}^{t_0+2\pi/\omega_0}$$

$$= \frac{1}{j(n-m)\omega_0} e^{i(n-m)\omega_0 t_0} [e^{i2\pi(n-m)} - 1]$$

Since both n and m are integers, $e^{j2\pi(n-m)}$ is equal to unity, and hence the integral is zero:

$$I = 0$$

Thus

$$\int_{t_0}^{t_0+2\pi/\omega_0} e^{jn\omega_0 t}(e^{jn\omega_0 t})^* \, dt = \begin{cases} \dfrac{2\pi}{\omega_0} & m = n \\ 0 & m \neq n \end{cases} \tag{1.54}$$

As before, let

$$\frac{2\pi}{\omega_0} = T$$

It is evident from Eq. 1.54 that the set of functions

$$\{e^{jn\omega_0 t}\} \qquad (n = 0, \pm 1, \pm 2, \ldots)$$

is orthogonal over the interval $(t_0, t_0 + T)$ where $T = 2\pi/\omega_0$. Further, it can be shown that this is a complete set. It is therefore possible to represent an arbitrary function $f(t)$ by a linear combination of exponential functions over an interval $(t_0, t_0 + T)$:

$$f(t) = F_0 + F_1 e^{j\omega_0 t} + F_2 e^{j2\omega_0 t} + \cdots + F_n e^{jn\omega_0 t} + \cdots$$

$$+ F_{-1} e^{-j\omega_0 t} + F_{-2} e^{-j2\omega_0 t} + \cdots + F_{-n} e^{-jn\omega_0 t} + \cdots$$

for

$$= \sum_{n=-\infty}^{\infty} F_n e^{jn\omega_0 t} \qquad (t_0 < t < t_0 + T) \tag{1.55}$$

where $\omega_0 = 2\pi/T$ and the summation in Eq. 1.55 is for integral values of n from $-\infty$ to ∞, including zero. Representation of $f(t)$ by exponential series, as shown in Eq. 1.55 is known as exponential Fourier series representation of $f(t)$ over the interval $(t_0, t_0 + T)$. The various coefficients in this series can be evaluated by using Eq. 1.43.

$$F_n = \frac{\int_{t_0}^{t_0+T} f(t)(e^{jn\omega_0 t})^* \, dt}{\int_{t_0}^{t_0+T} e^{jn\omega_0 t}(e^{jn\omega_0 t})^* \, dt}$$

$$= \frac{\int_{t_0}^{t_0+T} f(t)e^{-jn\omega t_0} \, dt}{\int_{t_0}^{t_0+T} e^{jn\omega_0 t}e^{-jn\omega_0 t} \, dt}$$

$$= \frac{1}{T}\int_{t_0}^{t_0+T} f(t)e^{-jn\omega_0 t} \, dt \tag{1.56}$$

We could also have obtained this result directly by multiplying both sides of Eq. 1.55 by $e^{-jn\omega_0 t}$ and integrating with respect to t over the interval $(t_0, t_0 + T)$. By virtue of orthogonality, all the terms except one, on the right-hand side, vanish and yield the expression for F_n as in Eq. 1.56.

Summarizing the results: Any given function $f(t)$ may be expressed as a discrete sum of exponential functions $\{e^{jn\omega_0 t}\}$, $(n = 0, \pm 1, \pm 2, \ldots)$ over an interval $t_0 < t < t_0 + T$, $(\omega_0 = 2\pi/T)$.

$$f(t) = \sum_{n=-\infty}^{\infty} F_n e^{jn\omega_0 t} \qquad (t_0 < t < t_0 + T) \tag{1.57}$$

where

$$F_n = \frac{1}{T}\int_{t_0}^{t_0+T} f(t)e^{-jn\omega_0 t} \, dt \tag{1.58}$$

It should be noted that the trigonometric and the exponential Fourier series are not two different types of series but two different ways of expressing the same series. The coefficients of one series can be obtained from those of the other. This can be seen from Eqs. 1.48 and 1.58. From these equations, it follows that

$$a_0 = F_0$$
$$a_n = F_n + F_{-n}$$
$$b_n = j(F_n - F_{-n}) \tag{1.59}$$

and

$$F_n = \tfrac{1}{2}(a_n - jb_n)$$

To illustrate, consider the function $f(t)$ in Example 1.4. This function has been represented by trigonometric Fourier series (Eq. 1.53). We can also represent it by exponential Fourier series by using Eqs. 1.57 and 1.58. Alternately, we may use Eq. 1.59 to obtain the coefficients of the exponential series from those of the trigonometric series. Substituting Eqs. 1.52 in Eq. 1.59, we get

$$F_0 = \frac{A}{2} \quad \text{and} \quad F_n = \frac{jA}{2\pi n} \tag{1.60a}$$

Hence

$$f(t) = \frac{A}{2} + \frac{jA}{2\pi} \sum_{n=-\infty}^{\infty} \frac{1}{n}\, e^{jn2\pi t} \tag{1.60b}$$

1.3　REPRESENTATION OF A PERIODIC FUNCTION BY THE FOURIER SERIES OVER THE ENTIRE INTERVAL $(-\infty < t < \infty)$

Thus far we have been able to represent a given function $f(t)$ by a Fourier series over a finite interval $(t_0, t_0 + T)$. Outside this interval the function $f(t)$ and the corresponding Fourier series need not be equal. If, however, the function $f(t)$ happens to be periodic, it can be shown that the series representation applies to the entire interval $(-\infty, \infty)$. This can be seen easily by considering some function $f(t)$ and its exponential Fourier series representation over an interval $(t_0, t_0 + T)$:

$$f(t) = \sum_{n=-\infty}^{\infty} F_n e^{jn\omega_0 t} \tag{1.61}$$

The equality holds over the interval $(t_0 < t < t_0 + T)$. The two sides of Eq. 1.61 need not be equal outside this interval. It is easy to see, however, that the right-hand side of Eq. 1.61 is periodic (with period $T = \omega_0/2\pi$). This follows from the fact that

$$e^{jn\omega_0 t} = e^{jn\omega_0(t+T)} \quad \text{for} \quad T = \frac{\omega_0}{2\pi}$$

It is therefore obvious that if $f(t)$ is periodic with period T, then the equality in Eq. 1.61 holds over the entire interval $(-\infty, \infty)$. Thus for a periodic function $f(t)$,

$$f(t) = \sum_{n=-\infty}^{\infty} F_n e^{jn\omega_0 t} \qquad (-\infty < t < \infty)$$

where

$$F_n = \frac{1}{T} \int_{t_0}^{t_0+T} f(t) e^{-jn\omega_0 t}\, dt$$

Note that the choice of t_0 is immaterial.

Example 1.5

Consider the rectified sine wave shown in Fig. 1.9. For this function

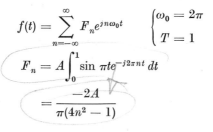

$$f(t) = \sum_{n=-\infty}^{\infty} F_n e^{jn\omega_0 t} \qquad \begin{cases} \omega_0 = 2\pi \\ T = 1 \end{cases}$$

$$F_n = A \int_0^1 \sin \pi t\, e^{-j2\pi n t}\, dt$$

$$= \frac{-2A}{\pi(4n^2 - 1)}$$

Hence

$$f(t) = \frac{-2A}{\pi} \sum_{n=-\infty}^{\infty} \frac{1}{4n^2 - 1}\, e^{j2\pi n t} \qquad (1.62)$$

Figure 1.9   Rectified sine wave.

1.4   THE COMPLEX FOURIER SPECTRUM

A Fourier series expansion of a periodic function is really equivalent to resolving the function in terms of its components of various frequencies. A periodic function with period T has frequency components of angular frequencies $\omega_0, 2\omega_0, 3\omega_0, \ldots, n\omega_0$, etc., where $\omega_0 = 2\pi/T$. Thus the periodic function $f(t)$ possesses its spectrum of frequencies. If we specify $f(t)$, we can find its spectrum. Conversely, if the spectrum is known, one can find the corresponding periodic function $f(t)$. We therefore have two ways of specifying a periodic function $f(t)$: the time domain representation where $f(t)$ is expressed as a function of time, and the frequency domain representation where the spectrum (that is, the amplitudes of various frequency components) is specified. Note that the spectrum exists only at $\omega = \omega_0,\ 2\omega_0,\ 3\omega_0, \ldots$, etc. Thus the spectrum is not a continuous curve but exists only at some discrete values of ω. It is therefore a *discrete spectrum*, sometimes referred to as a *line spectrum*. We may represent the spectrum graphically by drawing a vertical line at $\omega = \omega_0,\ 2\omega_0, \ldots$, etc., with their heights proportional to the amplitude of the corresponding frequency component. The discrete frequency spectrum thus appears on a graph as a series of equally spaced vertical lines with heights proportional to the amplitude of the corresponding frequency component.

We can use either the trigonometric or the exponential series to represent the spectrum. The exponential form, however, proves more useful for our purpose. In this series the periodic function is expressed as a sum of exponential functions of frequency $0,\ \pm\omega_0,\ \pm2\omega_0, \ldots$, etc. The significance of negative frequencies is not hard to understand. Both the signals $e^{j\omega t}$ and $e^{-j\omega t}$ oscillate at frequency ω. They may, however, be looked upon as two phasors rotating in opposite directions and, when added, yield a real function of time. Thus

$$e^{j\omega t} + e^{-j\omega t} = 2 \cos \omega t$$

For a periodic function of period T, the exponential series is given by

$$f(t) = F_0 + F_1 e^{j\omega_0 t} + F_2 e^{j2\omega_0 t} + \cdots + F_n e^{jn\omega_0 t} + \cdots$$
$$+ F_{-1} e^{-j\omega_0 t} + F_{-2} e^{-j2\omega_0 t} + \cdots + F_{-n} e^{-jn\omega_0 t} + \cdots$$

We thus have the frequencies $0,\ \omega_0,\ -\omega_0,\ 2\omega_0,\ -2\omega_0, \ldots,\ n\omega_0,\ -n\omega_0,$ $\ldots$, etc., and the amplitudes of these components are, respectively, $F_0,\ F_1,\ F_{-1},\ F_2,\ F_{-2}, \ldots,\ F_n,\ F_{-n}, \ldots$, etc.

The amplitudes F_n are usually complex and hence can be described by magnitude and phase. Therefore, in general, we need two line spectra: the *magnitude spectrum* and the *phase spectrum* for the frequency domain representation of a periodic function. In most of the cases, however, the amplitudes of frequency components are either real or imaginary, and thus it is possible to describe the function by only one spectrum.

Consider the periodic function in Example 1.5 (Fig. 1.9). This is a rectified sine wave, and the exponential Fourier series was found

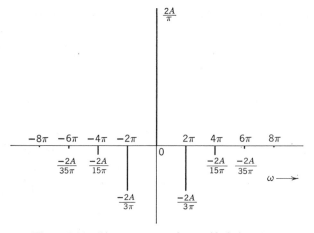

Figure 1.10   Line spectrum of a rectified sine wave.

to be (Eq. 1.62)

$$f(t) = \frac{2A}{\pi} - \frac{2A}{3\pi} e^{j2\pi t} - \frac{2A}{15\pi} e^{j4\pi t} - \frac{2A}{35\pi} e^{j6\pi t} - \cdots$$

$$- \frac{2A}{3\pi} e^{-j2\pi t} - \frac{2A}{15\pi} e^{-j4\pi t} - \frac{2A}{35\pi} e^{-j6\pi t} - \cdots$$

The spectrum exists at $\omega = 0$, $\pm 2\pi$, $\pm 4\pi$, $\pm 6\pi$, ..., etc., and the corresponding magnitudes are $2A/\pi$, $-2A/3\pi$, $-2A/15\pi$, $-2A/35\pi$, ..., etc. Note that all the amplitudes are real and, consequently, it is necessary to plot only one spectrum. The required frequency spectrum is shown in Fig. 1.10. It is evident from this figure that the spectrum is symmetrical about the vertical axis passing through the origin. This is not a coincidence. We shall presently show that the magnitude spectrum of every periodic function is symmetrical about the vertical axis passing through the origin. This can be demonstrated easily. The coefficient F_n is given by

$$F_n = \frac{1}{T} \int_{-T/2}^{T/2} f(t) e^{-jn\omega_0 t}\, dt$$

and

$$F_{-n} = \frac{1}{T} \int_{-T/2}^{T/2} f(t) e^{jn\omega_0 t}\, dt$$

It is evident from these equations that the coefficients F_n and F_{-n} are complex conjugates of each other, that is,

$$F_{-n} = F_n{}^* \qquad \text{hence} \quad |F_n| = |F_{-n}|$$

It therefore follows that the magnitude spectrum is symmetrical about the vertical axis passing through the origin and hence is an even function of ω.

If F_n is real, then F_{-n} is also real and F_n is equal to F_{-n}. If F_n is complex, let

$$F_n = |F_n| \, e^{j\theta_n} \tag{1.63a}$$

then

$$F_{-n} = |F_n| \, e^{-j\theta_n} \tag{1.63b}$$

The phase of F_n is θ_n; however, the phase of F_{-n} is $-\theta_n$. Hence it is obvious that the phase spectrum is antisymmetrical (an odd function) and the magnitude spectrum is symmetrical (an even function) about the vertical axis passing through the origin.

Example 1.6

Expand the periodic gate function shown in Fig. 1.11 by the exponential Fourier series and plot the frequency spectrum.

Solution. The gate function has a width δ and repeats every T seconds. The function may be described analytically over one period as follows.

$$f(t) = \begin{cases} A & (-\delta/2 < t < \delta/2) \\ 0 & (\delta/2 < t < T - \delta/2) \end{cases}$$

For convenience we shall choose as the limits of integration $-\delta/2$ to $(T - \delta/2)$.

$$
\begin{aligned}
F_n &= \frac{1}{T} \int_{-\delta/2}^{T-\delta/2} f(t) e^{-jn\omega_0 t} \, dt \\
&= \frac{1}{T} \int_{-\delta/2}^{\delta/2} A e^{-jn\omega_0 t} \, dt \\
&= \frac{-A}{jn\omega_0 T} e^{-jn\omega_0 t} \Big|_{-\delta/2}^{\delta/2} \\
&= \frac{2A}{n\omega_0 T} \frac{(e^{jn\omega_0\delta/2} - e^{-jn\omega_0\delta/2})}{2j} \\
&= \frac{2A}{n\omega_0 T} \sin (n\omega_0\delta/2) \\
&= \frac{A\delta}{T} \left[\frac{\sin (n\omega_0\delta/2)}{n\omega_0\delta/2} \right] \tag{1.64}
\end{aligned}
$$

The function in the bracket has a form $(\sin x)/x$. This function plays an important role in communication theory and is known as the *sampling*

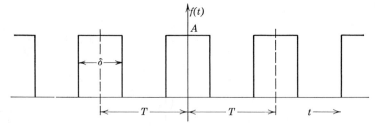

Figure 1.11   A periodic gate function.

function, abbreviated by $Sa(x)$.

$$Sa(x) = \frac{\sin x}{x} \tag{1.65}$$

The sampling function is shown in Fig. 1.12. Note that the function oscillates with a period 2π, with a decaying amplitude in either direction of x, and has zeros at $x = \pm\pi, \pm2\pi, \pm3\pi, \ldots$, etc. From Eq. 1.64, we have

$$F_n = \frac{A\delta}{T} Sa(n\omega_0\delta/2)$$

But

$$\omega_0 = \frac{2\pi}{T} \quad \text{and} \quad \frac{n\omega_0\delta}{2} = \frac{n\pi\delta}{T}$$

Hence

$$F_n = \frac{A\delta}{T} Sa\left(\frac{n\pi\delta}{T}\right) \tag{1.66a}$$

and

$$f(t) = \frac{A\delta}{T} \sum_{n=-\infty}^{\infty} Sa\left(\frac{n\pi\delta}{T}\right) e^{jn\omega_0 t} \tag{1.66b}$$

It is evident from Eq. 1.66 that F_n is real, and hence we need only one spectrum for frequency domain representation. Also, since $Sa(x)$ is an even function, it is obvious from Eq. 1.66 that $F_n = F_{-n}$.

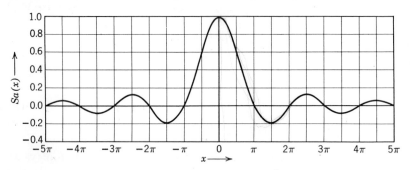

Figure 1.12   The sampling function $Sa(x)$.

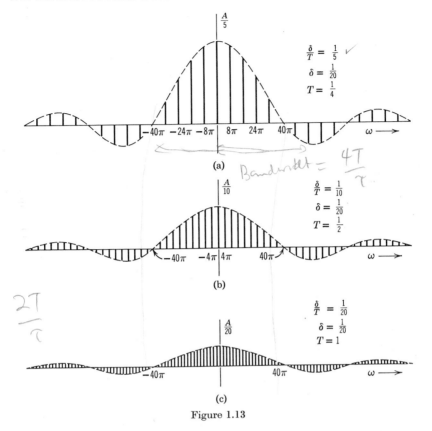

Figure 1.13

The fundamental frequency $\omega_0 = 2\pi/T$. The frequency spectrum is a discrete function and exists only at $\omega = 0, \pm 2\pi/T, \pm 4\pi/T, \pm 6\pi/T, \ldots$, etc., and has amplitudes $A\delta/T$, $(A\delta/T)Sa(\pi\delta/T)$, $(A\delta/T)Sa(2\pi\delta/T), \ldots$, etc., respectively. We shall consider the spectrum for some specific values of δ and T. The pulse width δ will be taken as $\frac{1}{20}$ second, and the period T will be chosen as $\frac{1}{4}$ second, $\frac{1}{2}$ second, and 1 second, successively.

For $\delta = \frac{1}{20}$ and $T = \frac{1}{4}$ second, Eq. 1.66 is given by

$$F_n = \frac{A}{5} Sa\left(\frac{n\pi}{5}\right)$$

The fundamental frequency $\omega_0 = 2\pi/T = 8\pi$. Thus the spectrum exists at $\omega = 0, \pm 8\pi, \pm 16\pi, \ldots$, etc., and is shown in Fig. 1.13a.

For $\delta = \frac{1}{20}$, $T = \frac{1}{2}$ second,

$$F_n = \frac{A}{10} Sa\left(\frac{n\pi}{10}\right)$$

The spectrum is shown in Fig. 1.13b and exists at frequencies $0, \pm 4\pi$, $\pm 8\pi, \ldots$, etc.

For $\delta = \frac{1}{20}$, $T = 1$ second,

$$F_n = \frac{A}{20} \, Sa\left(\frac{n\pi}{20}\right)$$

The spectrum exists at $\omega = 0, \pm 2\pi, \pm 4\pi, \ldots$, etc., and is shown in Fig. 1.13c.

It is evident that as the period T becomes larger and larger, the fundamental frequency $2\pi/T$ becomes smaller and smaller, and hence there are more and more frequency components in a given range of frequency. The spectrum therefore becomes denser as the period T becomes larger. However, the amplitudes of the frequency components become smaller and smaller as T is increased. In the limit as T is made infinity, we have a single rectangular pulse of width δ, and the fundamental frequency becomes zero. The spectrum now becomes continuous and exists at every frequency. Note, however, that the shape of the frequency spectrum does not change with the period T. Thus the envelope of the spectrum depends only upon the pulse shape that repeats at a period T but not upon the period (T) of repetition. In the limit as T is made infinite, the function $f(t)$ consists entirely of one nonrepetitive pulse, and the spectrum thus represents a nonperiodic function over the entire interval $(-\infty < t < \infty)$. We have thus extended the representation of a periodic function by summation of exponential functions to that for a nonperiodic function. This topic will be discussed in detail in Section 1.5.

1.5 REPRESENTATION OF AN ARBITRARY FUNCTION OVER THE ENTIRE INTERVAL $(-\infty, \infty)$: THE FOURIER TRANSFORM

We can now represent an arbitrary function in terms of exponential (or trigonometric) series over a finite interval. For the special case of a periodic function, we can extend this representation over the entire interval $(-\infty, \infty)$. It is, however, desirable to be able to represent any general function, periodic or not, over the entire interval $(-\infty, \infty)$ in terms of exponential signals. We shall presently see that this can be done. It will be shown that a nonperiodic signal generally can be expressed as a continuous sum (integral) of exponential signals, in contrast to the periodic signals, which can be represented by a discrete sum of exponential signals.

This problem may be approached in two ways. We can express a function $f(t)$ in terms of exponential functions over a finite interval $(-T/2 < t < T/2)$ and then let T go to infinity. Alternately, we may construct a periodic function of period T so that $f(t)$ represents the first cycle of this periodic waveform. In the limit we let the period T become infinity, and this periodic function then has only one cycle in the interval $(-\infty < t < \infty)$ and is represented by $f(t)$. There is virtually no difference between the two approaches. But the latter approach is more convenient, for it allows us to visualize the limiting process without altering the shape of the frequency spectrum. We have already discussed at some length this limiting process for a periodic gate function in Example 1.6 (Fig. 1.11). It was observed that as the period T is made larger, the fundamental frequency becomes smaller and the frequency spectrum becomes denser; that is, in a given frequency range there are more frequency components. But the amplitudes become smaller. The shape of the frequency spectrum, however, remains unaltered. This can be easily seen from Fig. 1.13 where three values of T ($T = \frac{1}{4}, \frac{1}{2}$, and 1 second) are considered.

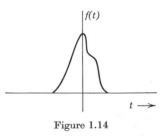

Figure 1.14

Let us consider a function $f(t)$, as shown in Fig. 1.14. We desire to represent this function as a sum of exponential functions over the entire interval $(-\infty < t < \infty)$. For this purpose we shall construct a new periodic function $f_T(t)$ with period T where the function $f(t)$ repeats itself every T seconds as shown in Fig. 1.15. The period T is made large enough so that there is no overlap between the pulses of the shape of $f(t)$. This new function $f_T(t)$ is a periodic function and, consequently, can be represented with an exponential Fourier series. In the limit, if we let T become infinite, then the pulses in the periodic function repeat after an infinite interval. Hence in the limit $T \to \infty$, $f_T(t)$ and $f(t)$ are identical. That is,

$$\lim_{T \to \infty} f_T(t) = f(t)$$

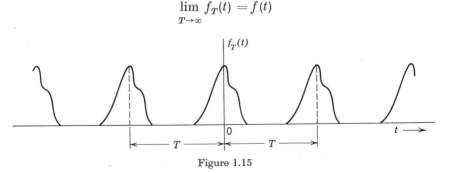

Figure 1.15

Thus the Fourier series representing $f_T(t)$ over the entire interval will also represent $f(t)$ over the entire interval if we let $T = \infty$ in this series.

The exponential Fourier series for $f_T(t)$ can be represented as

$$f_T(t) = \sum_{n=-\infty}^{\infty} F_n e^{jn\omega_0 t}$$

where

$$\omega_0 = \frac{2\pi}{T}$$

and

$$F_n = \frac{1}{T} \int_{-T/2}^{T/2} f_T(t) e^{-jn\omega_0 t}\, dt \tag{1.67}$$

The term F_n represents the amplitude of the component of frequency $n\omega_0$. We shall now let T become very large. As T becomes larger, ω_0 (the fundamental frequency) becomes smaller and the spectrum becomes denser. As seen from Eq. 1.67, the amplitudes of individual components become smaller too. The shape of the frequency spectrum, however, is unaltered (we have already observed this behavior for the case of periodic gate functions as shown in Fig. 1.13). In the limit when $T = \infty$, the magnitude of each component becomes infinitesimally small, but now there are also an infinite number of frequency components. The spectrum exists for every value of ω, and it is no longer a discrete but a continuous function of ω. To illustrate this point, let us make a slight change in notation here. Let

$$n\omega_0 = \omega_n \tag{1.68}$$

Then F_n is a function of ω_n, and we shall denote F_n by $F_n(\omega_n)$. Further, let

$$TF_n(\omega_n) = F(\omega_n) \tag{1.69}$$

Then

$$f_T(t) = \frac{1}{T} \sum_{n=-\infty}^{\infty} F(\omega_n) e^{j\omega_n t} \tag{1.70}$$

and, from Eqs. 1.67 and 1.69, we have

$$F(\omega_n) = TF_n = \int_{-T/2}^{T/2} f_T(t) e^{-j\omega_n t}\, dt \tag{1.71}$$

Substituting the value $T = 2\pi/\omega_0$ in Eq. 1.70, we get

$$f_T(t) = \frac{1}{2\pi} \sum_{n=-\infty}^{\infty} F(\omega_n) e^{j\omega_n t} \omega_0 \tag{1.72}$$

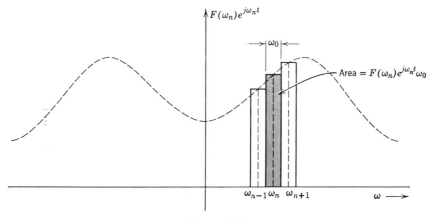

Figure 1.16

Equation 1.72 shows that $f_T(t)$ can be expressed as a sum of exponential signals of frequencies ω_1, ω_2, ω_3, . . . , ω_n, . . . , etc. The amplitude of the component of frequency ω_n is $F(\omega_n)\omega_0/2\pi$ (this is equal to F_n). We note, therefore, that the amplitude content in $f_T(t)$ of frequency ω_n is not $F(\omega_n)$ but is proportional to $F(\omega_n)$.

Let us try a graphical interpretation of Eq. 1.72, which represents a discrete sum or a sum of discrete frequency components. Actually, the quantity $F(\omega_n)e^{j\omega_n t}$ is complex in general, and a strict graphical representation will need two plots (real and imaginary plots or magnitude and phase plots). However, we shall assume that the quantity $F(\omega_n)e^{j\omega_n t}$ is real. This will sufficiently indicate our line of reasoning. Figure 1.16 shows such a plot of this quantity as a function of ω. This function exists only at discrete values of ω; that is, at $\omega = \omega_1$, ω_2, . . . , ω_n, etc., where $\omega_n = n\omega_0$.

Each frequency component is separated by distance ω_0. Therefore the area of the shaded rectangle in Fig. 1.16 is evidently $F(\omega_n)e^{jn\omega_n t}\omega_0$. Equation 1.72 represents the sum of areas under all such rectangles corresponding to $n = -\infty$ to ∞. The sum of rectangular areas represents approximately the area under the dotted curve. The approximation becomes better as ω_0 becomes smaller. In the limit when $T \to \infty$, ω_0 becomes infinitesimally small and may be represented by $d\omega$. The discrete sum in Eq. 1.72 becomes the integral or the area under this curve. The curve now is a continuous function of ω and is given by $F(\omega)e^{j\omega t}$. Also as $T \to \infty$, the function $f_T(t) \to f(t)$ and Eqs. 1.70 and

1.71 become

$$f(t) = \frac{1}{2\pi} \int_{-\infty}^{\infty} F(\omega)e^{j\omega t}\, d\omega \tag{1.73}$$

where

$$F(\omega) = \int_{-\infty}^{\infty} f(t)e^{-j\omega t}\, dt \tag{1.74}$$

We have thus succeeded in representing a nonperiodic function $f(t)$ in terms of exponential functions over an entire interval $(-\infty < t < \infty)$. Equation 1.73 represents $f(t)$ as a continuous sum of exponential functions with frequencies lying in the interval $(-\infty < \omega < \infty)$. The amplitude of the components of any frequency ω is proportional to $F(\omega)$. Therefore $F(\omega)$ represents the frequency spectrum of $f(t)$ and is called the *spectral-density function*. Note, however, that the frequency spectrum now is continuous and exists at all values of ω. The spectral-density function $F(\omega)$ can be evaluated from Eq. 1.74.

Equations 1.73 and 1.74 are usually referred to as the Fourier transform pair. Equation 1.74 is known as the direct Fourier transform of $f(t)$, and Eq. 1.73 is known as the inverse Fourier transform of $F(\omega)$. Symbolically, these transforms are also written as

$$F(\omega) = \mathscr{F}[f(t)] \quad \text{and} \quad f(t) = \mathscr{F}^{-1}[F(\omega)] \tag{1.75}$$

Thus $F(\omega)$ is the direct Fourier transform of $f(t)$, $f(t)$ is the inverse Fourier transform of $F(\omega)$, and

$$\mathscr{F}[f(t)] = \int_{-\infty}^{\infty} f(t)e^{-j\omega t}\, dt$$

$$\mathscr{F}^{-1}[F(\omega)] = \frac{1}{2\pi} \int_{-\infty}^{\infty} F(\omega)e^{j\omega t}\, d\omega$$

1.6 SOME REMARKS ABOUT THE CONTINUOUS SPECTRUM FUNCTION

We have expressed a nonperiodic function $f(t)$ as a continuous sum of exponential functions with frequencies in the interval $-\infty$ to ∞. The amplitude of a component of any frequency ω is infinitesimal, but it is proportional to $F(\omega)$ where $F(\omega)$ is the spectral density.

The concept of a continuous spectrum is sometimes confusing because we generally picture the spectrum as existing at discrete frequencies and

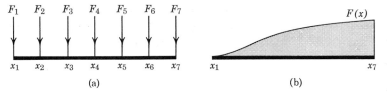

Figure 1.17   (a) A beam loaded at discrete points. (b) A continuously loaded beam.

with finite amplitudes in the manner of a periodic function. The continuous spectrum concept can be appreciated by considering the analogous concrete phenomenon. One familiar example of a continuous distribution is the loading of a beam. Consider a beam loaded with weights $F_1, F_2, F_3, \ldots, F_7$ units at uniformly spaced points $x_1, x_2, \ldots,$ x_7, as shown in Fig. 1.17a. The beam is loaded at 7 discrete points, and the total weight on the beam is given by the sum of these loads at 7 discrete points.

$$W_T = \sum_{r=1}^{7} F_r$$

Next consider the case of a continuously loaded beam as shown in Fig. 1.17b. The loading density is a function of x, and let it be given by $F(x)$ kg per meter. The total weight on the beam is now given by a continuous sum of the weights, that is, the integral of $F(x)$ over the entire length.

$$W_T = \int_{x_1}^{x_7} F(x)\, dx \quad \text{kg}$$

In the case of discrete loading, the weight existed only at discrete points. At other points there was no loading. On the other hand, in the continuously distributed case the loading exists at every point, but at any one point the loading is zero. However, the loading in a small distance dx is given by $F(x)\, dx$. Therefore $F(x)$ represents the relative magnitude of loading at a point x. An exactly analogous situation exists in the case of signals and their frequency spectrum. A periodic signal can be represented by a sum of discrete exponentials with finite amplitudes.

$$f(t) = \sum_{n=-\infty}^{\infty} F_n e^{j\omega_n t} \qquad (\omega_n = n\omega_0)$$

For a nonperiodic function, the distribution of exponentials becomes continuous; that is, the spectrum function exists at every value of ω. At any one frequency ω, the amplitude of that frequency component is zero. The total contribution in an infinitesimal interval $d\omega$ is given by $(1/2\pi)F(\omega)\,d\omega$, and the function $f(t)$ can be expressed in terms of the continuous sum of such infinitesimal components.

$$f(t) = \frac{1}{2\pi}\int_{-\infty}^{\infty} F(\omega)e^{j\omega t}\,d\omega \tag{1.76}$$

The factor 2π in Eq. 1.76 can be removed if the integration is performed with respect to variable f instead of ω. We have

$$\omega = 2\pi f$$

and

$$d\omega = 2\pi\,df$$

Equation 1.76 may now be expressed as

$$f(t) = \int_{-\infty}^{\infty} F(2\pi f)e^{j2\pi f t}\,df$$

1.7 TIME-DOMAIN AND FREQUENCY-DOMAIN REPRESENTATION OF A SIGNAL

The Fourier transform is a tool that resolves a given signal into its exponential components. The function $F(\omega)$ is the direct Fourier transform of $f(t)$ and represents relative amplitudes of various frequency components. Therefore $F(\omega)$ is the frequency-domain representation of $f(t)$. Time-domain representation specifies a function at each instant of time, whereas frequency-domain representation specifies the relative amplitudes of the frequency components of the function. Either representation uniquely specifies the function. However, the function $F(\omega)$ is complex in general and needs two plots for its graphical representation.

$$F(\omega) = |F(\omega)|\,e^{j\theta(\omega)}$$

Thus $F(\omega)$ may be represented by a magnitude plot $|F(\omega)|$ and a phase plot $\theta(\omega)$. In many cases, however, $F(\omega)$ is either real or imaginary and only one plot is necessary. $F(\omega)$, however, is generally a complex function of ω, and we shall now show that for a real function $f(t)$,

$$F^*(\omega) = F(-\omega)$$

We have

$$F(\omega) = \int_{-\infty}^{\infty} f(t)e^{-j\omega t}\,dt \tag{1.77a}$$

Similarly,

$$F(-\omega) = \int_{-\infty}^{\infty} f(t)e^{j\omega t}\, dt \qquad (1.77b)$$

From Eqs. 1.77a and 1.77b, it follows that if $f(t)$ is a real function of t, then

$$F^*(\omega) = F(-\omega)$$

Thus if

$$(1.78)$$

$$F(\omega) = |F(\omega)|\, e^{j\theta(\omega)}$$

then

$$F(-\omega) = |F(\omega)|\, e^{-j\theta(\omega)} \qquad (1.79)$$

It is evident from these equations that the magnitude spectrum $F(\omega)$ is an even function of ω, and the phase spectrum $\theta(\omega)$ is an odd function of ω.

1.8 EXISTENCE OF THE FOURIER TRANSFORM

From Eq. 1.74, defining the Fourier transform, it is evident that if $\int_{-\infty}^{\infty} f(t)e^{-j\omega t}\, dt$ is finite, then the Fourier transform exists. But since the magnitude of $e^{-j\omega t}$ is unity, a sufficient condition for the existence of a Fourier transform of a function $f(t)$ is that

$$\int_{-\infty}^{\infty} |f(t)|\, dt$$

must be finite. If, however, singularity functions (for example, impulse functions) are allowed, then this condition of absolute integrability is not always necessary. We shall see later that there are functions that are not absolutely integrable, but that do have transforms. Absolute integrability of $f(t)$ is thus a sufficient but not a necessary condition for the existence of the Fourier transform of $f(t)$.

Functions such as sin ωt, cos ωt, $u(t)$, etc., do not satisfy the above condition and, strictly speaking, do not possess the Fourier transform. These functions, however, do have Fourier transforms in the limit.

A function sin ωt may be assumed to exist only in the interval $-T/2 < t < T/2$. Under these conditions the function has a Fourier transform as long as T is finite. In the limit we make T very large but finite. It will be shown later that the Fourier transforms for such functions do exist in the limit.

I.9 FOURIER TRANSFORMS OF SOME USEFUL FUNCTIONS

I. Single-sided Exponential Signal $e^{-at}u(t)$

$$f(t) = e^{-at}u(t)$$

$$F(\omega) = \int_{-\infty}^{\infty} e^{-at}u(t)e^{-j\omega t}\, dt$$

$$= \int_{0}^{\infty} e^{-(a+j\omega)t}\, dt \qquad\qquad (1.80)$$

$$= \frac{1}{a + j\omega} \qquad\qquad (1.81)$$

$$= \frac{1}{\sqrt{a^2 + \omega^2}}\, e^{-j\tan^{-1}(\omega/a)}$$

Here

$$|F(\omega)| = \frac{1}{\sqrt{a^2 + \omega^2}} \quad\text{and}\quad \theta(\omega) = -\tan^{-1}\left(\frac{\omega}{a}\right)$$

The magnitude spectrum $|F(\omega)|$ and the phase spectrum $\theta(\omega)$ are shown in Fig. 1.18b.

Note that the integral Eq. 1.80 converges only for $a > 0$. For $a < 0$, the Fourier transform does not exist. This also follows from the fact that for $a < 0, f(t)$ is not absolutely integrable.

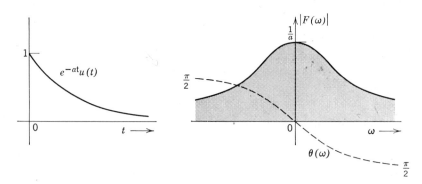

Figure 1.18   The time domain and the frequency domain representations of $e^{-at}u(t)$.

2. Double-sided Exponential Signal $e^{-a|t|}$

$$f(t) = e^{-a|t|}$$

$$F(\omega) = \int_{-\infty}^{\infty} e^{-a|t|} e^{-j\omega t}\, dt$$

$$= \int_{-\infty}^{0} e^{(a-j\omega)t}\, dt + \int_{\infty}^{0} e^{-(a+j\omega)t}\, dt$$

$$= \frac{2a}{a^2 + \omega^2} \tag{1.82}$$

Note that in this case the phase spectrum $\theta(\omega) = 0$. The magnitude spectrum $2a/(a^2 + \omega^2)$ is shown in Fig. 1.19.

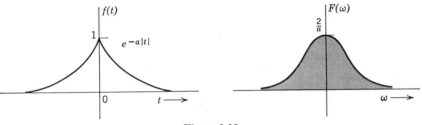

Figure 1.19

3. A Gate Function

A gate function $G_\tau(t)$ is a rectangular pulse, as shown in Fig. 1.20a, and is defined by

$$G_\tau(t) = \begin{cases} 1 & |t| < \tau/2 \\ 0 & |t| > \tau/2 \end{cases}$$

The Fourier transform of this function is given by

$$F(\omega) = \int_{-\tau/2}^{\tau/2} A e^{-j\omega t}\, dt$$

$$= \frac{A}{j\omega} \left(e^{j\omega\tau/2} - e^{-j\omega\tau/2} \right)$$

$$= A\tau \frac{\sin(\omega\tau/2)}{\omega\tau/2}$$

$$= A\tau Sa\left(\frac{\omega\tau}{2}\right) \tag{1.83}$$

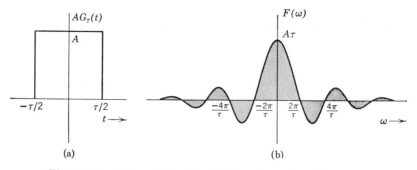

Figure 1.20   (a) A gate function. (b) Transform of a gate function.

Note that $F(\omega)$ is a real function and hence can be represented graphically by a single curve (Fig. 1.20b).

1.10 SINGULARITY FUNCTIONS

Consider a unit step voltage applied to a capacitor as shown in Fig. 1.21a. The current i through the capacitor is given by

$$i = C\frac{dv}{dt}$$

It is easy to see that dv/dt is zero for all values of t except at $t = 0$ where it is undefined. The derivative at $t = 0$ does not exist because the function $u(t)$ is discontinuous at this point. We are in a serious mathematical difficulty arising from the idealization of the source as well as the circuit element. If either the source or the capacitor were nonideal, the solution would exist. If, for example, the source voltage were as shown in Fig. 1.22a instead of that shown in Fig. 1.21a, the current through the capacitor would be a pulse of current as shown in Fig. 1.22b. The solution to an ideal unit step voltage does not exist, but it is possible to obtain a solution in the limit by assuming an unideal source as shown in Fig. 1.22a and then letting a go to zero in the limit.

Let us designate the unidealized voltage function by $u_a(t)$. In the limit when a goes to zero, this voltage function becomes a unit step function. The derivative of the function $u_a(t)$ is a rectangular pulse of height $1/a$ and width a. As a varies, the pulse shape varies, but the area of the pulse remains constant. Figure 1.23a shows the sequence of such

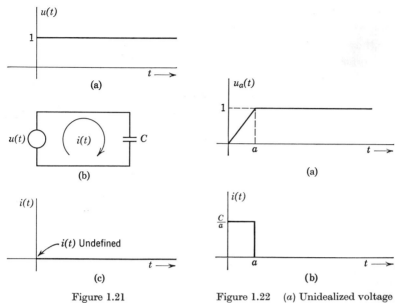

(a)

(b)

(c)

Figure 1.21

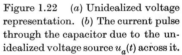

(a)

(b)

Figure 1.22 (a) Unidealized voltage representation. (b) The current pulse through the capacitor due to the un-idealized voltage source $u_a(t)$ across it.

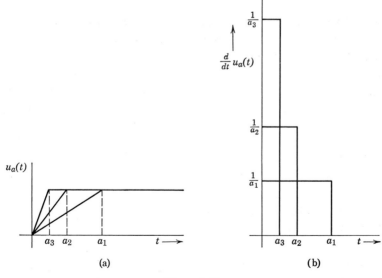

(a)

(b)

Figure 1.23

47

pulses as a varies. In the limit when a goes to zero, the height of the pulse goes to infinity and the width of the pulse is zero. The area of the pulse, however, remains unity. We define the unit impulse function as the derivative of a unit step function. Since the derivative of an idealized unit step function does not exist, we define the unit impulse function as the limit of the sequence of the derivatives of an unidealized function $u_a(t)$ as a goes to zero. The unit impulse function is denoted by $\delta(t)$. We therefore have

$$\delta(t) = \lim_{a \to 0} \frac{d}{dt} [u_a(t)]$$

$$= \lim_{a \to 0} \frac{1}{a} [u(t) - u(t - a)]$$

In the limit as $a \to 0$, the function $\delta(t)$ assumes the form of a pulse of infinite height and zero width. The area under the pulse, however, remains unity. The function $\delta(t)$ is zero everywhere except at $t = a$.

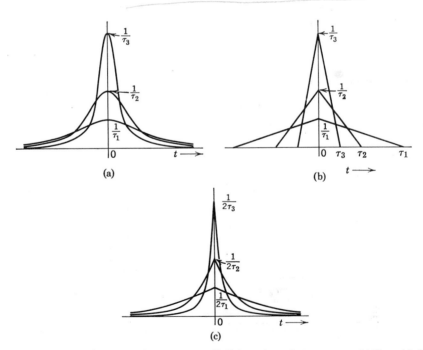

(a)

(b)

(c)

Figure 1.24   (a) Gaussian pulse sequence. (b) Triangular pulse sequence. (c) Two-sided exponential pulse sequence.

This fact can be expressed by the definition of impulse function as given by P. A. M. Dirac:

$$\left.\begin{aligned}\int_{-\infty}^{\infty} \delta(t)\, dt &= 1 \\[2mm] \delta(t) &= 0 \qquad t \neq 0\end{aligned}\right\} \tag{1.84}$$

and

The impulse function as seen from its definition, Eq. 1.84, is obviously not a true function in a mathematical sense, where a function is defined at every value of t. The impulse function, however, has been rigorously justified by the distribution theory of Schwartz.* A version of this theory (generalized functions) was given by Temple. In this approach, the impulse function is defined as a sequence of regular functions, and all operations on impulse function are viewed as operations on the sequence. One may define an impulse function accordingly by a number of sequences as shown in Fig. 1.24.

Figure 1.25 shows the impulse function as a limit of the sequence of sampling functions. All these sequences satisfy Eq. 1.84.

I. Gaussian Pulse.

$$\delta(t) = \lim_{\tau \to 0} \frac{1}{\tau} e^{-\pi t^2/\tau^2}$$

2. Triangular Pulse.

$$\delta(t) = \begin{cases} \lim_{\tau \to 0} \dfrac{1}{\tau}\left[1 - \dfrac{|t|}{\tau}\right] & |t| < \tau \\[4mm] 0 & |t| > \tau \end{cases}$$

3. Exponential Pulse.

$$\delta(t) = \lim_{\tau \to 0} \frac{1}{2\tau} e^{-|t|/\tau}$$

4. Sampling Function. It can be shown that

$$\int_{-\infty}^{\infty} \frac{k}{\pi} Sa(kt)\, dt = 1 \tag{1.85}$$

As k is made larger and larger, the amplitude of the function $(k/\pi)Sa(kt)$ becomes larger, and the function oscillates faster and decays very rapidly away from the origin (Fig. 1.25). In the limit as $k \to \infty$,

* M. J. Lighthill, *Fourier Analysis and Generalized Functions*, Cambridge University Press, 1959.

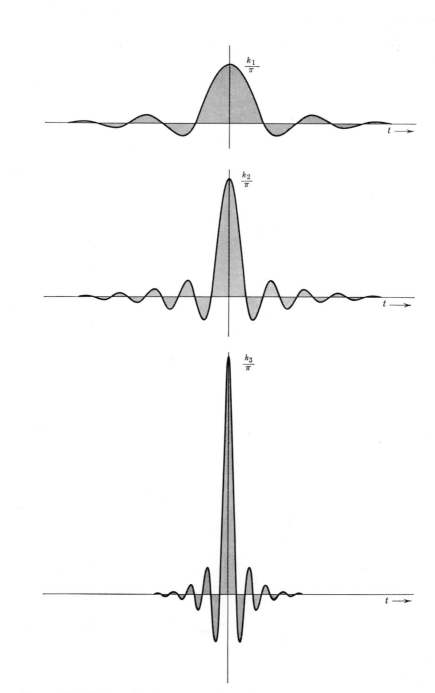

Figure 1.25 The impulse function as a limit of the sequence of a sampling function.

the function exists only at the origin, and the net area under the curve is still unity, as seen from Eq. 1.85. Hence

$$\delta(t) = \lim_{k \to \infty} \frac{k}{\pi} Sa(kt) \tag{1.86}$$

5. Sampling Square Function.

$$\delta(t) = \lim_{k \to \infty} \frac{k}{\pi} Sa^2(kt) \tag{1.87}$$

This follows from the arguments similar to those employed for the sampling function and the fact that

$$\int_{-\infty}^{\infty} \frac{k}{\pi} Sa^2(kx) \, dx = 1 \tag{1.88}$$

Returning to the impulse function $\delta(t)$, we observe that the area is concentrated at the origin $t = 0$. Hence we may write

$$\int_{-\infty}^{\infty} \delta(t) \, dt = \int_{0^-}^{0^+} \delta(t) \, dt = 1 \tag{1.89}$$

where 0^+ and 0^- denote arbitrarily small values of t approached from the right and the left side, respectively, of the origin. Also since $\delta(t) = 0$ everywhere except $t = 0$,

$$\int_{-\infty}^{\infty} f(t) \, \delta(t) \, dt = f(0) \int_{-\infty}^{\infty} \delta(t) \, dt = f(0) \tag{1.90a}$$

It also follows that

$$\int_{-\infty}^{\infty} f(t) \, \delta(t - t_0) \, dt = f(t_0) \tag{1.90b}$$

Equations 1.90 represent the sampling property (or sifting property) of the impulse function.*

It should be stressed again that the impulse function is not a true function in the usual mathematical sense where a function is defined for every value of t. Nevertheless, its formal use leads to results that

---

* We have shown here that Eqs. 1.90 follow from the definition of the impulse function in Eq. 1.84, Actually, in the rigorous approach, impulse function is defined by Eq. 1.90. This way we define the impulse function by its integral properties rather than as a function of time defined for every t. It can be shown that the definition (1.84) does not specify a unique function. See, for instance, A. Paponlis, *Fourier Integral and Its Applications*, McGraw-Hill, New York, 1962.

can be interpreted physically. The use of the impulse function is very common in the physical sciences and engineering to represent such entities as point masses, point charges, point sources, concentrated forces, line sources, and surface charges. Actually, such distributions do not exist in practice, but this idealization simplifies the derivation of results. Moreover, the measuring equipment now used, due to its finite resolution, cannot distinguish between the response of the idealized impulse function and a pulse of small but finite width.

The step function $u(t)$, the impulse function $\delta(t)$, and its higher derivatives are all known as singularity functions. The term singularity function is applied to all such functions which are discontinuous or have discontinuous derivatives. These functions may have continuous derivatives only of finite order. Thus a parabolic function $f(t) = at^2u(t)$ has a continuous derivative only up to first order, $[f'(t) = 2atu(t)]$. The second derivative is discontinuous. Hence it is a singularity function. Similarly, a ramp function and unit step function (also impulse function and its derivatives) are singularity functions. In fact any function represented by a polynomial in t is a singularity function.

1.11 FOURIER TRANSFORMS INVOLVING IMPULSE FUNCTIONS

1. The Fourier Transform of an Impulse Function

The Fourier transform of a unit impulse function $\delta(t)$ is given by

$$\mathscr{F}[\delta(t)] = \int_{-\infty}^{\infty} \delta(t)e^{-j\omega t}\, dt \qquad (1.91)$$

From the sampling property of an impulse function expressed in Eq. 1.90a, it is evident that the integral on the right-hand side of Eq. 1.91 is unity. Hence

$$\mathscr{F}[\delta(t)] = 1 \qquad (1.92)$$

Thus the Fourier transform of a unit impulse function is unity.

It is therefore evident that an impulse function has a uniform spectral density over the entire frequency interval. In other words, an impulse function contains all frequency components with the same relative amplitudes.

2. The Fourier Transform of a Constant

Let us now find the Fourier transform of a function:

$$f(t) = A$$

This function does not satisfy the condition of absolute integrability. Nevertheless, it has a Fourier transform in the limit. We shall consider the Fourier transform of a gate function of height A and width τ seconds

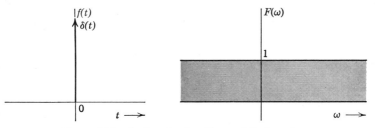

Figure 1.26   An impulse function and its transform.

as shown in Fig. 1.20a. In the limit as $\tau \to \infty$, the gate function tends to be a constant function A. The Fourier transform of a constant A is therefore the Fourier transform of a gate function $G_\tau(t)$ as $\tau \to \infty$. The Fourier transform of $AG_\tau(t)$ was found to be $A\tau Sa(\omega\tau/2)$. Hence

$$\mathscr{F}[A] = \lim_{\tau \to \infty} A\tau Sa \frac{\omega\tau}{2}$$

$$= 2\pi A \lim_{\tau \to \infty} \frac{\tau}{2\pi} Sa \frac{\omega\tau}{2}$$

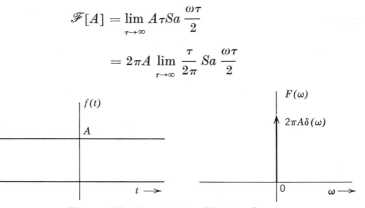

Figure 1.27   A constant and its transform.

From Eq. 1.86, it follows that the limit of the sampling function above defines an impulse function $\delta(\omega)$. Hence

$$\mathscr{F}[A] = 2\pi A \ \delta(\omega) \qquad\qquad (1.93a)$$

and

$$\mathscr{F}[1] = 2\pi \, \delta(\omega) \tag{1.93b}$$

Thus, when $f(t)$ equals a constant, it contains only a frequency component of $\omega = 0$. This is the logical result since a constant function is a d-c signal ($\omega = 0$), and it does not have any other frequency components.

3. Transform of sgn (t)

The signum function abbreviated as sgn (t) is defined by

$$\operatorname{sgn}\,(t) = \begin{cases} 1 & t > 0 \\ -1 & t < 0 \end{cases} \tag{1.94a}$$

It can be seen easily that

$$\operatorname{sgn}\,(t) = 2u(t) - 1 \tag{1.94b}$$

The Fourier transform of sgn (t) can be easily obtained when we observe that

$$\operatorname{sgn}\,(t) = \lim_{a \to 0} \ [e^{-at}u(t) - e^{at}u(-t)]$$

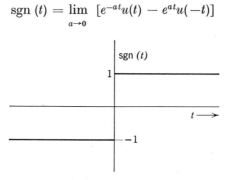

Figure 1.28

Hence

$$\mathscr{F}[\operatorname{sgn}\,(t)] = \lim_{a \to 0} \left[\int_0^{\infty} e^{-at}e^{-j\omega t}\,dt - \int_{-\infty}^{0} e^{at}e^{-j\omega t}\,dt \right]$$

$$= \lim_{a \to 0} \left[\frac{-2j\omega}{a^2 + \omega^2} \right]$$

$$= \frac{2}{j\omega} \tag{1.95}$$

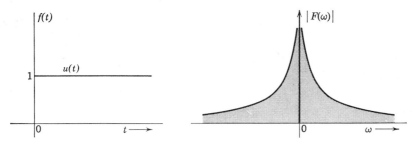

Figure 1.29   The unit step function and its spectral density function.

4. Transform of Unit Step Function $u(t)$

From Eq. 1.94b it follows that

$$u(t) = \tfrac{1}{2}[1 + \text{sgn }(t)]$$

Hence

$$\mathscr{F}[u(t)] = \tfrac{1}{2}\{\mathscr{F}[1] + \mathscr{F}[\text{sgn }(t)]\}$$

Using Eqs. 1.93b and 1.95, we obtain

$$\mathscr{F}[u(t)] = \pi \, \delta(\omega) + \frac{1}{j\omega} \tag{1.96}$$

The spectral density function contains an impulse at $\omega = 0$ (Fig. 1.29). Thus the function $u(t)$ contains a large d-c component as expected. In addition, it has other frequency components. The function $u(t)$ appears to be a pure d-c signal, and hence frequency components other than $\omega = 0$ may appear rather strange. However, the function $u(t)$ is not a true d-c signal. It is zero for $t < 0$, and there is an abrupt discontinuity at $t = 0$, giving rise to other frequency components. For a true d-c signal, $f(t)$ is constant for the entire interval $(-\infty, \infty)$. We have already seen (Eq. 1.93) that such a signal indeed has no frequency components other than d-c ($\omega = 0$).

5. Eternal Sinusoidal Signals $\cos \omega_0 t$ and $\sin \omega_0 t$

We shall now consider the sinusoidal signals $\cos \omega_0 t$ and $\sin \omega_0 t$ in the entire interval $(-\infty, \infty)$. These signals do not satisfy the condition of absolute integrability; yet their Fourier transforms exist and can be found by a limiting process analogous to that used for a constant function $f(t) = A$. We shall first assume these functions to exist in the interval $-\tau/2$ to $\tau/2$ and zero outside this interval. In the limit,

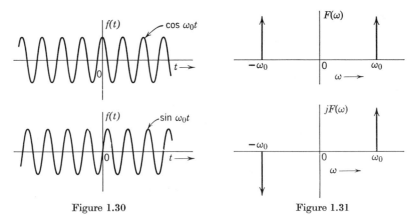

Figure 1.30   Figure 1.31

τ will be made infinity. The procedure is now demonstrated.

$$\mathscr{F}(\cos \omega_0 t) = \lim_{\tau \to \infty} \int_{-\tau/2}^{\tau/2} \cos \omega_0 t e^{-j\omega t}\, dt$$

$$= \lim_{\tau \to \infty} \frac{\tau}{2} \left\{ \frac{\sin\left[(\omega - \omega_0)\tau/2\right]}{(\omega - \omega_0)\tau/2} + \frac{\sin\left[(\omega + \omega_0)\tau/2\right]}{(\omega + \omega_0)\tau/2} \right\}$$

$$= \lim_{\tau \to \infty} \left\{ \frac{\tau}{2}\, Sa\left[\frac{\tau(\omega - \omega_0)}{2}\right] + \frac{\tau}{2}\, Sa\left[\frac{\tau(\omega + \omega_0)}{2}\right] \right\} \quad (1.97)$$

In the limit the sampling function becomes an impulse function according to Eq. 1.86, and we have

$$\mathscr{F}(\cos \omega_0 t) = \pi[\delta(\omega - \omega_0) + \delta(\omega + \omega_0)] \quad (1.98)$$

Similarly, it can be shown that

$$\mathscr{F}(\sin \omega_0 t) = j\pi[\delta(\omega + \omega_0) - \delta(\omega - \omega_0)] \quad (1.99)$$

Therefore the Fourier spectrum for these functions consists of two impulses at ω_0 and $-\omega_0$, respectively. It is interesting to see how the spectrum behaves in the limiting process as τ is made infinite. For a finite τ, the spectral-density function is given by Eq. 1.97. This spectral-density function is plotted in Fig. 1.32 for the case $\tau = 16\pi/\omega_0$. In other words, it represents the spectral-density function of a signal $\cos \omega_0 t$ which is trucated beyond 8 cycles.

$$f(t) = \begin{cases} \cos \omega_0 t & |t| < \dfrac{16\pi}{\omega_0} \\[2ex] 0 & |t| > \dfrac{16\pi}{\omega_0} \end{cases}$$

Note that there is a large concentration of energy at frequencies near $\pm\omega_0$. As we increase the interval τ, the spectral density concentrates more and more around frequencies $\pm\omega_0$. In the limit as $\tau \to \infty$, the spectral density is zero everywhere except at frequencies $\pm\omega_0$ where it is infinite in such a way that the area under the curve at each of these frequencies is π. In the limit the distribution therefore becomes two impulses of strength, π units each, located at frequencies $\pm\omega_0$ as shown in Fig. 1.30. It is evident that the spectral density functions for $\cos \omega_0 t$ and $\sin \omega_0 t$ exist only at $\omega = \omega_0$. This is quite logical since these signals do not contain components of any other frequency. On the other hand, functions $\cos \omega_0 t\, u(t)$ and $\sin \omega_0 t\, u(t)$ do contain components of frequencies other than ω_0. It can be shown that (see Eq. 1.116):

$$\mathscr{F}[(\cos \omega_0 t)u(t)] = \frac{\pi}{2}\left[\delta(\omega - \omega_0) + \delta(\omega + \omega_0)\right] + \frac{j\omega}{\omega_0{}^2 - \omega^2} \quad (1.100a)$$

and

$$\mathscr{F}[(\sin \omega_0 t)u(t)] = \frac{\pi}{2j}\left[\delta(\omega - \omega_0) - \delta(\omega + \omega_0)\right] + \frac{\omega_0}{\omega_0{}^2 - \omega^2} \quad (1.100b)$$

These functions evidently contain a large component of frequency ω_0, but they also contain other frequency components.

Apparently, the signals $\cos \omega_0 t\, u(t)$ and $\sin \omega_0 t\, u(t)$ are pure sinusoidal signals, and it may seem rather strange that these functions should contain components of frequencies other than ω_0. We must remember,

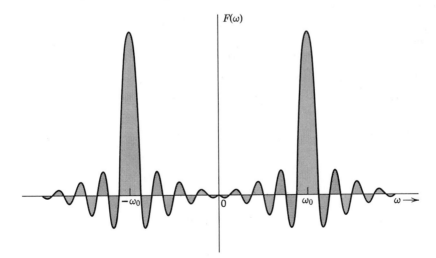

Figure 1.32   The spectral density function of 8 cycles of $\cos \omega_0 t$.

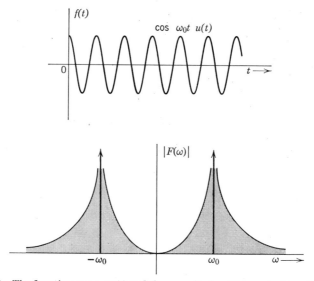

Figure 1.33   The function $\cos \omega_0 t\, u(t)$ and the magnitude of its spectral density function.

however, that we are expressing a function in terms of eternal exponential (or eternal sinusoidal) functions existing from $-\infty$ to ∞. The functions $\cos \omega_0 t\, u(t)$ and $\sin \omega_0 t\, u(t)$ are not eternal sinusoidal signals. These functions are zero for values of $t < 0$ and exist only for positive values of t. Hence in addition to ω_0, they also contain other frequency components. All these eternal frequency components in the spectrum of these functions (Fig. 1.33) add in such a way as to yield zero value for $t < 0$ and $\cos \omega_0 t$ (or $\sin \omega_0 t$) for $t > 0$. If the sinusoidal signals are eternal ($\cos \omega_0 t$ and $\sin \omega_0 t$ in the entire interval $-\infty$ to ∞), then we have already shown in Eqs. 1.98 and 1.99 that they indeed contain components of frequency ω_0 only.

6. Transform of an Eternal Exponential $e^{j\omega_0 t}$

We shall now find the Fourier transform of an eternal exponential signal $e^{j\omega_0 t}(-\infty < t < \infty)$. We have

$$e^{j\omega_0 t} = \cos \omega_0 t + j \sin \omega_0 t$$

Hence

$$\mathscr{F}[e^{j\omega_0 t}] = \mathscr{F}[\cos \omega_0 t + j \sin \omega_0 t]$$

Substituting Eqs. 1.98 and 1.99 in the preceding equation, we get

$$\mathscr{F}[e^{j\omega t}] = \pi[\delta(\omega - \omega_0) + \delta(\omega + \omega_0) - \delta(\omega + \omega_0) + \delta(\omega - \omega_0)]$$

$$= 2\pi\delta(\omega - \omega_0) \tag{1.101}$$

The Fourier transform of $e^{j\omega_0 t}$ is therefore a single impulse of strength 2π at $\omega = \omega_0$. Note that the signal $e^{j\omega_0 t}$ is not a real function of time, and hence it has a spectrum which exists at $\omega = \omega_0$ alone. We have shown previously that for any real function of time, the spectral density function $F(\omega)$ satisfies (see Eq. 1.78):

$$F^*(\omega) = F(-\omega)$$

and

$$|F(\omega)| = |F(-\omega)|$$

Hence for any real function of time the magnitude spectrum is an even function of ω, and if there is an impulse at $\omega = \omega_0$, there must exist an impulse at $\omega = -\omega_0$. This is the case for signals $\sin \omega_0 t$ and $\cos \omega_0 t$.

7. The Fourier Transform of a Periodic Function

We have developed a Fourier transform as a limiting case of the Fourier series by letting the period of a periodic function become infinite. We shall now proceed in the opposite direction and show that the Fourier series is just a limiting case of the Fourier transform. This point of view is very useful, since it permits a unified treatment of both the periodic and the nonperiodic functions.

Strictly speaking, the Fourier transform of a periodic function does not exist, since it fails to satisfy the condition of absolute integrability. For any periodic function $f(t)$:

$$\int_{\infty}^{-\infty} |f(t)| \, dt = \infty$$

But the transform does exist in the limit. We have already found the Fourier transform of $\cos \omega_0 t$, $\sin \omega_0 t$ in the limit. We use here exactly the same procedure by assuming that the periodic function exists only in a finite interval $(-\tau/2, \tau/2)$ and letting τ become infinite in the limit.

Alternatively, we may express a periodic function by its Fourier series. The Fourier transform of a periodic function is then the sum of Fourier transforms of its individual components. We can express a periodic function $f(t)$ with period T as

$$f(t) = \sum_{n=-\infty}^{\infty} F_n e^{jn\omega_0 t} \qquad \left(\omega_0 = \frac{2\pi}{T} \right)$$

Taking the Fourier transforms of both sides, we have

$$\mathscr{F}[f(t)] = \mathscr{F} \sum_{n=-\infty}^{\infty} F_n e^{jn\omega_0 t}$$

$$= \sum_{n=-\infty}^{\infty} F_n \mathscr{F}(e^{jn\omega_0 t})$$

Substituting the transform of $e^{j\omega_0 t}$ from Eq. 1.101, we get

$$\mathscr{F}[f(t)] = 2\pi \sum_{n=-\infty}^{\infty} F_n \, \delta(\omega - n\omega_0) \tag{1.102}$$

This is a significant result. Relation 1.102 states that *the spectral density function or the Fourier transform of a periodic signal consists of impulses located at the harmonic frequencies of the signal and that the strength of each impulse is the same as 2π times the value of the corresponding coefficient in the exponential Fourier series.* The sequence of equidistant impulses is just a limiting form of continuous density function. The result should, of course, be no surprise, since we know that a periodic function contains components only of discrete harmonic frequencies.

Example 1.7

Find the Fourier transform of a periodic gate function (rectangular pulse of width τ seconds and repeating every T seconds). The Fourier series for this function is given by (Eq. 1.66b):

$$f(t) = \sum_{n=-\infty}^{\infty} F_n e^{jn\omega_0 t}$$

where

$$F_n = \frac{A\tau}{T} \, Sa\!\left(\frac{n\pi\tau}{T}\right)$$

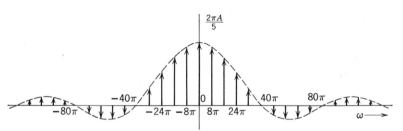

Figure 1.34   The spectral density function of a periodic gate pulse.

From Eq. 1.102, it follows that the Fourier transform of this function is given by

$$\mathscr{F}[f(t)] = \frac{2\pi A\tau}{T} \sum_{n=-\infty}^{\infty} Sa\left(\frac{n\pi\tau}{T}\right) \delta(\omega - n\omega_0) \qquad (1.103)$$

The transform of $f(t)$ therefore consists of impulses located at $\omega = 0, \pm\omega_0$, $\pm 2\omega_0, \ldots, \pm n\omega_0, \ldots$, etc. The magnitude of the impulse located at $\omega = n\omega_0$ is given by $2\pi(A\tau/T)Sa(n\pi\tau/T)$.

The spectrum for the case of $\tau = \frac{1}{20}$ second and $T = \frac{1}{4}$ second is shown in Fig. 1.34. Here $\omega_0 = 8\pi$.

Example 1.8

We shall find a Fourier transform of a sequence of equidistant impulses of unit strength and separated by T seconds as shown in Fig. 1.35. This

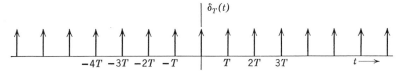

Figure 1.35 The sequence of a uniform equidistant impulse function.

function is very important in sampling theory, and hence it is convenient to denote it by a special symbol $\delta_T(t)$. Thus

$$\delta_T(t) = \delta(t) + \delta(t - T) + \delta(t - 2T) + \cdots + \delta(t - nT) + \cdots$$
$$+ \delta(t + T) + \delta(t + 2T) + \cdots + \delta(t + nT) + \cdots$$
$$= \sum_{n=-\infty}^{\infty} \delta(t - nT) \qquad (1.104)$$

This is obviously a periodic function with period T. We shall first find the Fourier series for this function.

$$\delta_T(t) = \sum_{n=-\infty}^{\infty} F_n e^{jn\omega_0 t}$$

where

$$F_n = \frac{1}{T} \int_{-T/2}^{T/2} \delta_T(t) e^{-jn\omega_0 t}\, dt$$

Function $\delta_T(t)$ in the interval $(-T/2, T/2)$ is simply $\delta(t)$. Hence

$$F_n = \frac{1}{T} \int_{-T/2}^{T/2} \delta(t) e^{-jn\omega_0 t}\, dt$$

From the sampling property of an impulse function as expressed in Eq. 1.90a the above equation reduces to

$$F_n = \frac{1}{T}$$

Consequently, F_n is a constant, $1/T$. It therefore follows that the impulse train function of period T contains components of frequencies $\omega = 0, \pm\omega_0, \pm 2\omega_0, \ldots, \pm n\omega_0, \ldots$, etc., ($\omega_0 = 2\pi/T$) in the same amount.

$$\delta_T(t) = \frac{1}{T} \sum_{n=-\infty}^{\infty} e^{jn\omega_0} \tag{1.105}$$

To find the Fourier transform of $\delta_T(t)$, we use Eq. 1.102. Since in this case $F_n = 1/T$, it is evident that

$$\mathscr{F}[\delta_T(t)] = 2\pi \sum_{n=-\infty}^{\infty} \frac{1}{T} \delta(\omega - n\omega_0)$$

$$= \frac{2\pi}{T} \sum_{n=-\infty}^{\infty} \delta(\omega - n\omega_0)$$

$$= \omega_0 \sum_{n=-\infty}^{\infty} \delta(\omega - n\omega_0)$$

$$= \omega_0 \delta_{\omega_0}(\omega) \tag{1.106}$$

Relation 1.106 is very significant. It states that the Fourier transform of a unit impulse train of period T is also a train of impulses of strength ω_0 and separated by ω_0 radians ($\omega_0 = 2\pi/T$). Therefore the impulse train function is its own transform. The sequence of impulses with periods $T = \frac{1}{2}$ and $T = 1$ second and their respective transforms are shown in Fig. 1.36. It is

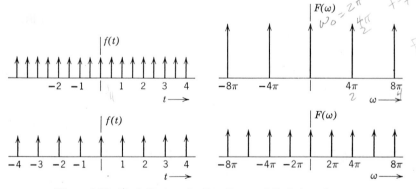

Figure 1.36   Periodic impulse functions and their transforms.

evident that as the periods of the impulses increase, the frequency spectrum becomes denser.

The various functions of time and their spectral-density functions are shown in Table 1.1A. Note that for a large number of signals in this table, $F(\omega)$ is real, and hence only one plot is necessary. Transforms of some functions are listed in Table 1.1B.

1.12 SOME PROPERTIES OF THE FOURIER TRANSFORM

The Fourier transform is a tool for expressing a function in terms of its exponential components of various frequencies. It has already been pointed out that the Fourier transform of a function is just another way of specifying the function. We therefore have two descriptions of the same function: the time-domain and frequency-domain descriptions. It is very illuminating to study the effect in one domain caused by certain operations over the function in the other domain. We may ask, for example: If a function is differentiated in the time domain, how is the spectrum of the derivative function related to the spectrum of the function itself? What happens to the spectrum of a function if the function is shifted in the time domain? We shall now seek to evaluate the effects on one domain caused by certain important operations on the function in the other domain.

It is important to point out at this stage that there is a certain amount of symmetry in the equations defining the two domains. This can be easily seen from the equations defining the Fourier transform.

$$F(\omega) = \int_{-\infty}^{\infty} f(t)e^{-j\omega t}\,dt$$

and
(1.107)

$$f(t) = \frac{1}{2\pi} \int_{-\infty}^{\infty} F(\omega)e^{j\omega t}\,d\omega$$

We should therefore expect this symmetry to be reflected in the properties. For example, we expect that the effect on the frequency domain due to differentiation in the time domain should be similar to the effect on the time domain due to differentiation in the frequency domain. We shall see that this indeed is the case.

For convenience, the correspondence between the two domains will be denoted by a double arrow. Thus the notation ·

$$f(t) \leftrightarrow F(\omega)$$

TABLE I.IA

Various Functions of Time and Their Spectral-Density Functions

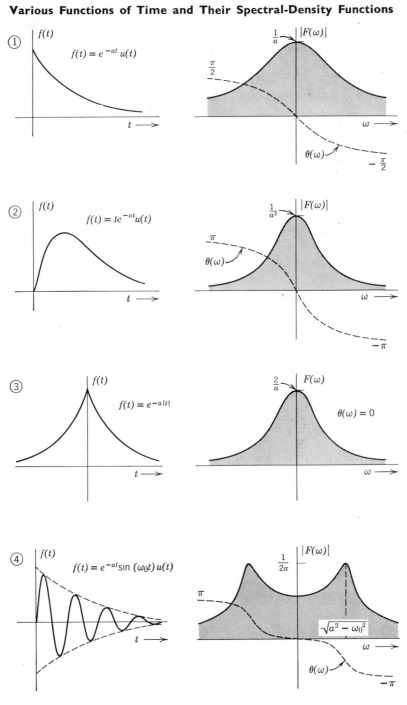

1. $f(t) = e^{-at} u(t)$

2. $f(t) = te^{-at}u(t)$

3. $f(t) = e^{-a|t|}$, $\theta(\omega) = 0$

4. $f(t) = e^{-at}\sin(\omega_0 t)u(t)$

TABLE I.IA (continued)

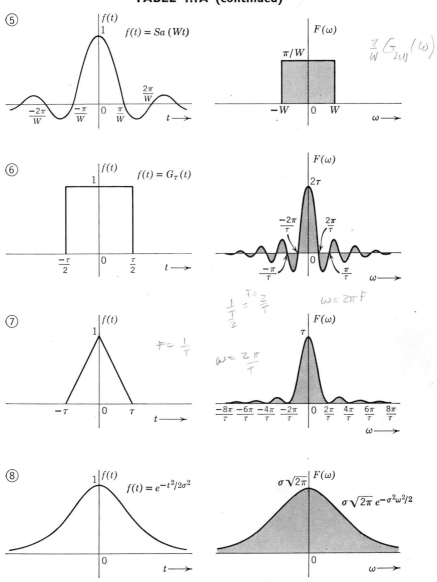

TABLE I.IA (continued)

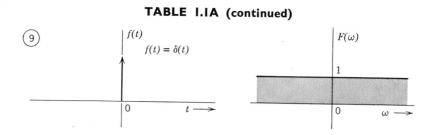

⑨ $f(t) = \delta(t)$

$F(\omega)$, value 1

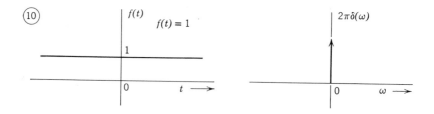

⑩ $f(t) = 1$

$2\pi\delta(\omega)$

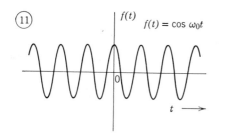

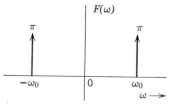

⑪ $f(t) = \cos \omega_0 t$

$F(\omega)$, π at $-\omega_0$ and π at ω_0

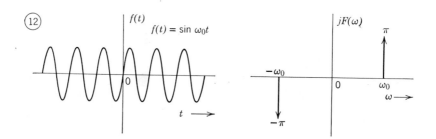

⑫ $f(t) = \sin \omega_0 t$

$jF(\omega)$, π at ω_0, $-\pi$ at $-\omega_0$

TABLE I.IA (continued)

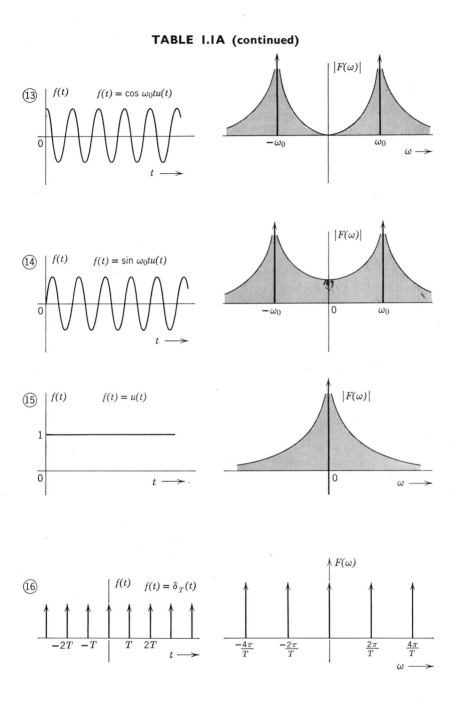

TABLE 1.IB
Fourier Transforms

$f(t)$	$F(\omega)$
1. $e^{-at}u(t)$	$\dfrac{1}{a + j\omega}$
2. $te^{-at}u(t)$	$\dfrac{1}{(a + j\omega)^2}$
3. $\lvert t \rvert$	$\dfrac{-2}{\omega^2}$
4. $\delta(t)$	1
5. 1	$2\pi\,\delta(\omega)$
6. $u(t)$	$\pi\,\delta(\omega) + \dfrac{1}{j\omega}$
7. $\cos \omega_0 t\, u(t)$	$\dfrac{\pi}{2}[\delta(\omega - \omega_0) + \delta(\omega + \omega_0)] + \dfrac{j\omega}{\omega_0^2 - \omega^2}$
8. $\sin \omega_0 t\, u(t)$	$\dfrac{\pi}{2j}[\delta(\omega - \omega_0) - \delta(\omega + \omega_0)] + \dfrac{\omega_0}{\omega_0^2 - \omega^2}$
9. $\cos \omega_0 t$	$\pi[\delta(\omega - \omega_0) + \delta(\omega + \omega_0)]$
10. $\sin \omega_0 t$	$j\pi[\delta(\omega + \omega_0) - \delta(\omega - \omega_0)]$
11. $e^{-at} \sin \omega_0 t\, u(t)$	$\dfrac{\omega_0}{(a + j\omega)^2 + \omega_0^2}$
12. $\dfrac{W}{2\pi} Sa\, \dfrac{(Wt)}{2}$	$G_W(\omega)$
13. $G_\tau(t)$	$\tau Sa\left(\dfrac{\omega\tau}{2}\right)$
14. $\begin{aligned} 1 - \dfrac{\lvert t \rvert}{\tau} &\cdots \lvert t \rvert < \tau \\ 0 &\cdots \lvert t \rvert > \tau \end{aligned}$	$\tau\left[Sa\left(\dfrac{\omega\tau}{2}\right)\right]^2$
15. $e^{-a\lvert t \rvert}$	$\dfrac{2a}{a^2 + \omega^2}$
16. $e^{-t^2/2\sigma^2}$	$\sigma\sqrt{2\pi}\,e^{-\sigma^2\omega^2/2}$
17. $\delta_T(t)$	$\omega_0\,\delta_{\omega_0}(\omega) \quad \left(\omega_0 = \dfrac{2\pi}{T}\right)$

68

denotes that $F(\omega)$ is the direct Fourier transform of $f(t)$ and that $f(t)$ is the inverse Fourier transform of $F(\omega)$ related by Eq. 1.107.

I. Symmetry Property

If

$$f(t) \leftrightarrow F(\omega)$$

then

$$F(t) \leftrightarrow 2\pi f(-\omega) \tag{1.108}$$

Proof. From Eq. 1.107 it follows that

$$2\pi f(-t) = \int_{-\infty}^{\infty} F(\omega)e^{-j\omega t}\, d\omega$$

Since ω is a dummy variable in this integral it may be replaced by another variable, x. Therefore

$$2\pi f(-t) = \int_{-\infty}^{\infty} F(x)e^{-jxt}\, dx$$

Hence

$$2\pi f(-\omega) = \int_{-\infty}^{\infty} F(x)e^{-jx\omega}\, dx$$

Replacing the dummy variable x by another variable t, we get

$$2\pi f(-\omega) = \int_{-\infty}^{\infty} F(t)e^{-j\omega t}\, dt$$

$$= \mathscr{F}[F(t)]$$

Hence

$$F(t) \leftrightarrow 2\pi f(-\omega) \tag{1.109}$$

The symmetry property holds perfectly if $f(t)$ is an even function. In that case $f(-\omega) = f(\omega)$ and Eq. 1.109 reduces to

$$F(t) \leftrightarrow 2\pi f(\omega)$$

This property is demonstrated in Fig. 1.37.

It can be seen easily that the Fourier transform of a gate function is a sampling function, and the Fourier transform of a sampling function is a gate function. The symmetry property holds for all, even $f(t)$. If $f(t)$ is not an even function, then the symmetry is not so perfect; nevertheless, there is some measure of symmetry, as seen from Eq. 1.109.

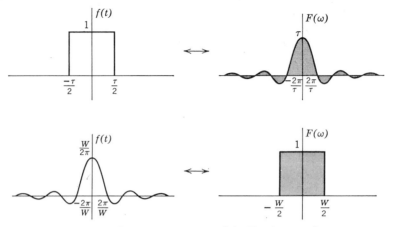

Figure 1.37   Symmetry property of the Fourier transform.

2. Linearity Property

If

$$f_1(t) \leftrightarrow F_1(\omega)$$

$$f_2(t) \leftrightarrow F_2(\omega)$$

then for any arbitrary constants a_1 and a_2,

$$a_1 f_1(t) + a_2 f_2(t) \leftrightarrow a_1 F_1(\omega) + a_2 F_2(\omega) \tag{1.110}$$

The proof is trivial. The linearity property is also valid for finite sums:

$$a_1 f_1(t) + a_2 f_2(t) + \cdots + a_n f_n(t) \leftrightarrow a_1 F_1(\omega) + a_2 F_2(\omega) + \cdots + a_n F_n(\omega)$$

3. Scaling Property

If

$$f(t) \leftrightarrow F(\omega)$$

then for a real constant a,

$$f(at) \leftrightarrow \frac{1}{|a|} F\left(\frac{\omega}{a}\right) \tag{1.111}$$

Proof. For a positive real constant a,

$$\mathscr{F}[f(at)] = \int_{-\infty}^{\infty} f(at) e^{-j\omega t} \, dt$$

Let $x = at$. Then for positive constant a,

$$\mathscr{F}[f(at)] = \frac{1}{a} \int_{-\infty}^{\infty} f(x) e^{(-j\omega/a)x} \, dx$$

$$= \frac{1}{a} F\left(\frac{\omega}{a}\right)$$

Hence

$$f(at) \leftrightarrow \frac{1}{a} F\left(\frac{\omega}{a}\right)$$

Similarly, it can be shown that if $a < 0$,

$$f(at) \leftrightarrow \frac{1}{-a} F\left(\frac{\omega}{a}\right)$$

Consequently,

$$f(at) \leftrightarrow \frac{1}{|a|} F\left(\frac{\omega}{a}\right)$$

Significance of the Scaling Property. The function $f(at)$ represents function $f(t)$ compressed in the time scale by a factor of a. Similarly, a function $F(\omega/a)$ represents a function $F(\omega)$ expanded in the frequency scale by the same factor a. The scaling-property therefore states that compression in the time domain is equivalent to expansion in the frequency domain and vice versa. This result is also obvious intuitively, since compression in the time scale by a factor a means that the function is varying rapidly by the same factor, and hence the frequencies of its components will be increased by the factor a. We therefore expect its frequency spectrum to be expanded by the factor a in the frequency scale. Similarly, if a function is expanded in the time scale, it varies slowly, and hence the frequencies of its components are lowered. Thus the frequency spectrum is compressed. As an example, consider the signal $\cos \omega_0 t$. This signal has frequency components at $\pm \omega_0$. The signal $\cos 2\omega_0 t$ represents compression of $\cos \omega_0 t$ by a factor of two, and its frequency components lie at $\pm 2\omega_0$. It is therefore evident that the frequency spectrum has been expanded by a factor of two. The effect of scaling is demonstrated in Fig. 1.38.

For a special case when $a = -1$, the scaling property yields

$$f(-t) \leftrightarrow F(-\omega) \tag{1.112}$$

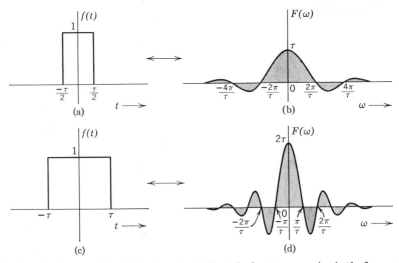

Figure 1.38   Compression in the time domain is equivalent to expansion in the frequency domain.

Example 1.9

We shall find the inverse Fourier transform of sgn (ω) and $u(\omega)$ using the symmetry and scaling properties. From Eq. 1.95, we have

$$\text{sgn } (t) \leftrightarrow \frac{2}{j\omega}$$

Applying the symmetry property (Eq. 1.109) to this equation, we get

$$\frac{2}{jt} \leftrightarrow 2\pi \text{ sgn } (-\omega)$$

But sgn $(-\omega) = -\text{sgn } (\omega)$. Hence

$$\frac{j}{\pi t} \leftrightarrow \text{sgn } (\omega) \tag{1.113}$$

Also

$$u(\omega) = \tfrac{1}{2}[1 + \text{sgn } (\omega)]$$

Hence

$$\mathscr{F}^{-1}[u(\omega)] = \tfrac{1}{2}\{\mathscr{F}^{-1}[1] + \mathscr{F}^{-1}[\text{sgn } (\omega)]\}$$

$$= \tfrac{1}{2}\delta(t) + \frac{j}{2\pi t}$$

Thus

$$\left[\frac{1}{2}\,\delta(t) - \frac{1}{2\pi j t}\right] \leftrightarrow u(\omega) \tag{1.114}$$

4. Frequency-Shifting Property

If

$$f(t) \leftrightarrow F(\omega)$$

then

$$f(t)e^{j\omega_0 t} \leftrightarrow F(\omega - \omega_0) \qquad (1.115)$$

Proof.

$$\mathscr{F}[f(t)e^{j\omega_0 t}] = \int_{-\infty}^{\infty} f(t)e^{j\omega_0 t}e^{-j\omega t} \, dt$$

$$= \int_{-\infty}^{\infty} f(t)e^{-j(\omega-\omega_0)t} \, dt$$

$$= F(\omega - \omega_0)$$

The theorem states that a shift of ω_0 in the frequency domain is equivalent to multiplication by $e^{j\omega_0 t}$ in the time domain. It is evident that multiplication by a factor $e^{j\omega_0 t}$ translates the whole frequency spectrum $F(\omega)$ by an amount ω_0. Hence this theorem is also known as the *frequency-translation theorem*.

In communication systems it is often desirable to translate the frequency spectrum. This is usually accomplished by multiplying a signal $f(t)$ by a sinusoidal signal. This process is known as *modulation*. Since a sinusoidal signal of frequency ω_0 can be expressed as the sum of exponentials, it is evident that multiplication of a signal $f(t)$ by a sinusoidal signal (modulation) will translate the whole frequency spectrum. This can be easily shown by observing the identity

$$f(t) \cos \omega_0 t = \tfrac{1}{2}[f(t)e^{j\omega_0 t} + f(t)e^{-j\omega_0 t}]$$

Using the frequency-shift theorem, it therefore follows that if

$$f(t) \leftrightarrow F(\omega)$$

then

$$f(t) \cos \omega_0 t \leftrightarrow \tfrac{1}{2}[F(\omega + \omega_0) + F(\omega - \omega_0)] \qquad (1.116a)$$

Similarly, it can be shown that

$$f(t) \sin \omega_0 t \leftrightarrow \frac{j}{2}[F(\omega + \omega_0) - F(\omega - \omega_0)] \qquad (1.116b)$$

Thus the process of modulation translates the frequency spectrum by the amount $\pm\omega_0$. This is a very useful result in communication theory. An example of frequency translation caused by modulation is shown in Fig. 1.39. This result is also known as the *modulation theorem*.

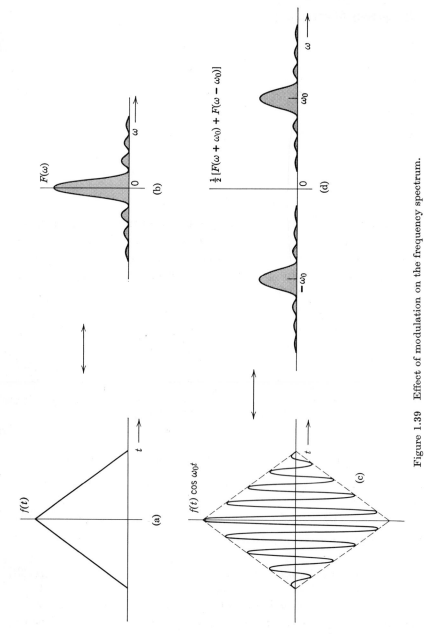

Figure 1.39 Effect of modulation on the frequency spectrum.

74

5. Time-Shifting Property

If
$$f(t) \leftrightarrow F(\omega)$$
then
$$f(t - t_0) \leftrightarrow F(\omega)e^{-j\omega t_0} \qquad (1.117)$$

Proof.

$$\mathscr{F}[f(t - t_0)] = \int_{-\infty}^{\infty} f(t - t_0)e^{-j\omega t}\, dt$$

Let
$$t - t_0 = x$$

Then

$$\mathscr{F}[f(t - t_0)] = \int_{-\infty}^{\infty} f(x)e^{-j\omega(x+t_0)}\, dx$$

$$= F(\omega)e^{-j\omega t_0}$$

This theorem states that if a function is shifted in the time domain by t_0 seconds, then its magnitude spectrum $|F(\omega)|$ remains unchanged, but the phase spectrum is changed by an amount $-\omega t_0$. This result is also obvious intuitively, since the shifting of a function in the time domain really does not change the frequency components of the signal, but each component is shifted by an amount t_0. A shift of time t_0 for a component of frequency ω is equivalent to a phase shift of $-\omega t_0$.

We may state that a shift of t_0 in the time domain is equivalent to multiplication by $e^{-j\omega t_0}$ in the frequency domain.

Example 1.10

Find the Fourier transform of a rectangular pulse shown in Fig. 1.40. The pulse in Fig. 1.40 is a gate function $G_\tau(t)$, shifted by $\tau/2$ seconds. Hence it can be expressed as $G_\tau(t - \tau/2)$. From Table 1.1B and the time-shifting property, we have

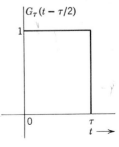

$$G_\tau\left(t - \frac{\tau}{2}\right) \leftrightarrow \tau Sa\left(\frac{\omega\tau}{2}\right)e^{-j\omega\tau/2}$$

Note the duality between the frequency-shifting property and the time-shifting property. There is also a dual of modulation theorem. It states that

$$\tfrac{1}{2}[f(t + T) + f(t - T)] \leftrightarrow F(\omega)\cos \omega T \qquad (1.118)$$

It is left as an exercise for the reader to prove Eq. 1.118.

Figure 1.40

6. Time Differentiation and Integration

If

$$f(t) \leftrightarrow F(\omega)$$

then*

$$\frac{df}{dt} \leftrightarrow (j\omega)F(\omega) \qquad (1.119a)$$

and

$$\int_{-\infty}^{t} f(\tau)\, d\tau \leftrightarrow \frac{1}{j\omega} F(\omega) \qquad (1.119b)$$

provided† that $F(\omega)/\omega$ is bounded at $\omega = 0$. This is equivalent to saying that $F(0) = 0$ or

$$\int_{-\infty}^{\infty} f(t)\, dt = 0$$

Proof.

$$f(t) = \frac{1}{2\pi} \int_{-\infty}^{\infty} F(\omega) e^{j\omega t}\, d\omega$$

Therefore

$$\frac{df}{dt} = \frac{1}{2\pi} \frac{d}{dt} \int_{-\infty}^{\infty} F(\omega) e^{j\omega t}\, d\omega$$

Changing the order of differentiation and integration, we get

$$\frac{df}{dt} = \frac{1}{2\pi} \int_{-\infty}^{\infty} j\omega F(\omega) e^{j\omega t}\, d\omega$$

From this equation it is evident that

$$\frac{df}{dt} \leftrightarrow j\omega F(\omega)$$

In a similar way the result can be extended to the nth derivative.

$$\frac{d^n f}{dt^n} \leftrightarrow (j\omega)^n F(\omega) \qquad (1.120)$$

Now consider the function

$$\varphi(t) = \int_{-\infty}^{t} f(\tau)\, d\tau$$

---

* Equation 1.119a does not guarantee the existence of the transform of df/dt. It merely says that if that transform exists, it is given by $j\omega F(\omega)$.

† If this condition is not satisfied, then Eq. 1.119b modified. See Problem 42 at the end of the chapter.

Then

$$\frac{d\varphi}{dt}(t) = f(t)$$

Hence if

$$\varphi(t) \longleftrightarrow \Phi(\omega)$$

then

$$f(t) \longleftrightarrow j\omega\Phi(\omega)$$

That is,

$$F(\omega) = j\omega\Phi(\omega)$$

Therefore

$$\Phi(\omega) = \frac{1}{j\omega} F(\omega)$$

and thus

$$\int_{-\infty}^{t} f(\tau)\, d\tau \longleftrightarrow \frac{1}{j\omega} F(\omega)$$

Note that this result is valid only if $\Phi(\omega)$ exists, that is, if $\varphi(t)$ is absolutely integrable. This is possible only if

$$\lim_{t \to \infty} \varphi(t) = 0$$

That is,

$$\int_{-\infty}^{\infty} f(t)\, dt = 0$$

This is equivalent to the condition that $F(0) = 0$ since

$$\int_{-\infty}^{\infty} f(t)\, dt = F(\omega)\big|_{\omega=0}$$

The time-differentiation and time-integration theorems as expressed in Eqs. 1.118 and 1.119 are also obvious intuitively. The Fourier transform actually expresses a function $f(t)$ in terms of a continuous sum of exponential functions of the form $e^{j\omega t}$. The derivative of $f(t)$ is therefore equal to the continuous sum of the derivatives of the individual exponential components. But the derivative of an exponential function $e^{j\omega t}$ is equal to $j\omega e^{j\omega t}$. Therefore the process of differentiation of $f(t)$ is equivalent to multiplication by $j\omega$ of each exponential component. Hence

$$\frac{df}{dt} \longleftrightarrow j\omega F(\omega)$$

A similar argument applies to the integration.

We conclude that differentiation in the time domain is equivalent to multiplication by $j\omega$ in the frequency domain and that integration in the time domain is equivalent to division by $j\omega$ in the frequency domain. The time-differentiation theorem proves convenient in deriving the Fourier transform of some piecewise continuous functions. This is illustrated by the next example.

Example 1.11

Evaluate the Fourier transform of a trapezoidal function $f(t)$ shown in Fig. 1.41. We differentiate this function twice to obtain a sequence of impulses. The transform of the impulses is readily found. It is evident from Fig. 1.41 that

$$\frac{d^2f}{dt^2} = \frac{A}{(b-a)} [\delta(t+b) - \delta(t+a) - \delta(t-a) + \delta(t-b)] \quad (1.121)$$

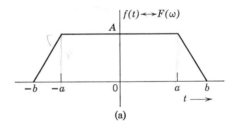

(a)

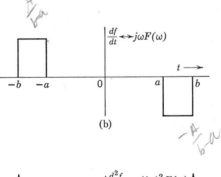

(b)

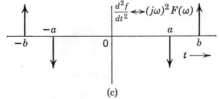

(c)

Figure 1.41

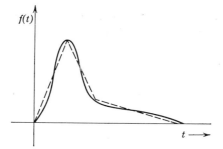

Figure 1.42   Approximation of a function with straight-line segments.

The Fourier transform of a unit impulse is 1. Therefore, using the time-shift theorem, we have

$$\delta(t - t_0) \longleftrightarrow e^{-j\omega t_0}$$

Using this result and the time-differentiation theorem, the transform of Eq. 1.121 can now be readily written as

$$(j\omega)^2 F(\omega) = \frac{A}{(b - a)} \left(e^{j\omega b} - e^{j\omega a} - e^{-j\omega a} + e^{-j\omega b}\right)$$

from which we get

$$F(\omega) = \frac{2A}{(b - a)} \left(\frac{\cos a\omega - \cos b\omega}{\omega^2}\right)$$

This problem suggests a numerical method of obtaining a Fourier transform of a function $f(t)$. Any function $f(t)$ may be approximated by straight-line segments as shown in Fig. 1.42. The approximation can be improved as much as desired by increasing the number of segments. The second derivative of the approximated function yields a train of impulses whose Fourier transform can be readily found. The transform $F(\omega)$ of the desired function is merely $1/(j\omega)^2$ times the transform of the second derivative.

7. Frequency Differentiation

If

$$f(t) \longleftrightarrow F(\omega)$$

then

$$-jtf(t) \longleftrightarrow \frac{dF}{d\omega} \tag{1.122}$$

Proof.

$$F(\omega) = \int_{-\infty}^{\infty} f(t)e^{-j\omega t}\,dt$$

Therefore

$$\frac{dF}{d\omega} = \frac{d}{d\omega} \int_{-\infty}^{\infty} f(t)e^{-j\omega t} \, dt$$

Changing the order of differentiation and integration, we get

$$\frac{dF}{d\omega} = \int_{-\infty}^{\infty} -jt f(t)e^{-j\omega t} \, dt$$

It is evident from the above equation that

$$-jtf(t) \longleftrightarrow \frac{dF}{d\omega}$$

The extension of this result to higher derivatives of $F(\omega)$ yields

$$(-jt)^n f(t) \Longleftrightarrow \frac{d^n F}{d\omega^n}$$

We therefore conclude that differentiation in the frequency domain is equivalent to multiplication by $-jt$ in the time domain.

8. The Convolution Theorem

The convolution theorem is perhaps one of the most powerful tools in frequency analysis. It permits the easy derivation of many important results and will be used often in this text.

Given two functions $f_1(t)$ and $f_2(t)$, we form the integral

$$f(t) = \int_{-\infty}^{\infty} f_1(\tau)f_2(t-\tau) \, d\tau \qquad (1.123)$$

This integral defines the convolution of functions $f_1(t)$ and $f_2(t)$. The convolution integral (Eq. 1.123) is also expressed symbolically as

$$f(t) = f_1(t) * f_2(t) \qquad (1.124)$$

The physical significance and the graphical interpretation of convolution will be considered later. We shall first state and prove the theorem. Again, as usual, we have two theorems: time convolution and frequency convolution.

Time-Convolution Theorem. If

$$f_1(t) \longleftrightarrow F_1(\omega)$$

and

$$f_2(t) \longleftrightarrow F_2(\omega)$$

then

$$\int_{-\infty}^{\infty} f_1(\tau) f_2(t - \tau) \, d\tau \leftrightarrow F_1(\omega) F_2(\omega) \qquad (1.125a)$$

that is,

$$f_1(t) * f_2(t) \leftrightarrow F_1(\omega) F_2(\omega) \qquad (1.125b)$$

Proof.

$$\mathscr{F}[f_1(t) * f_2(t)] = \int_{-\infty}^{\infty} e^{-j\omega t} \left[\int_{-\infty}^{\infty} f_1(\tau) f_2(t - \tau) \, d\tau \right] dt$$

$$= \int_{-\infty}^{\infty} f_1(\tau) \left[\int_{-\infty}^{\infty} e^{-j\omega t} f_2(t - \tau) \, dt \right] d\tau$$

From the time-shifting property (Eq. 1.117), it is evident that the integral inside the bracket, on the right-hand side, is equal to $F_2(\omega) e^{-j\omega \tau}$. Hence

$$\mathscr{F}[f_1(t) * f_2(t)] = \int_{-\infty}^{\infty} f_1(\tau) e^{-j\omega \tau} F_2(\omega) \, d\tau$$

$$= F_1(\omega) F_2(\omega)$$

Frequency Convolution. If

$$f_1(t) \leftrightarrow F_1(\omega)$$

and

$$f_2(t) \leftrightarrow F_2(\omega)$$

then

$$f_1(t) f_2(t) \leftrightarrow \frac{1}{2\pi} \int_{-\infty}^{\infty} F_1(u) F_2(\omega - u) \, du \qquad (1.126a)$$

That is,

$$f_1(t) f_2(t) \leftrightarrow \frac{1}{2\pi} [F_1(\omega) * F_2(\omega)] \qquad (1.126b)$$

This theorem can be proved in exactly the same way as the time-convolution theorem because of the symmetry in the direct and inverse Fourier transforms.

We therefore conclude that the convolution of two functions in the time domain is equivalent to multiplication of their spectra in the frequency domain and that multiplication of two functions in the time domain is equivalent to convolution of their spectra in the frequency domain.

Table 1.2 shows some of the important properties of the Fourier transform. Note the symmetry and correspondence between the time and frequency domains.

TABLE 1.2

Operation	$f(t)$	$F(\omega)$
1. Scaling	$f(at)$	$\dfrac{1}{|a|} F\left(\dfrac{\omega}{a}\right)$
2. Time shifting	$f(t - t_0)$	$F(\omega)e^{-j\omega t_0}$
3. Frequency shifting	$f(t)e^{j\omega_0 t}$	$F(\omega - \omega_0)$
4. Time differentiation	$\dfrac{d^n f}{dt^n}$	$(j\omega)^n F(\omega)$
5. Frequency differentiation	$(-jt)^n f(t)$	$\dfrac{d^n F}{d\omega^n}$
6. Time integration	$\displaystyle\int_{-\infty}^{t} f(\tau)\, d\tau$	$\dfrac{1}{(j\omega)} F(\omega)$
7. Time convolution	$f_1(t) * f_2(t)$	$F_1(\omega) F_2(\omega)$
8. Frequency convolution	$f_1(t)f_2(t)$	$\dfrac{1}{2\pi}[F_1(\omega) * F_2(\omega)]$

1.13 SOME CONVOLUTION RELATIONSHIPS

The symbolic representation of convolution suggests that convolution is a special kind of multiplication. Indeed, it is possible to write the laws of convolution algebra along lines that are similar to those for ordinary multiplication.

1. Commutative Law

$$f_1(t) * f_2(t) = f_2(t) * f_1(t) \qquad (1.127)$$

This relationship can be proved easily as follows:

$$f_1(t) * f_2(t) = \int_{-\infty}^{\infty} f_1(\tau)f_2(t - \tau)\, d\tau$$

Changing the variable τ to $t - x$, we get

$$f_1(t) * f_2(t) = \int_{-\infty}^{\infty} f_2(x)f_1(t - x)\, dx$$

$$= f_2(t) * f_1(t)$$

2. Distributive Law

$$f_1(t) * [f_2(t) + f_3(t)] = f_1(t) * f_2(t) + f_1(t) * f_3(t) \qquad (1.128)$$

The proof is trivial.

3. Associative Law

$$f_1(t) * [f_2(t) * f_3(t)] = [f_1(t) * f_2(t)] * f_3(t) \qquad (1.129)$$

This law follows from the convolution theorem and from the fact that

$$F_1(\omega)[F_2(\omega)F_3(\omega)] = [F_1(\omega)F_2(\omega)]F_3(\omega)$$

1.14 GRAPHICAL INTERPRETATION OF CONVOLUTION

The graphical interpretation of convolution is very useful in systems analysis as well as communication theory. It permits one to grasp visually the results of many abstract relationships. This is particularly true in communication theory. In linear systems, graphical convolution is very helpful in analysis if $f(t)$ and $h(t)$ are known only graphically. To illustrate, let us consider $f_1(t)$ and $f_2(t)$ as rectangular and triangular pulses as shown in Fig. 1.43a. We shall find the convolution $f_1(t) * f_2(t)$ graphically. By definition,

$$f_1(t) * f_2(t) = \int_{-\infty}^{\infty} f_1(\tau)f_2(t - \tau) \, d\tau \qquad (1.130)$$

The independent variable in the convolution integral is τ (Eq. 1.130). The functions $f_1(\tau)$ and $f_2(-\tau)$ are shown in Fig. 1.43b. Note that $f_2(-\tau)$ is obtained by folding $f_2(\tau)$ about the vertical axis passing through the origin. The term $f_2(t - \tau)$ represents the function $f_2(-\tau)$ shifted by t seconds along the positive τ axis. Figure 1.43c shows $f_2(t_1 - \tau)$. The value of the convolution integral at $t = t_1$ is given by the integral in Eq. 1.130 evaluated at $t = t_1$. This is clearly the area under the product curve of $f_1(\tau)$ and $f_2(t_1 - \tau)$. This area is shown shaded in Fig.1.43d. The value of $f_1(t) * f_2(t)$ at $t = t_1$ is equal to this shaded area and is plotted in Fig. 1.43f. We choose different values of t, shift the function $f_2(-\tau)$ accordingly, and find the area under the new product curve. These areas represent the value of the convolution function $f_1(t) * f_2(t)$

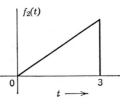
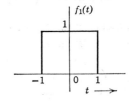

(a)

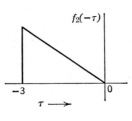
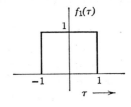

(b)

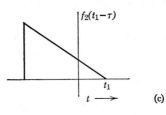
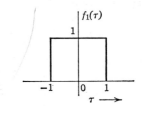

(c)

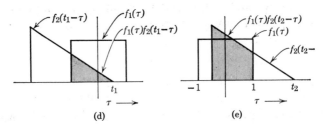

(d)   (e)

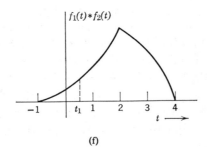

(f)

Figure 1.43

at the respective values of t. The plot of the area under the product curve as a function of t represents the desired convolution function $f_1(t) * f_2(t)$.

The graphical mechanism of convolution can be appreciated by visualizing the function $f_2(-\tau)$ as a rigid frame which is being progressed along the τ axis by t_1 seconds. The function represented by this frame is multiplied by $f_1(\tau)$, and the area under the product curve is the value of the convolution function at $t = t_1$. Therefore, to find the value of $f_1(t) * f_2(t)$ at any time, say $t = t_0$, we displace the rigid frame representing $f_2(-\tau)$ by t_0 seconds along the τ axis and multiply this function with $f_1(\tau)$. The area under the product curve is the desired value of $f_1(t) * f_2(t)$ at $t = t_0$. To find the function $f_1(t) * f_2(t)$, we progress the frame successively by different amounts and find the areas of the product curve at various positions. The plot of the area as a function of displacement of the frame represents the required convolution function $f_1(t) * f_2(t)$. To summarize:

1. Fold the function $f_2(\tau)$ about the vertical axis passing through the origin of the τ axis and obtain the function $f_2(-\tau)$.

2. Consider the folded function as a rigid frame and progress it along the τ axis by an amount, say t_0. The rigid frame now represents the function $f_2(t_0 - \tau)$.

3. The product of the function represented by this displaced rigid frame with $f_1(\tau)$ represents the function $f_1(\tau)f_2(t_0 - \tau)$, and the area under this curve is given by

$$\int_{-\infty}^{\infty} f_1(\tau)f_2(t_0 - \tau)\,d\tau = [f_1(t) * f_2(t)]_{t=t_0}$$

4. Repeat this procedure for different values of t by successively progressing the frame by different amounts and find the values of the convolution function $f_1(t) * f_2(t)$ at those values of t.

Note that to find the convolution function $f_1(t) * f_2(t)$ for the positive values of t, we progress the frame along the positive τ axis, whereas for the negative values of t, the frame is progressed along the negative τ axis.

It was shown in Eq. 1.127 that the convolution of $f_1(t)$ with $f_2(t)$ is equal to the convolution of $f_2(t)$ with $f_1(t)$. That is,

$$f_1(t) * f_2(t) = f_2(t) * f_1(t)$$

Thus we could have kept $f_2(\tau)$ fixed and taken the mirror image of $f_1(\tau)$ in the graphical convolution in Fig. 1.43. We get the same results either way.

I.15 CONVOLUTION OF A FUNCTION WITH A UNIT IMPULSE FUNCTION

The convolution of a function $f(t)$ with a unit impulse function $\delta(t)$ yields the function $f(t)$ itself. This can be proved easily by using the sampling property in Eq. 1.90b.

$$f(t) * \delta(t) = \int_{-\infty}^{\infty} f(\tau)\, \delta(t - \tau)\, d\tau$$

$$= f(t)$$

This result also follows from the time-convolution theorem and the fact that

$$f(t) \leftrightarrow F(\omega) \qquad \text{and} \qquad \delta(t) \leftrightarrow 1$$

Hence

$$f(t) * \delta(t) \leftrightarrow F(\omega)$$

Consequently,

$$f(t) * \delta(t) = f(t) \tag{1.131}$$

This result is also obvious graphically. Since the impulse is concentrated at one point and has an area of unity, the convolution integral in Eq. 1.130 yields the function $f(t)$. Thus the unit impulse function when convolved with a function $f(t)$ reproduces the function $f(t)$. A simple extension of Eq. 1.131 yields

$$f(t) * \delta(t - T) = f(t - T) \tag{1.132a}$$

$$f(t - t_1) * \delta(t - t_2) = f(t - t_1 - t_2) \tag{1.132b}$$

$$\delta(t - t_1) * \delta(t - t_2) = \delta(t - t_1 - t_2) \tag{1.132c}$$

Example 1.12

Find graphically the convolution of $f_1(t)$ (Fig. 1.44a) with a pair of impulses of strength k each, as shown in Fig. 1.44b.

Following the procedure of graphical convolution described in Section 1.14, we fold back $f_2(\tau)$ about the ordinate to obtain $f_2(-\tau)$. Since $f_2(\tau)$ is an even function of τ, $f_2(-\tau) = f_2(\tau)$. The convolution of $f_1(\tau)$ with $f_2(\tau)$ thus

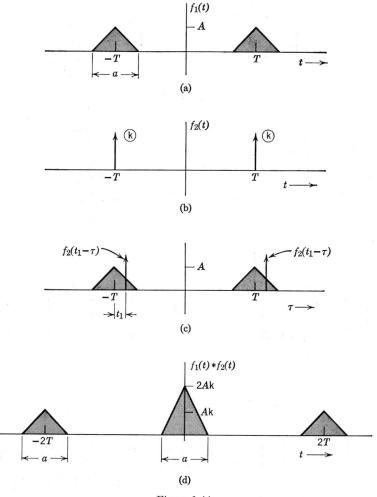

Figure 1.44

reduces to convolution of $f_1(\tau)$ with two impulses. From the property of an impulse function to reproduce the function by convolution (Eq. 1.131), it can be readily seen that each impulse produces a triangular pulse of height Ak at the origin ($t = 0$). Hence the net height of the triangular pulse is $2Ak$ at the origin. As the function $f_2(t - \tau)$ is moved farther in a positive direction, the impulse originally located at $-T$ encounters the triangular pulse at $\tau = T$ and reproduces the triangular pulse of height Ak at $t = 2T$. Similarly, the impulse originally located at T reproduces a triangular pulse of height Ak at $t = -2T$. The final result of convolution is shown in Fig. 1.44d.

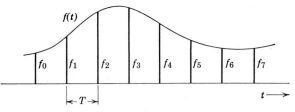

Figure 1.45

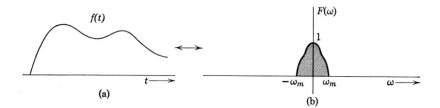

(a)

(b)

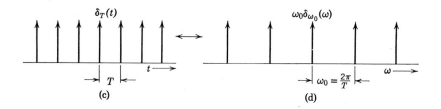

(c)

(d)

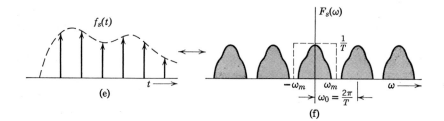

(e)

(f)

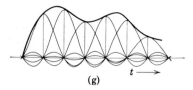

(g)

Figure 1.46

1.16 THE SAMPLING THEOREM

The sampling theorem has a deep significance in communication theory. It states the following:

A bandlimited signal which has no spectral components above a frequency f_m Hz is uniquely determined by its values at uniform intervals less than $1/2f_m$ seconds apart.

This theorem is known as the *uniform sampling theorem* since it pertains to the specification of a given signal by its samples at uniform intervals of $1/2f_m$ seconds.* This implies that if the Fourier transform of $f(t)$ is zero beyond a certain frequency $\omega_m = 2\pi f_m$, then the complete information about $f(t)$ is contained in its samples spaced uniformly at a distance less than $1/2f_m$ seconds. This is illustrated in Fig. 1.45. The function $f(t)$ is sampled once every T seconds ($T \leqslant 1/2f_m$) or at a rate greater than or equal to $2f_m$ samples per second. The successive samples are labeled as $f_0, f_1, f_2, \ldots$, etc. It follows from the sampling theorem that these samples contain the information about $f(t)$ at every value of t. The sampling rate, however, must be at least twice the highest frequency f_m present in the spectrum of $f(t)$. To say it another way, the signal must be sampled at least twice during each period or cycle of its highest frequency component.

The sampling theorem can be easily proved with the help of the frequency convolution theorem. Consider a bandlimited signal $f(t)$ which has no spectral components above f_m cycles per second. This means that $F(\omega)$, the Fourier transform of $f(t)$, is zero for $|\omega| > \omega_m$ ($\omega_m = 2\pi f_m$). Suppose we multiply the function $f(t)$ by a periodic impulse function $\delta_T(t)$ (Fig. 1.46c). The product function is a sequence of impulses located at regular intervals of T seconds and having strengths equal to the values of $f(t)$ at the corresponding instants. The

---

* This theorem is actually a special case of the general sampling theorem which is stated as follows:

If a signal is bandlimited and if the time interval is divided into equal parts forming subintervals such that each subdivision comprises an interval T seconds long where T is less than $1/2f_m$, and if one instantaneous sample is taken from each subinterval in any manner, then a knowledge of the instantaneous magnitude of each sample plus a knowledge of the instants within each subinterval at which the sample is taken contains all of the information of the original signal.

See, for instance, H. S. Black, *Modulation Theory*, Van Nostrand, New York, 1953, p. 41.

product $f(t)\ \delta_T(t)$ indeed represents the function $f(t)$ sampled at a uniform interval of T seconds. We shall denote this sampled function by $f_s(t)$ (see Fig. 1.46e).

$$f_s(t) = f(t)\ \delta_T(t)$$

The frequency spectrum of $f(t)$ is $F(\omega)$. We have shown (Eq. 1.106) that the Fourier transform of a uniform train of impulse function $\delta_T(t)$ is also a uniform train of impulse function $\omega_0\ \delta_{\omega_0}(\omega)$ (Fig. 1.46d). The impulses are separated by a uniform interval $\omega_0 = 2\pi/T$.

$$\delta_T(t) \leftrightarrow \omega_0\ \delta_{\omega_0}(\omega)$$

The Fourier transform of $f(t)\ \delta_T(t)$ will, according to the frequency convolution theorem, be given by the convolution of $F(\omega)$ with $\omega_0\ \delta_{\omega_0}(\omega)$.

$$f_s(t) \leftrightarrow \frac{1}{2\pi}\left[F(\omega) * \omega_0\ \delta_{\omega_0}(\omega)\right]$$

Substituting $\omega_0 = 2\pi/T$, we get

$$f_s(t) \leftrightarrow \frac{1}{T}\left[F(\omega) * \delta_{\omega_0}(\omega)\right] \qquad (1.134)$$

From Eq. 1.134 it is evident that the spectrum of the sampled signal $f_s(t)$ is given by the convolution of $F(\omega)$ with a train of impulses. Functions $F(\omega)$ and $\delta_{\omega_0}(\omega)$ (shown in Figs. 1.46b and d, respectively) can be convolved graphically by the procedure described in Section 1.14. In order to perform this operation, we fold back the function $\delta_{\omega_0}(\omega)$ about the vertical axis $\omega = 0$. Since $\delta_{\omega_0}(\omega)$ is an even function of ω, the folded function is the same as the original function $\delta_{\omega_0}(\omega)$. To perform the operation of convolution, we now progress the whole train of impulses $[\delta_{\omega_0}(\omega)]$ in a positive ω direction. As each impulse passes across $F(\omega)$, it reproduces $F(\omega)$ itself. Since the impulses are spaced at a distance $\omega_0 = 2\pi/T$, the operation of convolution yields $F(\omega)$ repeating itself every ω_0 radian per second as shown in Fig. 1.46f. The spectral density function (the Fourier transform) of $f_s(t)$ is therefore the same as $F(\omega)$ but repeating itself periodically, every ω_0 radian per second. This function will be designated as $F_s(\omega)$. Note that $F(\omega)$ will repeat periodically without overlap as long as $\omega_0 \geqslant 2\omega_m$, or

$$\frac{2\pi}{T} \geqslant 2(2\pi f_m)$$

That is,

$$T \leqslant \frac{1}{2f_m} \qquad (1.135)$$

Therefore, as long as we sample $f(t)$ at regular intervals less than $1/2f_m$ seconds apart, $F_s(\omega)$, the spectral density function of $f_s(t)$, will be a periodic replica of $F(\omega)$ and therefore contains all the information of $f(t)$. We can easily recover $F(\omega)$ from $F_s(\omega)$ by allowing the sampled signal to pass through a low-pass filter which will only allow frequency components below f_m and attenuate all the higher frequency components. Thus it is evident that the sampled function $f_s(t)$ contains all of the information of $f(t)$. To recover $f(t)$ from $f_s(t)$, we allow the sampled function $f_s(t)$ to pass through a low-pass filter which permits the transmission of all of the components of frequencies below f_m and attenuates all of the components of frequencies above f_m. The ideal filter characteristic to achieve this is shown dotted in Fig. 1.46f.

Note that if the sampling interval T becomes larger than $1/2f_m$, then the convolution of $F(\omega)$ with $\delta_{\omega_0}(\omega)$ yields $F(\omega)$ periodically. But now there is an overlap between successive cycles, and $F(\omega)$ cannot be recovered from $F_s(\omega)$. Therefore, if the sampling interval T is made too large, the information is partly lost, and the signal $f(t)$ cannot be recovered from the sampled signal $f_s(t)$. This conclusion is quite logical since it is reasonable to expect that the information will be lost if the sampling is too slow. The maximum interval of sampling $T = 1/2f_m$ is also called the *Nyquist interval*.

In the preceding discussion $F(\omega) * \delta_{\omega_0}(\omega)$ was obtained graphically. The same result can also be readily derived by analytical procedure. We have

$$\delta_{\omega_0}(\omega) = \delta(\omega) + \delta(\omega - \omega_0) + \cdots + \delta(\omega - n\omega_0) + \cdots$$
$$+ \delta(\omega + \omega_0) + \cdots + \delta(\omega + n\omega_0) + \cdots$$
$$= \sum_{n=-\infty}^{\infty} \delta(\omega - n\omega_0)$$

From Eq. 1.134, it follows that

$$F_s(\omega) = \frac{1}{T}[F(\omega) * \delta_{\omega_0}(\omega)] = \frac{1}{T}\left[F(\omega) * \sum_{n=-\infty}^{\infty} \delta(\omega - n\omega_0)\right]$$
$$= \frac{1}{T}\sum_{n=-\infty}^{\infty} F(\omega) * \delta(\omega - n\omega_0)$$

Use of Eq. 1.132a now yields

$$F_s(\omega) = \frac{1}{T} \sum_{n=-\infty}^{\infty} F(\omega - n\omega_0) \qquad (1.136)$$

The right-hand side of Eq. 1.136 represents function $F(\omega)$ repeating itself every ω_0 radian per second. This is exactly the same result as obtained by graphical convolution.

Recovering f(t) from Its Samples

As discussed earlier, the original function can be recovered by passing the sampled function through a low-pass filter with a cutoff frequency ω_m. This is obviously an operation in the frequency domain. Because of the duality in the frequency domain and the time domain, there is an equivalent operation in the time domain to recover $f(t)$ from its samples. We shall now explore this possibility.

Let us consider a signal $f(t)$ sampled at a minimum required rate ($2f_m$ samples per second). In this case

$$T = \frac{1}{2f_m} \quad \text{and} \quad \omega_0 = \frac{2\pi}{T} = 4\pi f_m = 2\omega_m$$

Hence Eq. 1.136 becomes

$$F_s(\omega) = \frac{1}{T} \sum_{n=-\infty}^{\infty} F(\omega - 2n\omega_m) \qquad (1.137)$$

As observed before, the spectrum $F(\omega)$ can be obtained by filtering $F_s(\omega)$ through a low-pass filter of cutoff frequency ω_m. It is obvious that such an operation of filtering is equivalent to multiplying $F_s(\omega)$ by a gate function $G_{2\omega_m}(\omega)$. Hence, from Eq. 1.137, we get

$$F_s(\omega)G_{2\omega_m}(\omega) = \frac{1}{T} F(\omega)$$

Therefore

$$F(\omega) = T F_s(\omega)G_{2\omega_m}(\omega) \qquad (1.138)$$

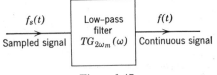

Figure 1.47

Thus transmitting the sampled signal $f_s(t)$ through a low-pass filter yields the signal $f(t)$. The filter has a cutoff frequency ω_m and gain of $T = 1/2f_m$. The transfer function $H(\omega)$ of this filter (Fig. 1.47) can be expressed as

$$H(\omega) = TG_{2\omega_m}(\omega)$$

$$= \frac{1}{2f_m} G_{2\omega_m}(\omega)$$

The application of the time convolution theorem to Eq. 1.138 yields

$$f(t) = Tf_s(t) * \frac{\omega_m}{\pi} Sa(\omega_m t)$$

$$= f_s(t) * Sa(\omega_m t) \qquad (1.139)$$

The sampled function $f_s(t)$ is given by

$$f_s(t) = \sum_n f_n \, \delta(t - nT)$$

where f_n is the nth sample of $f(t)$. Hence

$$f(t) = \sum_n f_n \, \delta(t - nT) * Sa(\omega_m t)$$

$$= \sum_n f_n Sa[\omega_m(t - nT)] \qquad (1.140a)$$

$$= \sum_n f_n Sa(\omega_m t - n\pi) \qquad (1.140b)$$

It is obvious that $f(t)$ can be constructed in the time domain from its samples according to Eq. 1.140. Graphically each sample is multiplied by a sampling function and all of the resulting waveforms are added to obtain $f(t)$. This is shown in Fig. 1.46g.

Most of the signals, in practice, closely approximate the bandlimited signals. It should be stated here that strictly speaking a bandlimited signal does not exist. It can be shown that if a signal exists over a finite interval of time it contains the components of all frequencies.* However, for all signals, in practice, the spectral density functions diminish

---

* This follows from the Paley-Wiener criterion discussed in Chapter 2 (Section 2.5). If $F(\omega)$ is bandlimited [$F(\omega) = 0$, for $|\omega| > \omega_m$], than $F(\omega)$ violates the Paley-Wiener condition, and hence its inverse transform $f(t)$ exists for all negative values of time. Therefore a bandlimited signal exists over an infinite time interval. Conversely, a signal which exists only over a finite interval of time cannot be bandlimited.

at higher frequencies. Most of the energy is carried by components lying within a certain frequency interval and, for all useful purposes, a signal may be considered to be bandlimited. The error introduced by ignoring high frequency components is negligible.

The sampling theorem is an important concept, for it allows us to replace a continuous bandlimited signal by a discrete sequence of its samples without the loss of any information. The information content of a continuous bandlimited signal is thus equivalent to discrete pieces of information. Since the sampling principle specifies the least number of discrete values necessary to reproduce a continuous signal, the problem of transmitting such a signal is reduced to that of transmitting a finite number of values. Such discrete information can be transmitted by a group of pulses whose amplitudes may be varied according to sample values (pulse amplitude modulation). Other forms of modulation are pulse position modulation where the position of the pulse is varied, the pulse width modulation (variation of pulse width in proportion to sample values), or pulse code modulation (where the samples are represented by a code formed by a group of pulses).

Sampling Theorem (Frequency Domain)

Sampling theorem in time domain has a dual which states that: A time limited signal which is zero for $|t| > T$ is uniquely determined by the samples of its frequency spectrum at uniform intervals less than $1/2T$ Hz apart (or π/T radians per second apart).

The proof of this theorem is similar to that in time domain with the roles of $f(t)$ and $F(\omega)$ reversed. It is left as an exercise for the reader to prove the dual of Eq. 1.140.

$$F(\omega) = \sum_{n=-\infty}^{\infty} F\left(\frac{n\pi}{T}\right) Sa(\omega T - n\pi) \qquad (1.141)$$

PROBLEMS

1. Show that over the interval $(0, 2\pi)$ the rectangular function in Fig. 1.3 is orthogonal to signals $\cos t, \cos 2t, \ldots, \cos nt$ for all integral values of n; that is, this function has a zero component of the waveform $\cos nt$ (n integral).

2. Show that if the two signals $f_1(t)$ and $f_2(t)$ are orthogonal over an interval t_1, t_2, then the energy of the signal $[f_1(t) + f_2(t)]$ is equal to the sum of the

energies of $f_1(t)$ and $f_2(t)$. The energy of a signal $f(t)$ over the interval (t_1, t_2) is defined as

$$\text{energy} = \int_{t_1}^{t_2} f^2(t)\, dt$$

Extend this result to n number of mutually orthogonal signals.

3. The rectangular function $f(t)$ in Fig. 1.3 is approximated by the signal $(4/\pi) \sin t$. Show that the error function

$$f_e(t) = f(t) - \frac{4}{\pi} \sin t$$

is orthogonal to the function $\sin t$ over the interval $(0, 2\pi)$. (Can you give the qualitative reason for this?) Now, show that the energy of $f(t)$ is the sum of energies of $f_e(t)$ and $(4/\pi) \sin t$.

4. Approximate the rectangular function in Fig. 1.3 by Legendre polynomials by the first two nonzero terms. Find the mean square error in the approximation when the approximation has only (a) the first term, and (b) the first and the second term. How does this approximation compare with that obtained by sinusoidal terms (in Eq. 1.38)?

5. Prove that if $f_1(t)$ and $f_2(t)$ are complex functions of a real variable t, then the component of $f_2(t)$ contained in $f_1(t)$ over the interval (t_1, t_2) is given by

$$C_{12} = \frac{\displaystyle\int_{t_1}^{t_2} f_1(t) f_2{}^*(t)\, dt}{\displaystyle\int_{t_1}^{t_2} f_2(t) f_2{}^*(t)\, dt}$$

The component is defined in the usual sense to minimize the magnitude of the mean square error. Now show that the signals $[f_1(t) - C_{12} f_2(t)]$ and $f_2(t)$ are mutually orthogonal.

6. Find the component of a waveform $\sin \omega_2 t$ contained in another waveform $\sin \omega_1 t$ over the interval $(-T, T)$ for all real values of ω_1 and ω_2 ($\omega_1 \neq \omega_2$). How does this component change with T? Show that as T is made infinite, the component vanishes. Show that this result holds for any pair of the functions $\sin \omega_1 t$, $\sin \omega_2 t$, $\cos \omega_1 t$, $\cos \omega_2 t$.

7. The two periodic functions $f_1(t)$ and $f_2(t)$ with zero d-c components have arbitrary waveforms with periods T and $\sqrt{2}\, T$, respectively. Show that the component in $f_1(t)$ of waveform $f_2(t)$ is zero over the interval $(-\infty < t < \infty)$. Show that this result is true for any two periodic functions if the ratio of their periods is an irrational number, and provided that either $f_1(t)$ or $f_2(t)$ or both have zero average values (zero d-c components).

8. Determine whether the following functions are periodic or nonperiodic. In the case of periodic functions, find the period.

(a) $a \sin t + b \sin 2t$

(b) $a \sin 5t + b \cos 8t$

(c) $a \sin 2t + b \cos \pi t$

(d) $a \cos 2t + b \sin 7t + c \sin 13t$

(e) $a \cos t + b \sin \sqrt{2}\, t$

(f) $a \sin (3t/2) + b \cos (16t/15) + c \sin (t/29)$

(g) $(a \sin t)^3$

(h) $(a \sin 2t + b \sin 5t)^2$

9. Represent the functions e^t and t^2 over the interval $(0 < t < 1)$ by the trigonometric Fourier series and the exponential Fourier series.

10. Represent each of the three functions in Fig. P-1.10 by a trigonometric Fourier series over the interval $(-\pi, \pi)$. We would now like to approximate these functions by finite terms of the Fourier series. Determine in each case

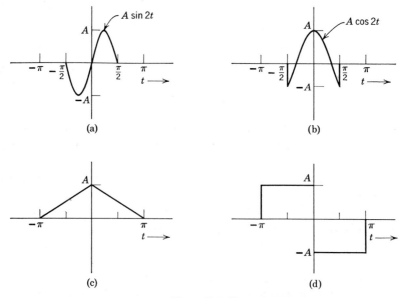

Figure P-1.10

the number of terms that should be taken in order to reduce the mean square error between the actual and the approximated function to less than 1% of the total energy of $f(t)$.

11. If a periodic signal satisfies certain symmetry conditions, the evaluation of Fourier coefficients is somewhat simplified. Show that the following are true.

(a) If $f(t) = f(-t)$ (even symmetry), then all the sine terms in the trigo-nometric Fourier series vanish.

(b) If $f(t) = -f(-t)$ (odd symmetry), then all the cosine terms in the trigonometric series vanish.

(c) If $f(t) = -f(t \pm T/2)$ (rotation symmetry), then all even harmonics vanish.

Further, show in each case the Fourier coefficients can be evaluated by integrating the periodic signal over the half cycle only.

12. A periodic waveform is formed by eliminating the alternate cycle of a sinusoidal waveform as shown in Fig. P-1.12.

(a) Find the Fourier series (trigonometric or exponential) by direct evaluation of the coefficients.

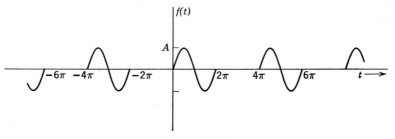

Figure P-1.12

(b) If the waveform $f(t)$ is shifted to the left by π seconds, the new waveform $f(t + \pi)$ is an odd function of time whose Fourier series contains only sine terms.

Find the Fourier series of $f(t + \pi)$. From this series, now write down the Fourier series for $f(t)$.

(c) Repeat (b) by shifting $f(t)$ to the right by π seconds.

13. A periodic waveform $f(t)$ is formed by inverting every other cycle of a sine wave as shown in Fig. P-1.13.

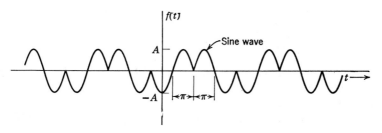

Figure P-1.13

(*a*) Find the Fourier series by direct evaluation of the coefficients.

(*b*) If this waveform $f(t)$ is shifted $\pi/2$ seconds to the right, the new waveform $f(t - \pi/2)$ is an even function of time whose Fourier series contains only cosine terms. Determine the Fourier series for $f(t - \pi/2)$ and from this series find the Fourier series for $f(t)$.

(*c*) The waveform $f(t + \pi/2)$ is an odd function of time. Determine the Fourier series for $f(t + \pi/2)$ and from this series find the Fourier series for $f(t)$.

(*d*) This waveform can also be expressed in terms of the waveform encountered in Problem 12 (Fig. P-1.12). Using the results in Problem 12, determine the Fourier series for $f(t)$.

14. Expand each of the functions $f(t)$ shown in Fig. P-1.14 by the trigonometric Fourier series, using the following terms.

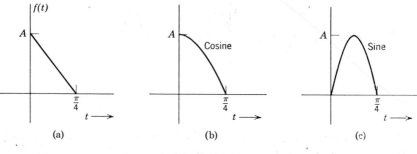

Figure P-1.14

(*a*) The sine and cosine terms of frequencies $\omega = 4, 8, 12, 16, \ldots$, etc., and a constant.

(*b*) Only the sine terms of frequencies $\omega = 2, 6, 10, 14, \ldots$, etc.

(*c*) A constant and only the cosine terms of frequencies $\omega = \frac{8}{3}, \frac{16}{3}, 8, \frac{32}{3}, \ldots$, etc.

(*d*) A constant and only the cosine terms of frequencies $\omega = 2, 4, 6, 8, \ldots$, etc.

(*e*) The sine and cosine terms of frequencies $\omega = 1, 2, 3, 4, \ldots$, etc., and a constant.

(*f*) Only the sine terms of frequencies $\omega = 1, 3, 5, 7, \ldots$, etc.

If you want to approximate $f(t)$ with a finite number of terms in these series, which of the above representations would you use? Give a qualitative and quantitative justification.

15. For each of the periodic waveforms shown in Fig. P-1.15, find the Fourier series and sketch the frequency spectrum.

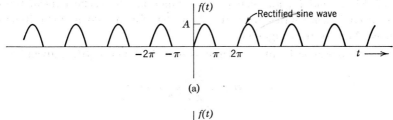

(a)

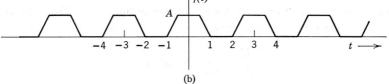

(b)

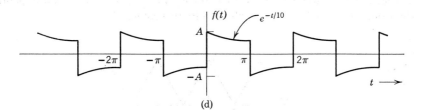

(c)

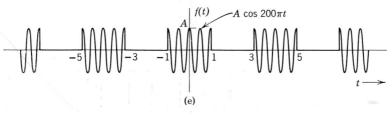

(d)

(e)

Figure P-1.15

16. Show that an arbitrary function $f(t)$ can always be expressed as a sum of an even function $f_e(t)$ and an odd function $f_o(t)$.

$$f(t) = f_e(t) + f_o(t)$$

Now find the even and odd components of the functions $u(t)$, $e^{-at}u(t)$, and e^{jt}. (*Hint:* $f(t) = \frac{1}{2}[f(t) + f(-t)] + \frac{1}{2}[f(t) - f(-t)]$.)

17. Show that for an even periodic function, the coefficients of the exponential Fourier series are real; show that for an odd periodic function, the coefficients are imaginary.

18. A Fourier series of a continuous periodic function $f(t)$ is given by

$$f(t) = \sum_{n=-\infty}^{\infty} F_n e^{jn\omega_0 t}$$

Show that the function df/dt is also a periodic function of the same period and may be expressed by a series

$$\frac{df}{dt} = \sum_{n=-\infty}^{\infty} (jn\omega_0 F_n) e^{jn\omega_0 t}$$

If the function $f(t)$ has a zero average value (that is, $F_0 = 0$), then show that its integral is also a periodic function and may be expressed by a series

$$\int f(t)\, dt = \sum_{n=-\infty}^{\infty} \frac{F_n}{jn\omega_0}\, e^{jn\omega_0 t}$$

19. Find the exponential Fourier series for the periodic functions shown in Fig. P-1.19. How do the coefficients F_n vary with n? Can you explain the results qualitatively, using the results in Problem 18?

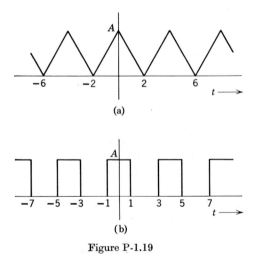

(a)

(b)

Figure P-1.19

20. A periodic function $f(t)$ is given with components of the first n harmonics only, and all the coefficients of the higher harmonics are zero. Such signals are known as bandlimited signals. Show that such a bandlimited

periodic signal is uniquely specified by its values at any $(2n + 1)$ instants in one period.

21. Show that the Fourier transform of $f(t)$ may also be expressed as

$$\mathscr{F}[f(t)] = F(\omega) = \int_{-\infty}^{\infty} f(t) \cos \omega t - j \int_{-\infty}^{\infty} f(t) \sin \omega t \, dt$$

Show also that if $f(t)$ is an even function of t, then

$$F(\omega) = 2 \int_{0}^{\infty} f(t) \cos \omega t \, dt$$

and if $f(t)$ is an odd function of t, then

$$F(\omega) = -2j \int_{0}^{\infty} f(t) \sin \omega t \, dt$$

Hence, prove that if $f(t)$ is a: Then $F(\omega)$ is a:

    Real and even function of t Real and even function of ω

    Real and odd   Imaginary and odd

    Imaginary and even   Imaginary and even

    Complex and even   Complex and even

    Complex and odd   Complex and odd

22. A function $f(t)$ can be expressed as a sum of even function and odd function (see Problem 16):

$$f(t) = f_e(t) + f_o(t)$$

where $f_e(t)$ is an even function of t and $f_o(t)$ is an odd function of t.

(a) Show that if $F(\omega)$ is the Fourier transform of a real signal $f(t)$, then $\operatorname{Re}[F(\omega)]$ is the Fourier transform of $f_e(t)$ and $j \operatorname{Im}[F(\omega)]$ is the Fourier transform of $f_o(t)$.

(b) Show that if $f(t)$ is complex,

$$f(t) = f_r(t) + jf_i(t)$$

and if $F(\omega)$ is the Fourier transform of $f(t)$, then

$$\mathscr{F}[f_r(t)] = \tfrac{1}{2}[F(\omega) + F^* (-\omega)]$$
$$\mathscr{F}[f_i(t)] = \tfrac{1}{2}[F(\omega) - F^* (-\omega)]$$

Hint: $f^*(t) = f_r(t) - jf_i(t)$

and

$$\mathscr{F}[f^* (t)] = F^*(-\omega)$$

23. Find the Fourier transforms of functions $f(t)$ shown in Fig. P-1.23.

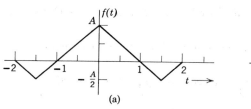

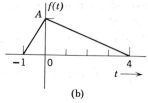

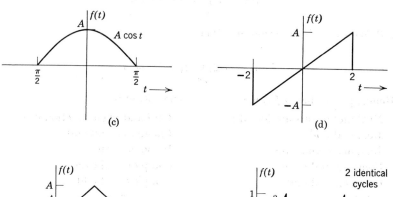

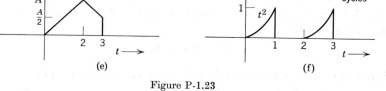

Figure P-1.23

24. Derive Eqs. 1.100a and 1.100b from the modulation theorem and Eq. 1.96.

25. Using the sampling property of the impulse function, evaluate the following integrals:

$$\int_{-\infty}^{\infty} \delta(t-2)\sin t\, dt$$

$$\int_{-\infty}^{\infty} \delta(t+3)e^{-t}\, dt$$

$$\int_{-\infty}^{\infty} \delta(1-t)(t^3+4)\, dt$$

26. Determine the functions $f(t)$ whose Fourier transforms are shown in Fig. P-1.26.

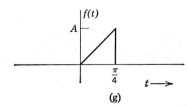

(g)

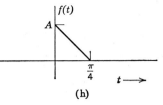

(h)

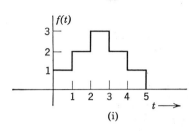

(i)

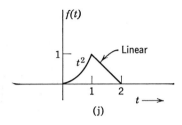

(j)

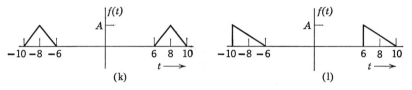

(k)   (l)

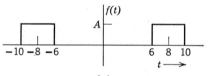

(m)

Figure P-1.23 (continued)

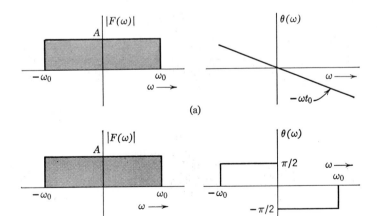

(a)

(b)

Figure P-1.26

103

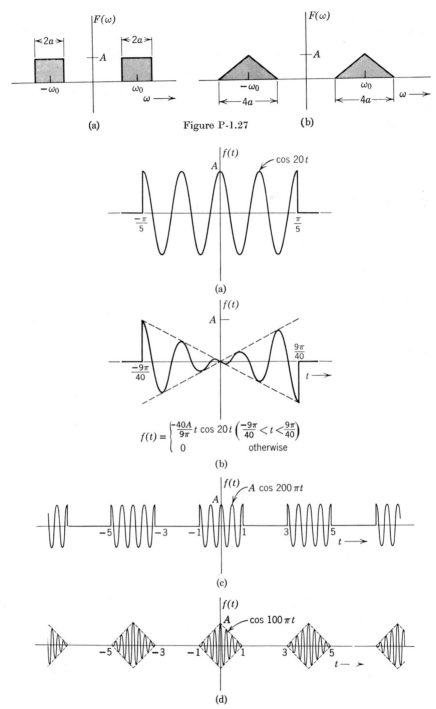

Figure P-1.27

(a)

$$f(t) = \begin{cases} \dfrac{-40A}{9\pi} t \cos 20t \left(\dfrac{-9\pi}{40} < t < \dfrac{9\pi}{40} \right) \\ 0 \qquad\qquad \text{otherwise} \end{cases}$$

(b)

(c)

(d)

Figure P-1.28

27. Find the function $f(t)$ whose Fourier transforms are shown in Fig. P-1.27. (*Hint:* Use the modulation theorem.)

28. Find the Fourier transform of the functions shown in Fig. P-1.28 using the modulation theorem. Sketch the frequency spectrum in each case.

29. If

$$f(t) \longleftrightarrow F(\omega)$$

determine the Fourier transforms of the following:

(a) $tf(2t)$

(b) $(t - 2)f(t)$

(c) $(t - 2)f(-2t)$

(d) $t\dfrac{df}{dt}$

(e) $f(1 - t)$

(f) $(1 - t)f(1 - t)$

30. Find the Fourier transforms of functions $f(t)$ in Fig. P-1.23 using the frequency-differentiation property, the time-shifting property, and Table 1.1B.

31. Find the Fourier transform of $f(t)$ in Fig. P-1.23 k, l, and m by using the dual of modulation theorem (Eq. 1.118).

32. Find the Fourier transform of a function shown in Fig. P-1.32 by:

(a) Straightforward integration.

(b) Using only the time-integration property and transform Table 1.1B.

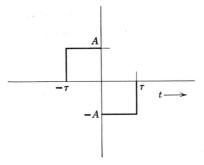

Figure P-1.32

(c) Using only the time-differentiation property, the time-shifting property, and transform Table 1.1B.

(d) Using only the time-shifting property and transform Table 1.1B.

33. The nth moment m_n of a function $f(t)$ is defined by

$$m_n = \int_{-\infty}^{\infty} t^n f(t)\, dt$$

Using the frequency-differentiation theorem, show that

$$m_n = (j)^n \frac{d^n F(0)}{d\omega^n}$$

Using this result, show that Taylor's series expansion of $F(\omega)$ can be expressed as

$$F(\omega) = m_0 - jm_1\omega - \frac{m_2\omega^2}{2!} + \frac{jm_3\omega^3}{3!} + \frac{m_4\omega^4}{4!} + \cdots$$

$$= \sum_{n=0}^{\infty} (-j)^n m_n \frac{\omega^n}{n!}$$

Determine the various moments of a gate function and, using the above equation, find its Fourier transform.

34. Show that if

$$f(t) \longleftrightarrow F(\omega)$$

then

$$|F(\omega)| \leqslant \int_{-\infty}^{\infty} |f(t)| \, dt$$

$$|F(\omega)| \leqslant \frac{1}{|\omega|} \int_{-\infty}^{\infty} \left|\frac{df}{dt}\right| \, dt$$

$$|F(\omega)| \leqslant \frac{1}{\omega^2} \int_{-\infty}^{\infty} \left|\frac{d^2f}{dt^2}\right| \, dt$$

These inequalities determine the upper bounds of $|F(\omega)|$.

35. Evaluate the following convolution integrals:

(a) $u(t) * u(t)$
(b) $u(t) * e^{-t}u(t)$
(c) $e^{-t}u(t) * e^{-2t}u(t)$
(d) $u(t) * tu(t)$
(e) $e^{-t}u(t) * tu(t)$
(f) $e^{-2t}u(t) * e^{-t}$

Verify your results for parts (b) through (f) by using Fourier transforms.

36. If $f(t)$ is a continuous signal bandlimited to ω_m radians per second, then show that

$$\frac{k}{\pi} [f(t) * Sa(kt)] = f(t) \qquad \text{for} \quad k \geqslant \omega_m$$

Hence show that

$$\frac{\omega_n}{\pi} [Sa(\omega_m t) * Sa(\omega_n t)] = Sa(\omega_m t) \qquad \text{for} \quad \omega_n \geqslant \omega_m$$

37. Evaluate the inverse Fourier transform of $Sa^2(Wt)$ by using time convolution theorem. Evaluate the convolution integral graphically.

38. Determine the minimum sampling rate and the Nyquist interval for the following signals:

   (a) $Sa(100t)$

   (b) $Sa^2(100t)$

   (c) $Sa(100t) + Sa(50t)$

   (d) $Sa(100t) + Sa^2(60t)$

39. Evaluate $f_1 * f_2$ and $f_2 * f_1$ for the functions shown in Fig. P-1.39.

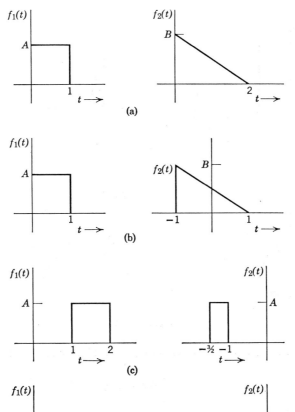

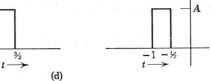

Figure P-1.39

40. Show that a causal function $f(t)$ (a function which is zero for $t < 0$) can be expressed as a continuous sum of unit ramp function as

$$f(t) = \int_0^t \frac{d^2 f}{d\tau^2} (t - \tau) \, d\tau$$

(*Hint:* Use the convolution theorem.)

41. Evaluate

$$h(t) * f_1(t), \qquad h(t) * f_2(t) \qquad \text{and} \quad h(t) * f_3(t)$$

in Fig. P-1.41.

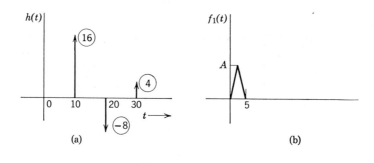

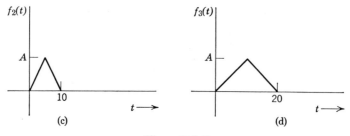

Figure P-1.41

42. The time-integration theorem (Eq. 1.119b) holds only if

$$\int_{-\infty}^{\infty} f(t) \, dt = 0$$

If this condition is not satisfied, then show that

$$\int_{-\infty}^{t} f(\tau) \, d\tau \longleftrightarrow \pi F(\omega) \, \delta(\omega) + \frac{1}{j\omega} \, F(\omega)$$

$\left[\textit{Hint:} \text{ Express } \int_{-\infty}^{t} f(\tau) \, d\tau \text{ as a convolution of } f(t) \text{ and } u(t). \right]$

43. Let the function $f(t)$ shown in Fig. 1.14 have a Fourier transform $F(\omega)$.

(a) Show that the Fourier transform of the periodic function $f_T(t)$, formed by repetition of $f(t)$ every T seconds as shown in Fig. 1.15, is given by

$$f_T(t) \leftrightarrow \frac{2\pi}{T} F(\omega)\delta_{\omega_0}(\omega) \qquad \left(\omega_0 = \frac{2\pi}{T}\right)$$

[Hint: Expand $f_T(t)$ by the Fourier series and note that the nth coefficient F_n of this series is $(1/T)F(\omega)|_{\omega=n\omega_0}$; or express $f_T(t)$ as a convolution of $f(t)$ and $\delta_T(t)$.]

(b) Sketch the Fourier transform of a gate function $G_\tau(t)$ for $\tau = \frac{1}{20}$. Using the result in (a), sketch the Fourier transform of a periodic gate function repeating every $\frac{1}{4}$ second.

Compare these results with those obtained in Example 1.7 in the text.

44. Let the function $f(t)$, shown in Fig. 1.14, have a Fourier transform $F(\omega)$. Form a new function as shown in Fig. 1.15, but with n pulses only, spaced T seconds apart. The new function $f_n(t)$ thus formed exists in the interval $(-nT/2 < t < nT/2)$ and is zero outside this interval. Show that

$$f_n(t) \leftrightarrow F(\omega) \frac{\sin(n\omega T/2)}{\sin(\omega T/2)}$$

Sketch the function $[\sin(n\omega T/2)]/[\sin(\omega T/2)]$ for $n = 15$. Show that as $n \to \infty$, this function tends to a sequence of impulse functions; that is,

$$\lim_{n \to \infty} \frac{\sin(n\omega T/2)}{\sin(\omega T/2)} = \omega_0 \delta_{\omega_0}(\omega)$$

where

$$\omega_0 = \frac{2\pi}{T}$$

Hint: Use

$$\sum_{k=-m}^{m} e^{jkx} = \frac{\sin(nx/2)}{\sin(x/2)} \qquad n = 2m + 1$$

45. A function $h(t)$ is given so that it is zero for all $t < 0$ (such a function is called a causal function). $H(\omega)$ is the Fourier transform of $h(t)$. If $R(\omega)$ and $X(\omega)$ are the real and imaginary parts, respectively, of $H(\omega)$, and if $h(t)$ contains no impulse function at the origin, then show that

$$R(\omega) = \frac{1}{\pi} \int_{-\infty}^{\infty} \frac{X(y)}{\omega - y} \, dy$$

and

$$X(\omega) = \frac{1}{\pi} \int_{-\infty}^{\infty} \frac{R(y)}{\omega - y} \, dy$$

This pair of equations defines the Hilbert transform. Modify this result if $h(t)$ is a negative time function, that is, $h(t) = 0$, for all $t > 0$. [*Hint:* Express $h(t)$ in terms of even and odd components $h_e(t)$ and $h_o(t)$ (see Problem 16)].

It follows from the results of Problem 22a that

$$h_e(t) \leftrightarrow R(\omega) \qquad \text{and} \quad h_o(t) \leftrightarrow jX(\omega)$$

Also note that for a causal function $h(t)$

$$h_e(t) = h_o(t) \text{ sgn } t$$

and

$$h_o(t) = h_e(t) \text{ sgn } t$$

Now use the result of Eq. 1.95 and the convolution theorem.

46. Show that

$$\frac{d^n \delta(t)}{dt^n} \leftrightarrow (j\omega)^n$$

$$t^n \leftrightarrow 2\pi j^n \frac{d^n \delta(\omega)}{d\omega^n}$$

chapter 2

Transmission of Signals and Power Density Spectra

2.1 SIGNAL TRANSMISSION THROUGH LINEAR SYSTEMS

Linear systems are characterized by the principle of superposition. This implies that if $r_1(t)$ is the response to a driving function $f_1(t)$ and $r_2(t)$ is the response to another driving function $f_2(t)$, then for the driving function $f_1(t) + f_2(t)$ the response will be $r_1(t) + r_2(t)$. This is the statement of the principle of superposition. More generally, the response to a driving function $\alpha f_1(t) + \beta f_2(t)$ is given by $\alpha r_1(t) + \beta r_2(t)$ for arbitrary constants α and β.

In determining a response of a linear system to a given driving function, one can take advantage of the principle of superposition. A driving function may be expressed as a sum of simpler functions for which the response can be evaluated easily. We have already seen in Chapter 1 that an arbitrary driving function $f(t)$ can be expressed as a (continuous) sum of exponentials by Fourier transform. We can utilize this fact to obtain the response of a system by Fourier (or Laplace) methods. Here, however, we shall consider another class of elementary functions, the impulse function. We shall first express a signal $f(t)$ as a continuous sum of impulse functions. From the sampling property of the impulse function (or use of Eq. 1.131), we obtain

$$f(t) = f(t) * \delta(t) = \int_{-\infty}^{\infty} f(\tau)\, \delta(t - \tau)\, d\tau \qquad (2.1)$$

This equation can be viewed as a representation of $f(t)$ in terms of impulse components. The right-hand side of Eq. 2.1 represents a continuous sum (integral) of impulse functions. In order to bring out this fact clearly, we shall express the continuous sum (integral) as a limiting form of the discrete sum. Equation 2.1 can be expressed as

$$f(t) = \lim_{\Delta\tau \to 0} \sum_{\tau=-\infty}^{\infty} [f(\tau) \, \Delta\tau] \, \delta(t - \tau) \tag{2.2}$$

Here $f(t)$ is expressed as a sum of impulses. A typical impulse located at $t = \tau$ has a strength $f(\tau) \, \Delta\tau$.

If $h(t)$ is the response of the system to a unit impulse $\delta(t)$, then the response of the system to $[f(\tau) \, \Delta\tau] \, \delta(t - \tau)$ will be* $f(t) \, \Delta\tau \, h(t - \tau)$, and the total response $r(t)$ to the driving function $f(t)$ is given by (using the principle of superposition in Eq. 2.2):

$$r(t) = \lim_{\Delta\tau \to 0} \sum_{\tau=-\infty}^{\infty} [f(t) \, \Delta\tau] h(t - \tau)$$

$$= \int_{-\infty}^{\infty} f(\tau) h(t - \tau) \, d\tau \tag{2.3}$$

$$= f(t) * h(t) \tag{2.4}$$

Use of time convolution theorem now yields

$$R(\omega) = F(\omega) H(\omega) \tag{2.5a}$$

where

$$r(t) \leftrightarrow R(\omega), \qquad f(t) \leftrightarrow F(\omega) \quad \text{and} \quad h(t) \leftrightarrow H(\omega)$$

$H(\omega)$ is called the transfer function of the system†.

If the signal $f(t)$ starts at $t = 0$ and is zero for $t < 0$, then the lower limit of integration in Eq. 2.3 can be replaced by zero. If, in addition, $h(t) = 0$ for $t < 0$ (this is true for all physical systems), then $h(t - \tau) = 0$ for $\tau > t$. Hence the upper limit of the integration in Eq. 2.3 may be

---

* Here we are assuming implicitly that the system is linear and time-invariant. For such systems, the system parameters do not change with time. Hence if the response of the system to $\delta(t)$ is $h(t)$, then the response to $\delta(t - \tau)$ must be $h(t - \tau)$. There is a class of linear systems (time-varying linear systems) where one or more of the system parameters may change with time and the response to $\delta(t - \tau)$ is not necessarily $h(t - \tau)$. If the response of a linear system is $h(t - \tau)$ to the driving function $\delta(t - \tau)$ for all values of τ, then one can infer that the system is time-invariant.

† It can be shown that for a driving function $e^{j\omega t}$ the, system response is $H(\omega) e^{j\omega t}$. For further discussion see B. P. Lathi, *Signals, Systems, and Communication*, John Wiley and Sons, New York, 1965.

replaced by t. Thus for physical systems and when $f(t) = 0$ for $t < 0$,

$$r(t) = \int_0^t f(\tau)h(t - \tau)\,d\tau \qquad (2.5b)$$

Note that Eq. 2.3 is a general form and Eq. 2.5b is a special case when both $f(t)$ and $h(t)$ vanish for $t < 0$.

2.2 THE FILTER CHARACTERISTIC OF LINEAR SYSTEMS

For a given system, an input signal $f(t)$ gives rise to a response signal $r(t)$, thus processing the signal $f(t)$ in a way that is characteristic of the system. The spectral density function of the input signal is given by $F(\omega)$ whereas the spectral density function of the response is given by $F(\omega)H(\omega)$. The system therefore modifies the spectral density function of the input signal. It is evident that the system acts as a kind of filter to various frequency components. Some frequency components are boosted in strength, some are attenuated, and some may remain unaffected. Similarly, each frequency component undergoes a different amount of phase shift in the process of transmission. Thus the system modifies the spectral density function according to its filter characteristics. The modification is carried out according to the transfer function $H(\omega)$, which represents the response of the system to various frequency components. Therefore $H(\omega)$ acts as a weighting function to different frequencies. The resultant response has the spectral density $F(\omega)H(\omega)$ (Fig. 2.1). The input signal has a spectral density $F(\omega)$, and the system response is given by $H(\omega)$. The spectral density of the response is evidently $F(\omega)H(\omega)$.

Consider the simple R-C network shown in Fig. 2.2a. A square pulse shown in Fig. 2.2c is applied at the input terminals aa' of this network. The response is the output voltage $v_o(t)$ observed across the output terminals of this network. The spectral density function of the input signal (rectangular pulse) is shown in Fig. 2.2d. The transfer function $H(\omega)$ of the network, relating the output voltage to the input voltage, is obviously $1/(j\omega + 1)$. Hence $H(\omega) = 1/(j\omega + 1)$.

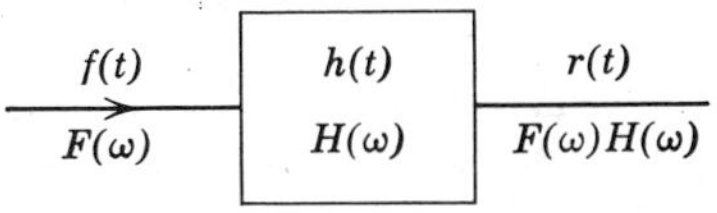

$$\begin{array}{c|c|c} f(t) & h(t) & r(t) \\ \hline F(\omega) & H(\omega) & F(\omega)H(\omega) \end{array}$$

Figure 2.1

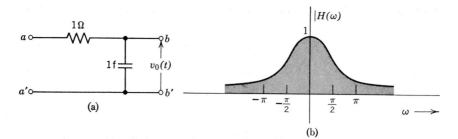

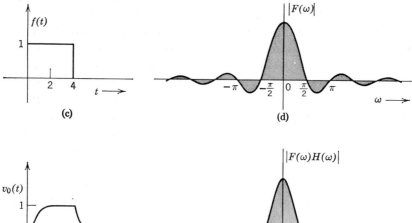

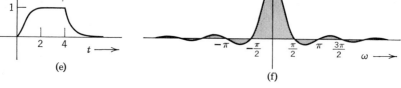

Figure 2.2

The magnitude plot $|H(\omega)|$ of the network filter characteristics is shown in Fig. 2.2b. At the present time we shall ignore the phase characteristics. Observe that this network attenuates high frequencies and allows lower frequencies to pass with a relatively small attenuation. Thus this network is the simplest form of a low-pass filter. The high frequency components in the input spectrum experience severe attenuation compared to the low frequency components. The spectral density function of the response is the product of $F(\omega)$ and $H(\omega)$. The magnitude $|F(\omega)H(\omega)|$ of the response spectral density function is shown in Fig. 2.2f. A comparison of Figs. 2.2d and 2.2f shows clearly the attenuation of high frequency components caused by the network. The response function $v_o(t)$ (Fig. 2.2e) is obviously a distorted replica of the input signal, the distortion being caused by the fact that the network does not allow equal access to the transmission of all the frequency components of the input signal. In particular, the high frequency components suffer most. This is manifested in the rising and falling characteristic of the response voltage. The input signal rises sharply at $t = 0$. The sharp rise means a rapid change, implying very high frequency components. Since the network does not allow the high frequency components, the output voltage cannot change at a rapid rate, and hence it rises and falls sluggishly compared to the input signal.

2.3 DISTORTIONLESS TRANSMISSION

The preceding discussion immediately suggests the requirement to be met by a system in order to allow the distortionless transmission of a signal. A system must attenuate all the frequency components equally; that is, $H(\omega)$ should have constant magnitude for all frequencies. Even this requirement is not sufficient to guarantee the distortionless transmission. The phase shift of each component must also satisfy certain relationships. Thus far we have ignored the effect of phase shift. It is conceivable that even if all of the frequency components of a signal are transmitted through the system with equal attenuation, if they acquire different phase shifts in the process of transmission, they may add up to an entirely different signal. We shall now investigate the requirement of the relative phase shifts of various components for a distortionless transmission.

For a distortionless transmission we require that the response be an exact replica of the input signal. This replica may, of course, have a different magnitude. The important thing is the waveform and not its relative magnitude. In general, there may also be some time delay associated with this replica. We may therefore say that a signal $f(t)$ is transmitted without distortion if the response is $kf(t - t_0)$. It is evident that the response is the exact replica of the input with a magnitude k times the original signal and delayed by t_0 seconds.

Thus if $f(t)$ is the input signal, for distortionless transmission we need the response $r(t)$ to be

$$r(t) = kf(t - t_0)$$

From the time-shifting property (Eq. 1.117) we have

$$R(\omega) = kF(\omega)e^{-j\omega t_0}$$

From Eq. 2.5a we have

$$R(\omega) = F(\omega)H(\omega) = kF(\omega)e^{-j\omega t_0}$$

Hence for a distortionless system,

$$H(\omega) = ke^{-j\omega t_0} \qquad (2.6)$$

Therefore, to achieve distortionless transmission through a system, the transfer function of the system must be of the form shown in Eq. 2.6 (Fig. 2.3a). It is evident that $|H(\omega)|$, the magnitude of the transfer function, is k, and it is constant for all frequencies. The phase shift, on the other hand, is proportional to frequency; that is,

$$\theta(\omega) = -\omega t_0 \qquad (2.7)$$

The reason for this is obvious: If two different frequency components are shifted by the same time interval, their corresponding phase changes are proportional to the frequency. For example, if a signal $\cos \omega t$ is shifted by t_0 seconds, the resulting signal $\cos \omega(t - t_0)$ can be expressed as

$$\cos \omega(t - t_0) = \cos (\omega t - \omega t_0)$$

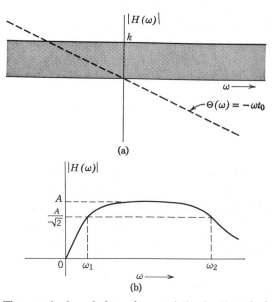

(a)

(b)

Figure 2.3   (a) The magnitude and phase characteristic of a distortionless transmission system.

It is evident that the phase shift of the new signal is $-\omega t_0$, proportional to frequency ω.

Strictly speaking, Eq. 2.7 should be

$$\theta(\omega) = n\pi - \omega t_0 \qquad (n \text{ positive or negative integer}) \qquad (2.8)$$

since the addition of excess phase of $n\pi$ radians may, at most, change the sign of the signal. Hence the phase function of a distortionless system must be of the form shown in Eq. 2.8 (Fig.2.3a).

Bandwidth of a System

The constancy of the magnitude $|H(\omega)|$ in a system is usually specified by its bandwidth. The bandwidth of a system is arbitrarily defined as the interval of frequencies over which the magnitude $|H(\omega)|$ remains within $1/\sqrt{2}$ times (within 3 db) its value at the midband. The bandwidth of a system whose $|H(\omega)|$ plot is shown in Fig. 2.3b is $\omega_2 - \omega_1$.

For distortionless transmission, we obviously need a system with infinite bandwidth. Because of physical limitations, it is impossible to construct such a system. Actually, a satisfactory distortionless transmission can be achieved by systems with finite but fairly large bandwidths. For any physical signal, the energy content decreases with frequency. Hence it is necessary only to construct a system that will transmit the frequency components which contain most of the energy of the signal. Attenuation of extremely high frequency components would tend to introduce very little distortion, since these components carry very little energy.

2.4 IDEAL FILTERS

An ideal low-pass filter transmits, without any distortion, all of the signals of frequencies below a certain frequency W radians per second. The signals of frequencies above W radians per second are completely attenuated (Fig. 2.4a). The frequency response (magnitude characteristic) of a low-pass filter is thus a gate function $G_{2W}(\omega)$. The corresponding phase function for distortionless transmission is $-\omega t_0$. The transfer function of such a filter is evidently given by

$$H(\omega) = |H(\omega)|\, e^{j\theta(\omega)}$$
$$= G_{2W}(\omega)e^{-j\omega t}$$

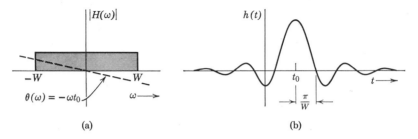

Figure 2.4   An ideal low-pass filter characteristic and its impulse response.

The unit impulse response $h(t)$ of this filter can be found by taking the inverse Fourier transform of $H(\omega)$.

$$h(t) = \mathscr{F}^{-1}[H(\omega)]$$
$$= \mathscr{F}^{-1}[G_{2W}(\omega)e^{-j\omega t_0}]$$

Using pair 12 in Table 1.1B and the time-shifting property, we get

$$h(t) = \frac{W}{\pi}\, Sa[W(t - t_0)]$$

A glance at Fig. 2.4b shows that the impulse response exists for negative values of t. This is certainly a strange result in view of the fact that the driving function (unit impulse) was applied at $t = 0$. The response therefore appears even before the driving function is applied. The system seems to anticipate the driving function. Unfortunately, it is impossible in practice to build a system with such foresight. Hence we must conclude that although an ideal low-pass filter is very desirable, it cannot be physically realizable. One can show similarly that other ideal filters (such as ideal high-pass or ideal band-pass filters shown in Fig. 2.5) are also physically unrealizable.

In practice we are satisfied with filters having characteristics close to ideal filters. A simple low-pass filter is shown in Fig. 2.6a. The transfer function of this filter is given by

$$H(\omega) = \frac{1/(1/R + j\omega C)}{j\omega L + 1/(1/R + j\omega C)}$$
$$= \frac{1}{1 - \omega^2 LC + j\omega L/R}$$

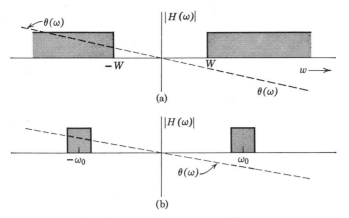

(a)

(b)

Figure 2.5   Ideal high-pass and bandpass filter characteristics.

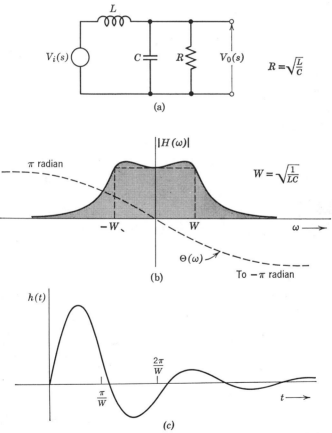

$$R = \sqrt{\frac{L}{C}}$$

(a)

$$W = \sqrt{\frac{1}{LC}}$$

(b)

(c)

Figure 2.6   (a) and (b) A simple realizable low-pass filter and its transfer function. (c) Impulse response of the low-pass filter shown in a.

119

Since

$$\frac{1}{\sqrt{LC}} = W \quad \text{and} \quad R = \sqrt{\frac{L}{C}}$$

$$H(\omega) = \frac{2W}{\sqrt{3}} \frac{\frac{\sqrt{3}}{2} W}{\left(\frac{W}{2} + j\omega\right)^2 + \left(\frac{\sqrt{3}}{2} W\right)^2}$$

The impulse response $h(t)$ is given by (pair 11, Table 1.1B)

$$h(t) = \mathscr{F}^{-1}[H(\omega)] = \frac{2W}{\sqrt{3}} e^{-Wt/2} \sin\left(\frac{\sqrt{3}}{2} Wt\right)$$

The magnitude and phase of the frequency response $H(\omega)$ are sketched in Fig. 2.6b. The impulse response $h(t)$ is shown in Fig. 2.6c. Compare the magnitude and phase characteristics of this filter with those of the ideal filter. The impulse response is very similar to that of an ideal filter except that it starts at $t = 0$.

We would like to find a test which can distinguish a physically realizable characteristic from an unrealizable one. Such a test is the famous *Paley-Wiener* criterion, which will now be introduced.

2.5 CAUSALITY AND PHYSICAL REALIZABILITY: THE PALEY-WIENER CRITERION

Physical realizability has been defined in the literature in different ways. Here we shall use the least restrictive definition that will distinguish the systems that are physically possible from those that are not. It is intuitively evident that a physically realizable system cannot have a response before the driving function is applied. This is known as the *causality condition*. This condition may be expressed alternatively. A unit impulse response $h(t)$ of a physically realizable system must be causal. A signal is said to be causal if it is zero for $t < 0$. Thus the impulse response $h(t)$ of a physically realizable system must be zero for $t < 0$. This is the time domain criterion of physical realizability. In the frequency domain this criterion implies that a necessary and sufficient condition for a magnitude function $|H(\omega)|$ to be physically realizable is that*

$$\int_{-\infty}^{\infty} \frac{|\ln |H(\omega)||}{1 + \omega^2} \, d\omega < \infty \qquad (2.9)$$

The magnitude function $|H(\omega)|$ must, however, be square-integrable before the Paley-Wiener criterion is valid†; that is,

$$\int_{-\infty}^{\infty} |H(\omega)|^2 \, d\omega < \infty$$

A system whose magnitude function violates the Paley-Weiner criterion (Eq. 2.9) has a noncausal impulse response, that is, the response exists forever in the past, prior to the application of the driving function.

We can draw some significant conclusions from the Paley-Wiener criterion. It is evident that the magnitude function $|H(\omega)|$ may be zero at some discrete frequencies, but it cannot be zero over a finite band of frequencies since this will cause the integral in Eq. 2.9 to become infinite. It is therefore evident that the ideal filters shown in Figs. 2.4 and 2.5 are not physically realizable. We can conclude from Eq. 2.9 that the amplitude function cannot fall off to zero faster than a function of exponential order. Thus

$$|H(\omega)| = ke^{-\alpha|\omega|}$$

is permissible. But the gaussian function

$$|H(\omega)| = ke^{-\alpha\omega^2}$$

is not realizable, since it violates Eq. 2.9. In short, a realizable magnitude characteristic cannot have too great a total attenuation. It is interesting to note that although the ideal filter characteristics shown in Figs. 2.4 and 2.5 are not realizable, it is possible to approach these characteristics as closely as desired. Thus the low-pass filter characteristic shown in

---

* Raymond E. A. C. Paley and Norbert Wiener, *Fourier Transforms in the Complex Domain*, American Mathematical Society Colloquium Publication 19, New York, 1934.

† If the magnitude of function $H(\omega)$ satisfies the Paley-Wiener criterion (Eq. 2.9), it does not follow that the system is physically realizable. It merely says that a suitable phase function $\theta(\omega)$ may be associated with $H(\omega)$, so that the resulting transfer function is physically realizable.

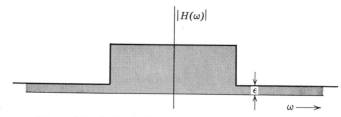

Figure 2.7   A physically realizable filter characteristic.

Fig. 2.7 is physically realizable for arbitrarily small values of ε. The reader can verify that this characteristic does not violate the Paley-Wiener criterion.

2.6 RELATIONSHIP BETWEEN THE BANDWIDTH AND THE RISE TIME

A sharp change in signal amplitude implies rapid variations of the signal with time. This gives rise to high frequency components in the signal. A signal which is relatively smooth contains predominantly low frequencies. We have observed (Fig. 2.2) that if a signal with a jump discontinuity is transmitted through a low-pass filter, the discontinuity is rounded off in the output. The change occurs gradually. If a unit step function $u(t)$ is applied to an ideal low-pass filter, the output will show a gradual rise (instead of a sharp rise in the input). The rise time of the output will depend upon the cutoff frequency of the filter. We shall now show that the rise time is inversely proportional to the cutoff frequency of the filter. The smaller the cutoff frequency, the more gradually will the signal rise at the output.

This can be seen easily by considering the response of the low-pass filter to a unit step function $u(t)$. If $H(\omega)$ is the transfer function of the ideal low-pass filter, then (Fig. 2.4a)

$$H(\omega) = G_{2W}(\omega)e^{-j\omega t_0}$$

Also

$$u(t) \leftrightarrow \pi\,\delta(\omega) + \frac{1}{j\omega}$$

If $r(t)$ is the response of the low-pass filter to $u(t)$, then

$$R(\omega) = \left[\pi\,\delta(\omega) + \frac{1}{j\omega}\right] G_{2W}(\omega)e^{-j\omega t_0}$$

Note that $H(0) = 1$. Hence $\delta(\omega)H(\omega) = H(0)\,\delta(\omega)$ and

$$R(\omega) = \pi\,\delta(\omega) + \frac{1}{j\omega}\,G_{2W}(\omega)e^{-j\omega t_0}$$

$$r(t) = \mathscr{F}^{-1}\left[\pi\,\delta(\omega) + \frac{1}{j\omega}\,G_{2W}(\omega)e^{-j\omega t_0}\right]$$

$$= \frac{1}{2} + \mathscr{F}^{-1}\left[\frac{1}{j\omega}\,G_{2W}(\omega)e^{-j\omega t_0}\right]$$

$$= \frac{1}{2} + \frac{1}{2\pi}\int_{-\infty}^{\infty}\frac{G_{2W}(\omega)}{j\omega}\,e^{j\omega(t-t_0)}\,d\omega$$

$$= \frac{1}{2} + \frac{1}{2\pi}\int_{-W}^{W}\frac{e^{j\omega(t-t_0)}}{j\omega}\,d\omega$$

$$= \frac{1}{2} + \frac{1}{2\pi}\int_{-W}^{W}\frac{\cos\,[\omega(t-t_0)]}{j\omega}\,d\omega + \frac{1}{2\pi}\int_{-W}^{W}\frac{\sin\,[\omega(t-t_0)]}{\omega}\,d\omega$$

Note that the first integral vanishes because the integrand

$$\frac{\{\cos\,[\omega(t-t_0)]\}}{\omega}$$

is an odd function of ω. The integrand in the second integral is an even function of ω. Hence we have

$$r(t) = \frac{1}{2} + \frac{1}{\pi}\int_0^W\frac{\sin\,[\omega(t-t_0)]}{\omega}\,d\omega$$

$$= \frac{1}{2} + \frac{1}{\pi}\int_0^{W(t-t_0)}\frac{\sin x}{x}\,dx$$

$$= \frac{1}{2} + \frac{1}{\pi}\int_0^{W(t-t_0)}Sa(x)\,dx$$

The integral that appears on the right-hand side cannot be evaluated in a closed form but must be obtained by expanding the sampling function in a power series. This integral is tabulated in standard tables* and is given a special name *sine integral* denoted by $Si(x)$

$$Si(x) = \int_0^x Sa(\tau)\,d\tau \qquad (2.10)$$

* E. Jahnke and F. Emde, *Tables of Functions*, Dover Publications, New York, 1945.

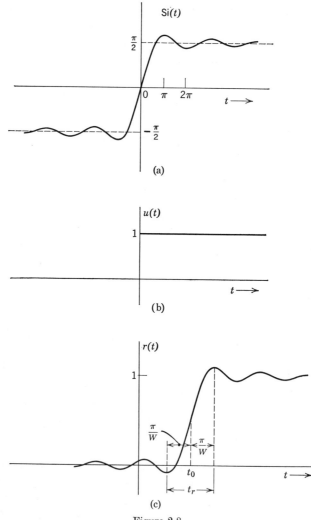

Figure 2.8

The function $Si(x)$ is plotted in Fig. 2.8a. The response $r(t)$ can now be expressed as

$$r(t) = \frac{1}{2} + \frac{1}{\pi} Si[W(t - t_0)]$$

The unit step function $u(t)$ and its response $r(t)$ are shown in Figs. 2.8b and c, respectively. The cutoff frequency of the low-pass filter is W (Fig. 2.4a).

It is obvious from Fig. 2.8c that as the cutoff frequency W is lowered, the output $r(t)$ rises more slowly. If we define the rise time t_r as the time required for the output to reach from the minimum to the maximum* (Fig. 2.8c), then it is evident from Fig. 2.8c that

$$t_r = \frac{2\pi}{W} = \frac{1}{B} \qquad (2.11)$$

where B is the filter bandwidth in Hz. Thus the rise time is inversely proportional to the cutoff frequency (in Hz) of the filter.

From these results, one can easily obtain the response of a low-pass filter to a rectangular pulse $p(t)$ shown in Fig. 2.9a:

$$p(t) = u(t) - u(t - \tau)$$

The response $r_p(t)$ of the filter to this pulse can be obtained from Eq. 2.10b by using principle of superposition:

$$r_p(t) = \frac{1}{\pi} \{Si\,[W(t - t_0)] - Si\,[W(t - t_0 - \tau)]\}$$

This response is shown in Fig. 2.9b. A reasonable approximation of this response by a trapezoidal pulse is shown in Fig. 2.9c. It can be readily seen that the transmission of a pulse through a low-pass filter causes the dispersion (spreading out) of the pulse.

2.7 THE ENERGY DENSITY SPECTRUM

A useful parameter of a signal $f(t)$ is its normalized energy. We define the normalized energy (or simply the energy) E of a signal $f(t)$ as the energy dissipated by a voltage $f(t)$ applied across a 1-ohm resistor (or by a current $f(t)$ passing through a 1-ohm resistor). Thus

$$E = \int_{-\infty}^{\infty} f^2(t)\, dt \qquad (2.12)$$

---

* There are varied definitions of the rise time t_r in the literature. Some define it as the time required for the response to rise from its zero value (nearest to the minimum) to the maximum value. This can be shown to be $0.8/B$. Alternatively, t_r is defined as the reciprocal of the slope of $r(t)$ at $t = t_0$. This yields $t_r = 0.5/B$. In electronics circuits, the rise time t_r is defined as the time required for the output to rise from 10% to 90% of its final value. In Fig. 2.8c, this is about $0.44/B$. In all these definitions, it can be seen that t_r is inversely proportional to the cutoff frequency.

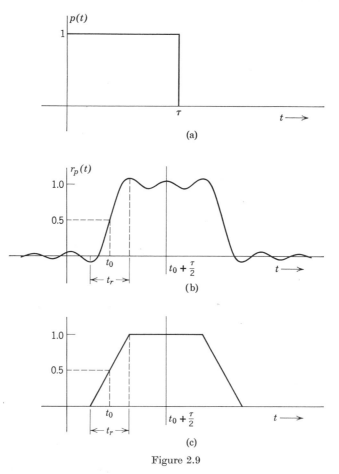

Figure 2.9

The concept of signal energy is meaningful only if the integral in Eq. 2.12 is finite. The signals for which the energy E is finite are known as *energy signals* (also known as pulse signals). With some signals, for example, periodic signals, the integral (2.12) is obviously infinite and the concept of energy is meaningless. In such cases we consider the time average of the energy, which is obviously the average power of the signal. Such signals are known as *power signals* and will be discussed later.

If $F(\omega)$ is the Fourier transform of $f(t)$,

$$f(t) = \frac{1}{2\pi} \int_{-\infty}^{\infty} F(\omega)e^{j\omega t}\, d\omega$$

and the energy E of $f(t)$ is given by

$$E = \int_{-\infty}^{\infty} f^2(t)\, dt = \int_{-\infty}^{\infty} f(t)\left[\frac{1}{2\pi}\int_{-\infty}^{\infty} F(\omega)e^{j\omega t}\, d\omega\right] dt$$

Interchanging the order of integration on the right-hand side, we get

$$E = \int_{-\infty}^{\infty} f^2(t)\, dt = \frac{1}{2\pi}\int_{-\infty}^{\infty} F(\omega)\left[\int_{-\infty}^{\infty} f(t)e^{j\omega t}\, dt\right] d\omega$$

The inner integral on the right-hand side is obviously $F(-\omega)$. Hence we have

$$\int_{-\infty}^{\infty} f^2(t)\, dt = \frac{1}{2\pi}\int_{-\infty}^{\infty} F(\omega)F(-\omega)\, d\omega$$

We have already proved (see Eq. 1.78) that for real $f(t)$

$$F(\omega)F(-\omega) = |F(\omega)|^2$$

and

$$\int_{-\infty}^{\infty} f^2(t)\, dt = \frac{1}{2\pi}\int_{-\infty}^{\infty} |F(\omega)|^2\, d\omega \qquad (2.13a)$$

$$= \int_{-\infty}^{\infty} |F(\omega)|^2\, df \qquad (2.13b)$$

This equation* states that the energy of a signal is given by the area under the $|F(\omega)|^2$ curve (integrated with respect to the frequency variable $f = \omega/2\pi$).

Interpretation of Energy Density

The energy density of a signal has a very interesting physical interpretation. We have

$$E = \frac{1}{2\pi}\int_{-\infty}^{\infty} |F(\omega)|^2\, d\omega$$

Let us consider the signal $f(t)$ applied at the input of an ideal band-pass filter whose transfer function $H(\omega)$ is shown in Fig. 2.10. This filter suppresses all frequencies except a narrow band $\Delta\omega$ ($\Delta\omega \to 0$) centered at frequency ω_0. If $R(\omega)$ is the Fourier transform of the response $r(t)$ of this filter, then

$$R(\omega) = F(\omega)H(\omega)$$

* Equation 2.13, which corresponds to Parseval's theorem (Eq. 1.34) for nonperiodic signals, is called Parseval's theorem or Plancharel's theorem.

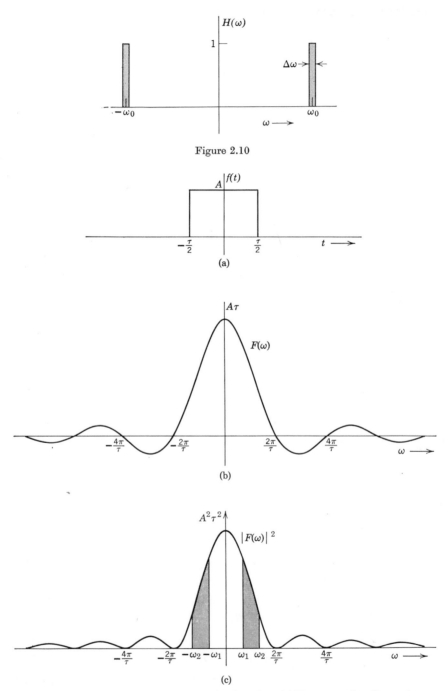

Figure 2.10

(a)

(b)

(c)

Figure 2.11   (a) $f(t)$.   (b) The spectral density function.   (c) The energy density spectrum.

and the energy E_o of the output signal $r(t)$ is given by Eq. 2.13,

$$E_o = \frac{1}{2\pi} \int_{-\infty}^{\infty} |F(\omega)H(\omega)|^2 \, d\omega$$

Since $H(\omega) = 0$ everywhere except over a narrowband $\Delta\omega$ where it is unity, we have (for $\Delta\omega \to 0$)

$$E_o = 2 \frac{1}{2\pi} |F(\omega_0)|^2 \, \Delta\omega$$

The energy of the output signal thus is $2 |F(\omega_0)|^2 \, \Delta f$. As can be seen from Fig. 2.10, only the frequency components of $f(t)$ which lie in the narrowband $\Delta\omega$ are transmitted intact through the filter. The remaining frequency components are completely suppressed. Obviously, $2 |F(\omega_0)|^2 \, \Delta f$ represents the contribution to the energy of $f(t)$ by frequency components of $f(t)$ lying in the narrowband Δf centered at ω_0. Hence $2 |F(\omega)|^2$ is the energy per unit bandwidth (in Hz) contributed by frequency components centered at ω. Note that the units of the energy density spectrum are joules per Hz.

It should be noted that we have the energy contribution from negative as well as positive frequency components. Moreover, the contribution by negative and positive frequency components is equal since

$$|F(\omega)|^2 = |F(-\omega)|^2$$

Hence we may interpret that the amount $|F(\omega)|^2$ [half of the energy $2 |F(\omega)|^2$] was contributed by positive frequency components, and the remaining $|F(\omega)|^2$ was contributed by the negative frequency components.* For this reason $|F(\omega)|^2$ is called the energy density spectrum. This represents the energy per unit bandwidth (either positive or negative). The energy density spectrum $\Psi_f(\omega)$ is thus defined† as

$$\Psi_f(\omega) = |F(\omega)|^2 \tag{2.14}$$

The energy density spectrum provides us with the relative contribution of energy by various frequency components. Figure 2.11 shows the

---

* This distinction is more for convenience than natural. Actually, a combination of negative and positive frequencies contributes the energy associated with any particular frequency band.

† In the literature $\Psi_f(\omega)$ is defined in several ways. One definition is $\Psi_f(\omega) = 2 |F(\omega)|^2$. In this case the energies of positive and negative frequencies are lumped together. The total energy E is found by integrating $\Psi_f(\omega)$ over 0 to ∞ or 0 to $-\infty$ only. Alternately $\Psi_f(\omega)$ is defined as $(1/\pi) |F(\omega)|^2$. Here the energy density is defined as the energy per radian bandwidth instead of cycle bandwidth.

gate function, its Fourier transform and the energy density spectrum $|F(\omega)|^2$.

The total energy E is given by

$$E = \frac{1}{2\pi} \int_{-\infty}^{\infty} |F(\omega)|^2 \, d\omega \qquad (2.15)$$

$$= \int_{-\infty}^{\infty} |F(\omega)|^2 \, df \qquad (2.16)$$

$$= \int_{-\infty}^{\infty} \Psi_f(\omega) \, df \qquad (2.17)$$

It is evident from the fact that

$$|F(\omega)|^2 = |F(-\omega)|^2$$

the energy density function is a real and even function of ω. Hence Eq. 2.15 may be expressed as

$$E = \frac{1}{\pi} \int_{0}^{\infty} F(\omega)^2 \, d\omega \qquad (2.18)$$

$$= 2 \int_{0}^{\infty} |F(\omega)|^2 \, df \qquad (2.19)$$

Energy Densities of the Input and the Response

If $f(t)$ and $r(t)$ are the driving function and the corresponding response of a linear system with transfer function $H(\omega)$, then

$$R(\omega) = H(\omega)F(\omega)$$

The energy density spectrum of the driving function is $|F(\omega)|^2$ and that of the response is $|R(\omega)|^2$. It is evident that

$$|R(\omega)|^2 = |H(\omega)|^2 \, |F(\omega)|^2 = |H(\omega)|^2 \Psi_f(\omega) \qquad (2.20)$$

Hence the energy density spectrum of the response is given by the energy density spectrum of the driving signal multiplied by $|H(\omega)|^2$.

2.8 THE POWER DENSITY SPECTRUM

It was mentioned earlier that some signals (for example, periodic signals) have infinite energy. Such signals are called power signals. The meaningful parameter for a power signal $f(t)$ is the average power

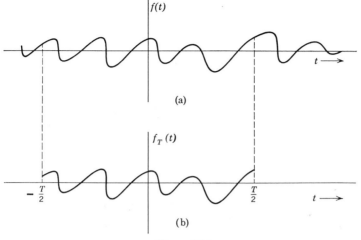

Figure 2.12

P. We define the average power (or simply power) or a signal $f(t)$ as the average power dissipated by a voltage $f(t)$ applied across a 1-ohm resistor (or by a current $f(t)$ passing through a 1-ohm resistor). Thus the average power P of a signal $f(t)$ is given by

$$P = \lim_{T \to \infty} \frac{1}{T} \int_{-T/2}^{T/2} f^2(t)\, dt \tag{2.21a}$$

Note that the average power as defined in Eq. 2.21a is also the mean square value of $f(t)$. If we denote the mean square value of $f(t)$ by $\overline{f^2(t)}$, then

$$P = \overline{f^2(t)} = \lim_{T \to \infty} \frac{1}{T} \int_{-T/2}^{T/2} f^2(t)\, dt \tag{2.21b}$$

We can now continue with the same procedure as that used for obtaining energy density.

Let us form a new function $f_T(t)$ by truncating $f(t)$ outside the interval $|t| > T/2$. The truncated function (Fig. 2.12b) can be expressed as

$$f_T(t) = \begin{cases} f(t) & |t| < \dfrac{T}{2} \\ 0 & \text{otherwise} \end{cases}$$

As long as T is finite, $f_T(t)$ has finite energy. Let

$$f_T(t) \leftrightarrow F_T(\omega)$$

Then the energy E_T of $f_T(t)$ is given by

$$E_T = \int_{-\infty}^{\infty} f_T{}^2(t)\, dt = \int_{-\infty}^{\infty} |F_T(\omega)|^2\, df$$

But

$$\int_{-\infty}^{\infty} f_T{}^2(t)\, dt = \int_{-T/2}^{T/2} f^2(t)\, dt$$

Hence the average power P is given by

$$P = \lim_{T \to \infty} \frac{1}{T} \int_{-T/2}^{T/2} f^2(t)\, dt = \int_{-\infty}^{\infty} \lim_{T \to \infty} \frac{|F_T(\omega)|^2}{T}\, df$$

As T increases, the energy of $f_T(t)$ also increases. Thus $|F_T(\omega)|^2$ increases with T. In the limit as $T \to \infty$, the quantity $|F_T(\omega)|^2/T$ may approach a limit. Assuming that such a limit exists, we define $S_f(\omega)$, the power density spectrum of $f(t)$, as

$$S_f(\omega) = \lim_{T \to \infty} \frac{|F_T(\omega)|^2}{T} \tag{2.22}$$

$S_f(\omega)$ is called the average power density spectrum, or simply the *power density spectrum* (also spectral power density) of $f(t)$. Hence

$$\text{Average power } P = \overline{f^2(t)} = \lim_{T \to \infty} \frac{1}{T} \int_{-T/2}^{T/2} f^2(t)\, dt$$

$$= \int_{-\infty}^{\infty} S_f(\omega)\, df \tag{2.23a}$$

$$= \frac{1}{2\pi} \int_{-\infty}^{\infty} S_f(\omega)\, d\omega \tag{2.23b}$$

Note that

$$|F_T(\omega)|^2 = F_T(\omega) F_T(-\omega)$$

It is obvious from this equation and Eq. 2.22 that the power density is an even function of ω. Hence Eqs. 2.23a and b may be expressed as

$$\text{Average power} = \overline{f^2(t)} = 2 \int_0^{\infty} S_f(\omega)\, df \tag{2.23c}$$

$$= \frac{1}{\pi} \int_0^{\infty} S_f(\omega)\, d\omega \tag{2.23d}$$

From Eq. 2.22 it is evident that the power density spectrum of a signal retains only the information of magnitude of the frequency spectrum $F_T(\omega)$. The phase information is lost. It follows that all signals with identical frequency spectrum magnitude but different phase functions will have identical power density spectra. Thus for a given signal there is a unique power density spectrum. But the converse is not true; there may be a large number of signals (in fact, infinite) which can have the same power density spectrum.

We have shown that for energy signals, if

$$f(t) \longleftrightarrow F(\omega)$$

then

$$f(t) \cos \omega_0 t \longleftrightarrow \tfrac{1}{2}[F(\omega + \omega_0) + F(\omega - \omega_0)]$$

and

$$f(t) \sin \omega_0 t \longleftrightarrow \frac{j}{2}[F(\omega + \omega_0) - F(\omega - \omega_0)]$$

We can extend these results for power signals.

Consider a power signal $f(t)$ with a power density spectrum $S_f(\omega)$:

$$S_f(\omega) = \lim_{T \to \infty} \frac{|F_T(\omega)|^2}{T}$$

Consider the signal $\varphi(t)$ given by

$$\varphi(t) = f(t) \cos \omega_0 t$$

If $S_\varphi(\omega)$ is the power density spectrum of $\varphi(t)$, then by definition

$$S_\varphi(\omega) = \lim_{T \to \infty} \frac{|\Phi_T(\omega)|^2}{T}$$

where

$$f_T(t) \cos \omega_0 t = \varphi_T(t) \longleftrightarrow \Phi_T(\omega)$$

From the modulation theorem (Eq. 1.116a), it follows that

$$\Phi_T(\omega) = \tfrac{1}{2}[F_T(\omega + \omega_0) + F_T(\omega - \omega_0)]$$

and

$$\begin{aligned} S_\varphi(\omega) &= \frac{1}{4} \lim_{T \to \infty} \frac{|F_T(\omega + \omega_0) + F_T(\omega - \omega_0)|^2}{T} \\ &= \frac{1}{4} \lim_{T \to \infty} \frac{|F_T(\omega + \omega_0)|^2 + |F_T(\omega - \omega_0)|^2}{T} \end{aligned} \tag{2.24a}$$

Note that the cross-product term $F_T(\omega + \omega_0)F_T(\omega - \omega_0)$ vanishes because the two spectra are nonoverlapping (see Fig. 2.13).

Equation 2.22 now yields*

$$S_\varphi(\omega) = \tfrac{1}{4}[S_f(\omega + \omega_0) + S_f(\omega - \omega_0)] \qquad (2.24b)$$

It can be easily seen that if

$$\varphi(t) = f(t) \sin \omega_0 t$$

then

$$S_\varphi(\omega) = \lim_{T \to \infty} \frac{\left| \dfrac{j}{2}[F_T(\omega + \omega_0) - F(\omega - \omega_0)] \right|^2}{T}$$

$$= \frac{1}{4} \lim_{T \to \infty} \frac{|F_T(\omega + \omega_0)|^2 + |F_T(\omega - \omega_0)|^2}{T}$$

$$= \tfrac{1}{4}[S_f(\omega + \omega_0) + S_f(\omega - \omega_0)] \qquad (2.24c)$$

It is evident that $f(t) \cos \omega_0 t$ and $f(t) \sin \omega_0 t$ have identical power spectra. The power density spectra of $f(t)$ and $\varphi(t)$ are shown in Fig. 2.13. Equations 2.24b and 2.24c represent the extension of the modulation theorem (Eq. 1.116) to power signals.

Example 2.1   *Power of Amplitude Modulated Signal*

Calculate the power (the mean square value) of the modulated signal $\varphi(t) = f(t) \cos \omega_0 t$ where $f(t)$ is a power signal with the mean square value $\overline{f^2(t)}$.

The power of a signal is $1/2\pi$ times the area under its power density spectrum:

$$\overline{\varphi^2(t)} = \frac{1}{2\pi} \int_{-\infty}^{\infty} S_\varphi(\omega)\, d\omega$$

and

$$\overline{f^2(t)} = \frac{1}{2\pi} \int_{-\infty}^{\infty} S_f(\omega)\, d\omega$$

* The results derived here apply to low-pass signals $f(t)$ where $F_T(\omega + \omega_0)$ and $F_T(\omega - \omega_0)$ are nonoverlapping. If $f(t)$ is a bandpass signal, it can be easily seen that $F_T(\omega + (\omega_0))$ and $F_T(\omega - \omega_0)$ overlap around the origin and, strictly speaking, these results Eqs. 2.24b and 2.24c) do not apply. If, however, $f(t)$ is a random signal, these results do apply. If a source generates random signal, every time it is turned on it generates a different waveform. For this reason the power density spectrum of a random signal (random process) is defined as the mean of the power density spectra of all possible waveforms generated by the source. In such a case it can be shown that, although $F_T(\omega + \omega_0)$ and $F_T(\omega - \omega_0)$ are overlapping, the mean of their product is zero and the results in Eqs. 2.24b and 2.24c are valid. This point can be fully appreciated only with some background in random processes.

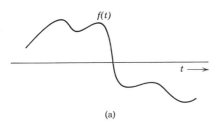

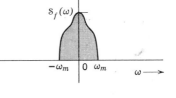

(a)

(b)

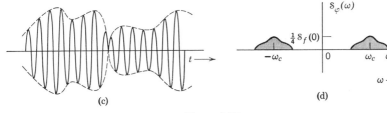

$\varphi(t) = f(t) \cos \omega_c t$ or $f(t) \sin \omega_c t$

(c)

(d)

Figure 2.13

The area under $S_f(\omega)$ is $2\pi\overline{f^2(t)}$ (Fig. 2.13). But from Eq. 2.24b, it follows that the area under $S_\varphi(\omega)$ is half the area under $S_f(\omega)$. This is also obvious from Fig. 2.13. Hence

$$\overline{\varphi^2(t)} = \tfrac{1}{2}\overline{f^2(t)} \tag{2.25a}$$

Thus the mean square value of the modulated signal $f(t) \cos \omega_c t$ is half the mean square value of the modulating signal $f(t)$:

$$\overline{[f(t) \cos \omega_c t]^2} = \tfrac{1}{2}\overline{f^2(t)} \tag{2.25b}$$

Following along similar lines, the reader can easily show that

$$\overline{[f(t) \sin \omega_c t]^2} = \tfrac{1}{2}\overline{f^2(t)} \tag{2.25c}$$

In deriving these results, we have implicitly assumed that $\omega_c > \omega_m$. If $\omega_c < \omega_m$, then $S_f(\omega + \omega_c)$ and $S_f(\omega - \omega_c)$ in Fig. 2.13d will overlap and Eq. 2.24 will be invalid. Thus Eq. 2.25 is valid only if $\omega_c > \omega_m$ [the highest frequency of $f(t)$].

Example 2.2   *Mean Square Value of a Bandlimited Signal in Terms of its Sample Values*

Determine the power (the mean square value) of a bandlimited signal in terms of its sample values.

A signal $f(t)$, bandlimited to ω_m radians per second, can be expressed in terms of its samples as (Eq. 1.140)

$$f(t) = \sum_k f_k Sa[\omega_m(t - kT)] \tag{2.26}$$

where f_k is the kth sample, $f_k = f(kT)$, and T is the sampling interval

$$T = \frac{1}{2f_m} = \frac{\pi}{\omega_m} \tag{2.27}$$

The mean square value of $f(t)$ is given by

$$\overline{f^2(t)} = \lim_{\tau \to \infty} \frac{1}{\tau} \int_{-\tau/2}^{\tau/2} f^2(t)\, dt$$

Substituting Eq. 2.26 in the preceding equation and interchanging the summation with integration, we obtain

$$\overline{f^2(t)} = \lim_{\tau \to \infty} \frac{1}{\tau} \int_{-\tau/2}^{\tau/2} \left(\sum_k f_k Sa[\omega_m(t - kT)] \right)^2 dt \tag{2.28}$$

Various sampling functions appearing in Eq. 2.28 can be shown to be orthogonal, that is,

$$\int_{-\infty}^{\infty} Sa[\omega_m(t - nT)]Sa[\omega_m(t - mT)]\, dt = \begin{cases} \dfrac{\pi}{\omega_m} & m = n \\ 0 & m \neq n \end{cases} \tag{2.29}$$

Hence the integral of all cross-product terms in Eq. 2.28 vanishes, and (after changing the operation of integration and summation) we get

$$\overline{f^2(t)} = \sum_k \lim_{\tau \to \infty} \frac{1}{\tau} \int_{-\tau/2}^{\tau/2} f_k^2 Sa^2[\omega_m(t - kT)]\, dt$$

$$= \sum_k \frac{f_k^2}{\tau} \lim_{\tau \to \infty} \int_{-\tau/2}^{\tau/2} Sa^2[\omega_m(t - kT)]\, dt$$

Using the result of Eq. 2.29 in the above equation yields

$$\overline{f^2(t)} = \sum_k \frac{\pi}{\omega_m \tau} f_k^2 = \frac{1}{2f_m \tau} \sum_k f_k^2 \tag{2.30}$$

Note that τ is the interval over which averaging is performed. Since the sampling rate is $2f_m$ samples per second, $2f_m\tau$ is the total number of samples in the interval τ. The right-hand side of Eq. 2.30 is obviously the mean of the square of the samples (the mean square of the sample amplitudes). It is therefore evident from Eq. 2.30 that the mean square value of a band-limited signal is equal to the mean square value of its samples. This result (Eq. 2.30) may also be expressed as

$$\overline{f^2(t)} = \overline{f_k^2} \tag{2.31}$$

Power Density Spectrum of a Periodic Signal

Consider a periodic signal $f(t)$ and its Fourier series representation

$$f(t) = \sum_{n=-\infty}^{\infty} F_n e^{jn\omega_0 t}$$

$F(\omega)$, the Fourier transform of $f(t)$, is given by Eq. 1.102,

$$F(\omega) = 2\pi \sum_{n=-\infty}^{\infty} F_n \delta(\omega - n\omega_0)$$

The truncated function $f_T(t)$ can be obtained by multiplying $f(t)$ by a gate function

$$f_T(t) = G_T(t) f(t)$$

Using the frequency convolution theorem, we get

$$F_T(\omega) = \frac{1}{2\pi} T\, Sa\left(\frac{\omega T}{2}\right) * F(\omega)$$

$$= T\, Sa\left(\frac{\omega T}{2}\right) * \sum_{n=-\infty}^{\infty} F_n \delta(\omega - n\omega_0)$$

$$= T \sum_{n=-\infty}^{\infty} F_n Sa\left(\frac{\omega T}{2}\right) * \delta(\omega - n\omega_0)$$

$$= T \sum_{n=-\infty}^{\infty} F_n Sa\left[\frac{(\omega - n\omega_0)T}{2}\right]$$

Therefore

$$\lim_{T \to \infty} \frac{|F_T(\omega)|^2}{T} = \lim_{T \to \infty} T \sum_{n=-\infty}^{\infty} |F_n|^2 Sa^2\left[\frac{(\omega - n\omega_0)T}{2}\right] \qquad (2.32)$$

Note that as $T \to \infty$, the function $Sa\{[(\omega - n\omega_0)T]/2\}$ tends to be concentrated at $\omega = n\omega_0$ (see Eq. 1.86). Hence the expression for $|F_T(\omega)|^2$ in Eq. 2.32 does not have any cross-product terms since each component exists where all the other components are zero. Use of Eq. 1.87 in Eq. 2.26 now yields

$$S_f(\omega) = \lim_{T \to \infty} \frac{|F_T(\omega)|^2}{T} = 2\pi \sum_{n=-\infty}^{\infty} |F_n|^2 \delta(\omega - n\omega_0) \qquad (2.33)$$

It follows that if

$$f(t) = a \cos (\omega_0 t + \theta)$$

$$= \frac{a}{2} [e^{j(\omega_0 t + \theta)} + e^{-j(\omega_0 t + \theta)}]$$

$$= \left(\frac{a}{2} e^{j\theta}\right) e^{j\omega_0 t} + \left(\frac{a}{2} e^{-j\theta}\right) e^{-j\omega_0 t}$$

then from Eq. 2.33,

$$S_f(\omega) = 2\pi \left[\frac{a^2}{4} \delta(\omega - \omega_0) + \frac{a^2}{4} \delta(\omega + \omega_0)\right]$$

$$= \frac{\pi a^2}{2} [\delta(\omega - \omega_0) + \delta(\omega + \omega_0)] \tag{2.34}$$

Thus the power density spectrum of a sinusoidal signal $a \cos (\omega_0 t + \theta)$ is given by two impulses at $\pm \omega_0$ of strength $\pi a^2/2$ each. Note that the power density spectrum is independent of θ.

The power of a signal is its mean square value. Hence P, the power of a sinusoidal signal

$$f(t) = a \cos (\omega_0 t + \theta)$$

is given by

$$P = \overline{f^2(t)} = \frac{a^2}{2} \tag{2.35}$$

This follows from the fact that the mean square value of any sinusoidal signal of amplitude a is $a^2/2$. Alternatively, the power P is $1/2\pi$ times the area under $S_f(\omega)$ in Eq. 2.34. This is easily seen to be $a^2/2$.

Example 2.3

Find the power of a signal $A + f(t)$, where A is a constant and the signal $f(t)$ is a power signal with zero mean value.

Let P be the power of $A + f(t)$. Then by definition

$$P = \lim_{T \to \infty} \frac{1}{T} \int_{-T/2}^{T/2} [A + f(t)]^2 \, dt$$

$$= \lim_{T \to \infty} \frac{1}{T} \left[\int_{-T/2}^{T/2} A^2 \, dt + \int_{-T/2}^{T/2} f^2(t) \, dt + 2A \int_{-T/2}^{T/2} f(t) \, dt\right]$$

$$= A^2 + \overline{f^2(t)} + \lim_{T \to \infty} \frac{1}{T} \int_{-T/2}^{T/2} f(t) \, dt$$

The integral on the right-hand side is the mean value of $f(t)$, which is given to be zero. Hence

$$P = \overline{[A + f(t)]^2} = A^2 + \overline{f^2(t)} \tag{2.36}$$

Power Densities of the Input and the Response

Let us apply a power signal $f(t)$ at the input of a linear system with transfer function $H(\omega)$, and let the output signal be $r(t)$. We shall express this fact symbolically as

$$f(t) \to r(t)$$

The signals $f_T(t)$ and $r_T(t)$ represent signals $f(t)$ and $r(t)$ respectively, truncated beyond $|t| = T/2$.

Let us now apply at the input the truncated signal $f_T(t)$. This is the same as applying the signal $f(t)$ over the interval $|t| < T/2$ and no signal beyond this interval. The response generally will not be $r_T(t)$; it will extend beyond $t = T/2$. However, since the input is zero for $t > T/2$, for a stable system the response for $t > T/2$ must decay with time. In the limit as $T \to \infty$, this contribution (beyond $t = T/2$) will be of no significance when viewed in a proper perspective of a signal of infinite time duration.*

Hence for $T \to \infty$, the response for $f_T(t)$ may be considered to be $r_T(t)$ without much error. Thus

$$\lim_{T \to \infty} f_T(t) \to r_T(t)$$

and

$$\lim_{T \to \infty} R_T(\omega) = H(\omega) F_T(\omega)$$

Furthermore, by definition $S_r(\omega)$, the power density spectrum of the output signal $r(t)$, is given by

$$S_r(\omega) = \lim_{T \to \infty} \frac{1}{T} |R_T(\omega)|^2$$

$$= \lim_{T \to \infty} \frac{1}{T} |H(\omega) F_T(\omega)|^2$$

$$= |H(\omega)|^2 \lim_{T \to \infty} \frac{1}{T} |F_T(\omega)|^2$$

$$= |H(\omega)|^2 \, S_f(\omega) \tag{2.37}$$

---

* Similar argument applies for $t < -T/2$. The response for $t > -T/2$ due to the input signal $f(t)$ applied before $t = -T/2$ will become insignificant in the perspective as $T \to \infty$.

It is therefore evident that the output signal power density is given by $|H(\omega)|^2$ times the power density of the input signal.

The mean square value of a signal is given by $1/2\pi$ times the area under its power density spectrum. Hence the mean square value of the response $r(t)$ is given by

$$\overline{r^2(t)} = \frac{1}{2\pi} \int_{-\infty}^{\infty} |H(\omega)|^2 \, S_f(\omega) \, d\omega$$

Note that $\overline{r^2(t)}$, the mean square value of $r(t)$, by definition is the power of $r(t)$.

Interpretation of Power Density Spectrum

Power density spectrum has a physical interpretation very similar to that of energy density. Let us consider a power signal $f(t)$ applied at the input of an ideal bandpass filter whose transfer function $H(\omega)$ is shown in Fig. 2.10. This filter suppresses all frequencies except a narrowband $\Delta\omega(\Delta\omega \to 0)$ centered at frequency ω_0. The power density spectrum of the output signal will be given by $S_f(\omega_0)$ over the narrowband $\Delta\omega$ centered at $\pm\omega_0$ as shown in Fig. 2.14.

The power P_o of the output signal according to Eq. 2.23 is given by

$$P_o = 2S_f(\omega_0) \, \Delta f \tag{2.38}$$

It can be seen that this is the power contribution of frequency components of $f(t)$ lying in the band Δf centered at ω_0. As usual, we have negative and positive frequency components. We therefore attribute the $S_f(\omega_0) \, \Delta f$ contribution to positive frequencies and an equal amount

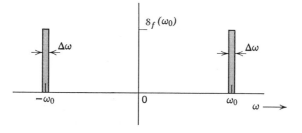

Figure 2.14

to negative frequencies. Thus $S_f(\omega)$ is the power per unit bandwidth (in Hz) contributed by frequency components centered at frequency ω. The units of the power density spectrum are watts per Hz.

Example 2.4 *Power Density of df/dt.*

A power signal $f(t)$ has a power $S_f(\omega)$. Find the power density spectrum of the signal df/dt.

If a signal $f(t)$ is transmitted through an ideal differentiator the output will be df/dt. The transfer function of the differentiator is $j\omega$. This follows from the time differentiation property (Eq. 1.119a), which states that if

$$f(t) \leftrightarrow F(\omega), \qquad \text{then} \quad \frac{df}{dt} \leftrightarrow j\omega F(\omega)$$

Thus, for an ideal differentiator

$$|H(\omega)|^2 = |j\omega|^2 = \omega^2$$

and $S_{\dot{f}}(\omega)$, the power density spectrum of df/dt, is given by

$$S_{\dot{f}}(\omega) = \omega^2 S_f(\omega) \tag{2.39}$$

PROBLEMS

1. Consider an ideal low-pass filter as in Fig. 2.4a. Show that the response of this filter to signals $(\pi/W)\,\delta(t)$ and $Sa(Wt)$ is identical. Comment upon this result.

2. A resistive network formed by two resistors R_1 and R_2 is used as an attenuator to reduce the voltage applied at terminals ab. Resistors R_1 and R_2 have stray capacitances across them of magnitude C_1 and C_2, respectively, as shown in Fig. P-2.2. What should be the relationship between R's and C's in order to have a distortionless attenuation?

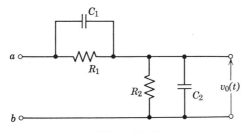

Figure P-2.2

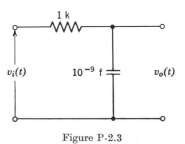

Figure P-2.3

3. Find the transfer function relating the output voltage to the input voltage for the R-C network shown in Fig. P-2.3. Sketch the magnitude and the phase characteristics of the transfer function. What kind of signals can be transmitted with reasonable fidelity through this network. What is the amount of delay in the output waveform.

4. An amplifier has a gain characteristic (transfer function) $H(\omega)$ given by

$$H(\omega) = \frac{K}{j\omega + \omega_0}$$

(a) Sketch the magnitude and phase characteristics of the amplifier.

(b) State what type of signals can be amplified by this amplifier with a reasonable fidelity and give the reasons. Find the delay time in the transmission of such signals.

5. Find the transfer function of the lattice network shown in Fig. P-2.5.

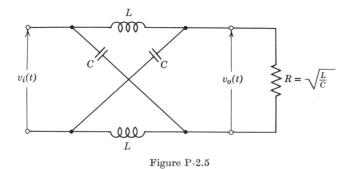

Figure P-2.5

(a) For this network find the transfer function relating $v_o(t)$ to $v_i(t)$.

(b) Sketch the magnitude and the phase functions.

(c) Can this network be used in general to transmit signals without distortion? State the conditions on signals which may be transmitted through this network with reasonable fidelity.

(d) Find the input impedance of this network.

(e) Can you form a delay line using large numbers of such lattice sections in cascade? What is the condition on signals which may be delayed (with reasonable fidelity) by such sections? Find the amount of delay obtained by using n sections in cascade.

6. Find the transfer function relating the output voltage $v_o(t)$ to the input current $i(t)$ in Fig. P-2.6. The output voltage should have the same waveform as $i(t)$ (distortionless transmission). Find the desired element values. Is there any delay involved in the transmission?

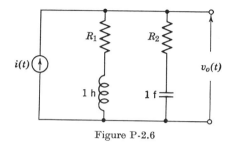

Figure P-2.6

7. The transfer function of an ideal bandpass filter is given by

$$H(\omega) = k[G_W(\omega - \omega_0) + G_W(\omega + \omega_0)]e^{-j\omega t_0}$$

(a) Sketch the magnitude and the phase function of this transfer function.
(b) Evaluate the impulse response of this filter.
(c) Sketch this response and state whether the filter is physically realizable.

8. Find the response of the bandpass filter in Problem 7 to the input signal $\cos \omega_0 t \, u(t)$. (Assume the filter is narrowband so that the function $1/j(\omega + \omega_0) \simeq 1/2j\omega_0$ over the entire passband at $\omega = \omega_0$, and $1/j(\omega - \omega_0) \simeq -1/2j\omega_0$ over the entire passband at $\omega = -\omega_0$.)

9. Find and sketch the response of an ideal low-pass filter (Fig. 2.4a) to a signal $f(t)$ shown in Fig. P-2.9. The cutoff frequency of the filter is 10 kHz. and the delay time t_0 is 1 μs.

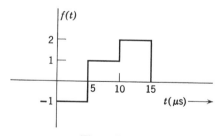

Figure P-2.9

10. Find the unit impulse and the unit step response of the ideal high-pass filter whose transfer function is given by

$$H(\omega) = [1 - G_{2W}(\omega)]e^{-j\omega t_0}$$

11. An equivalent circuit of a vacuum tube amplifier is shown in Fig. P-2.11.

(a) Find the transfer function $H(\omega)$ relating the output voltage to the input voltage, and determine the bandwidth of voltage gain.

(b) Find and sketch the output voltage $v_0(t)$ when a unit voltage is applied at the input terminals gk.

(c) If the rise time is defined as the time required for the response in b to rise from 10 to 90 percent of its final value, show that the bandwidt

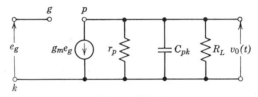

Figure P-2.11

(in radians per second) multiplied by the rise time is a constant, independent of the tube parameters. Show that this constant is 2.2.

(d) Show that the product of the d-c gain $[H(0)]$ and the bandwidth found in (a) is equal to g_m/C_{pk}.

12. Assume a signal $f(t)$ to be bandlimited; that is, the spectral density function $F(\omega)$ has no frequency component beyond a certain frequency W. In other words, $F(\omega) = 0$ for $|\omega| > W$. Such a signal can be amplified without distortion by an amplifier whose transfer function $H(\omega)$ has an ideal low-pass filter characteristic

$$H(\omega) = kG_{2W}(\omega)e^{-j\omega t_0}$$

Deviation of either magnitude or phase characteristic from that given in this equation introduces what is known as paired echo type of distortion. Assume that the phase characteristic of such an amplifier is ideal; that is, $\theta(\omega) = -\omega t_0$, but the magnitude characteristic drops at higher frequencies

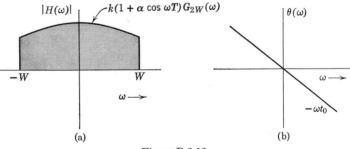

(a)

(b)

Figure P-2.12

as shown in Fig. P-2.12. This may be expressed as

$$H(\omega) = k(1 + \alpha \cos \omega T)G_{2W}(\omega)$$

Find the output of the amplifier when a pulse signal $f(t)$ bandlimited to W radians per second is applied at the input. (*Hint:* Use Eq. 1.118.)

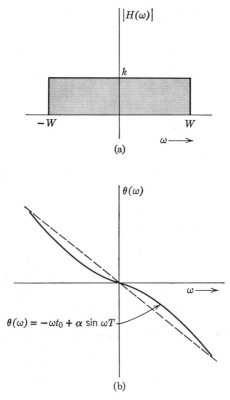

(a)

$\theta(\omega) = -\omega t_0 + \alpha \sin \omega T$

(b)

Figure P-2.13

13. In the amplifier discussed in Problem 12, assume that the magnitude characteristic of the amplifier is ideal; that is, $H(\omega) = kG_{2W}(\omega)$, but the phase characteristic is nonideal (Fig. P-2.13) and is given by

$$\theta(\omega) = -\omega t_0 + \alpha \sin \omega T$$

Find the output of the amplifier when the same bandlimited pulse signal $f(t)$ is applied at the input. (*Hint:* Assume α and T to be very small, and expand $e^{j\alpha \sin \omega T}$ by the first two terms of the Taylor series:

$$e^{j\alpha \sin T\omega} \simeq 1 + j\alpha \sin \omega T$$

$$= 1 + \frac{\alpha}{2} (e^{j\omega T} - e^{-j\omega T})$$

14. A signal $f(t) = 2e^{-t}u(t)$ is passed through an ideal low-pass filter with cutoff frequency 1 radian per second. Find the energy density spectrum of the output of the filter. Determine energies of the input signal and the output signal.

15. For an energy signal $f(t)$, show that the energy density function $\Psi_f(\omega)$ is the Fourier transform of function $\varphi(\tau)$, where

$$\varphi(\tau) = \int_{-\infty}^{\infty} f(t)f(t - \tau)\, dt$$

$$= \int_{-\infty}^{\infty} f(t)f(t + \tau)\, dt$$

16. Derive Parseval's theorem (Eq. 2.13b) from the time convolution theorem. [*Hint:* If $f(t) \leftrightarrow F(\omega)$, then $f(-t) \leftrightarrow F(-\omega)$ and $f(t) * f(-t) \leftrightarrow |F(\omega)|^2$.]

17. For a real power signal $f(t)$, show that the power density $S_f(\omega)$ is the Fourier transform of $\mathcal{R}_f(\tau)$, defined by

$$\mathcal{R}_f(\tau) = \lim_{T \to \infty} \frac{1}{T} \int_{-T/2}^{T/2} f(t)f(t - \tau)\, dt$$

$$= \lim_{T \to \infty} \frac{1}{T} \int_{-T/2}^{T/2} f(t)f(t + \tau)\, dt$$

The function $\mathcal{R}_f(\tau)$ is known as the *time-autocorrelation* function of $f(t)$.

18. For a power signal the power is defined as the power dissipated through a 1-ohm resistor. Equation 2.21a defines power for a real power signal $f(t)$. Modify this equation for a complex power signal $f(t)$.

19. Find the power (mean square values) of the following signals and sketch their power density spectra.
    (a) $A \cos (2000\pi t) + B \sin (200\pi t)$.
    (b) $[A + \sin (200\pi t)] \cos (2000\pi t)$.
    (c) $A \cos (200\pi t) \cos (2000\pi t)$.
    (d) $A \sin (200\pi t) \cos (2000\pi t)$.
    (e) $A \sin (300\pi t) \cos (2000\pi t)$.
    (f) $A \sin^2 (200\pi t) \cos (2000\pi t)$.

20. A periodic signal $f(t)$ shown in Fig. P-2.20a is transmitted through a system with transfer function $H(\omega)$. For three different values of T ($T = 2\pi/3$, $\pi/3$, and $\pi/6$), find the power density spectrum and the power (mean

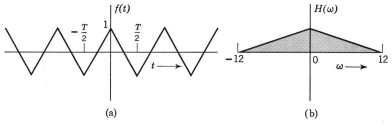

(a)　　　　　　　　　　(b)

Figure P-2.20

square value) of the output signal. Calculate the power of the input signal $f(t)$.

21. Find the mean square value of the output voltage $v_o(t)$ of an R-C network shown in Fig. P-2.21 if the input voltage has a power density

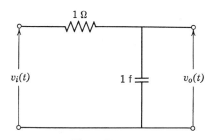

Figure P-2.21

spectrum $S_i(\omega)$ given by:
  (a) $S_i(\omega) = K$.
  (b) $S_i(\omega) = G_2(\omega)$ (gate function with cutoff at $\omega = 1$).
  (c) $S_i(\omega) = \pi[\delta(\omega + 1) + \delta(\omega - 1)]$.

In each case, calculate the power (the mean square value) of the input signal.

chapter 3

Communication Systems: Amplitude Modulation

In communication engineering one is concerned with the transmission of various signals from one point to another. We encounter this problem in radio and television broadcasting, long distance communication on telephone lines, satellite communication, remote control systems, telemetering, etc. In this chapter we shall study some systems of communication.

The signals are transmitted from one point to the other through a channel which may be in the form of a transmission line (such as a telephone channel) or merely an open space in which the signals bearing the desired information are radiated (such as radio and television broadcasting, satellite communication, etc.). Each of the signals to be transmitted generally has a small finite bandwidth compared to the bandwidth of the channel itself. It is therefore wasteful to transmit one signal at a time on the channel. The channel is being operated very much below its capacity to transmit information. We cannot, however, directly transmit more than one signal at a time, because this will cause an interference between the signals, and it will be impossible to recover the individual signals at the receiving end. This means that it is not possible by a direct method to transmit more than one conversation on a telephone line or to broadcast more than one radio or television station at a time. We shall presently see that by using *frequency division multiplexing* or *time division multiplexing* techniques it is possible to transmit several signals simultaneously on a channel.

3.1 FREQUENCY DIVISION MULTIPLEXING AND TIME DIVISION MULTIPLEXING

As mentioned before, the transmission of one signal at a time on a channel is a highly wasteful situation. We can surmount this difficulty, however, if we can shift the frequency spectra of various signals so that they occupy different frequency ranges without overlapping. We have already seen in Chapter 1 (frequency shifting) that it is possible to shift the frequency spectrum of a signal by modulating it (that is, multiplying the signal by a sinusoidal signal). Therefore it should be possible to transmit a large number of signals at the same time on one channel by using modulation techniques.

For the case of several signals, the spectrum of each individual signal is translated by a proper amount so that there is no overlap between the spectra of various signals. At the receiving end, the various signals can be separated by using appropriate filters. The individual spectra which are thus separated, however, do not represent the original signal because they have been translated from their original position. Thus, to obtain the original signal, each individual spectrum is retranslated by a proper amount to bring it back to its original form.

Modulation, however, serves another very useful purpose for systems which transmit signals by radiation in space. It can be shown from the theory of electromagnetic waves that a signal can be radiated effectively only if the radiating antenna is of the order of one-tenth or more of the wavelength corresponding to the frequencies of signals to be radiated. For human speech, the maximum frequency is about 10,000 Hz, which corresponds to a minimum wavelength of 30,000 meters. Thus, to radiate electromagnetic waves corresponding to the frequency range of the human voice, one would need an antenna of several miles in length. This is rather impractical. The process of modulation shifts the frequency spectrum to any desired higher frequency range, making it easier to radiate by electromagnetic waves. In practice, all the radio and television signals are modulated, thus in effect shifting the frequency spectrum of the desired signal to a very high frequency range. Modulation therefore not only allows the simultaneous transmission of several signals without interfering with each other, but it also makes it possible to transmit (radiate) these signals effectively.

The frequency translation method just discussed is not the only way of transmitting several signals simultaneously on a channel. We have shown in Chapter 1 that a bandlimited signal (a signal which has no spectral components beyond a certain frequency f_m Hz) is uniquely specified by its values at intervals $1/(2f_m)$ seconds (the uniform sampling theorem). It has been shown that the complete signal can be reconstructed from the knowledge of the signal at these instants alone. We therefore need to transmit only the samples of the signals at these finite number of instants. The channel is thus occupied only at these instants and conveys no signals for the rest of the time. During this idle period we may transmit the samples of other signals. We can thus interweave the samples of several signals on the channel. At the receiving end, these samples can be separated by a proper synchronous detector.

We can therefore transmit several signals simultaneously on a channel, provided that these signals can be separated at the receiving end. Each signal can be specified in the time domain or the frequency domain. Therefore, at the receiving end, we may recover the individual signals either in the time domain or the frequency domain. In the method of frequency translation, all of the signals are mixed in the time domain, but their spectra are so separated that they occupy different frequency bands. At the receiving end, we can recover the various individual signals by using proper filters. Here we have recovered the spectrum of individual signals, and hence this method actually separates at the receiving end the various signals in the frequency domain. This approach, where different signals share the different frequency intervals, is known as *frequency division multiplexing*. In the latter approach the samples of various signals are interweaved, and the samples of individual signals can be separated at the receiving end of the proper synchronous detector. In this method we actually recover the various signals in the time domain. In this case the frequency spectra of all the sampled signals occupy the same frequency range and are actually mixed. This approach, where all of the signals share the different time intervals, is known as *time division multiplexing*. In this chapter and the following two chapters (chapters 4 and 5), we shall study various types of communication systems using the multiplexing techniques mentioned above.

3.2 AMPLITUDE MODULATION: SUPPRESSED CARRIER SYSTEMS (AM-SC)

This technique essentially translates the frequency spectrum of the signal to be transmitted by multiplying it by a sinusoidal signal of the

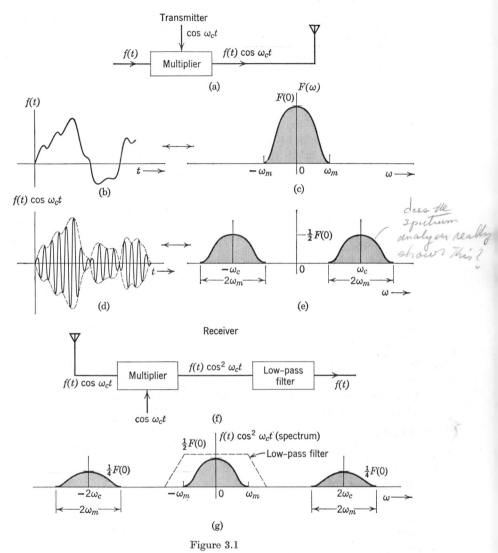

does the spectrum analyzer really show this?

Figure 3.1

frequency of the desired translation. From the modulation theorem (Eq. 1.116a) it is evident that the spectrum of $f(t) \cos \omega_c t$ is the same as that of $f(t)$, but shifted by $\pm \omega_c$ radians per second (Fig. 3.1e); that is, if

$$f(t) \leftrightarrow F(\omega)$$

then

$$f(t) \cos \omega_c t \leftrightarrow \tfrac{1}{2}[F(\omega + \omega_c) + F(\omega - \omega_c)] \qquad (3.1)$$

Concept.

The signal $\cos \omega_c t$ is called the carrier. The multiplication of $\cos \omega_c t$ by $f(t)$ is really equivalent to varying the carrier amplitude in proportion to $f(t)$. The carrier signal $\cos \omega_c t$ is said to be modulated by the signal $f(t)$. The signal $f(t)$ is thus a modulating signal, and the carrier signal $\cos \omega_c t$ is the modulated signal. This mode of transmission is called amplitude modulation with *suppressed carrier* (AM-SC). It is called a suppressed carrier because the modulated signal $f(t) \cos \omega_c t$ has no extra carrier signal with it. The free carrier is thus eliminated or suppressed. Later we shall study amplitude modulation with an additional free carrier, where the modulated signal is $f(t) \cos \omega_c t + A \cos \omega_c t$. Here an additional carrier is also transmitted for certain advantages which will be discussed. Such systems are called merely amplitude modulated systems (AM).

The amplitude modulation (AM-SC) therefore translates the frequency spectrum by $\pm\omega_c$ radians per second as seen from Eq. 3.1. To recover the original signal $f(t)$ from the modulated signals, it is necessary to retranslate the spectrum to its original position. The process of retranslation of the spectrum to its original position is referred to as demodulation or detection.

The spectrum of the modulated waveform (Fig. 3.1e) can be conveniently retranslated to the original position by multiplying the modulated signal by $\cos \omega_c t$ at the receiving end. Since multiplication in the time domain is equivalent to convolving the spectra in the frequency domain, it is evident that the spectrum of the resultant signal $[f(t) \cos^2 \omega_c t]$ will be obtained by convolving the spectrum of the received signal (Fig. 3.1e) with the spectrum of $\cos \omega_c t$ (two impulses at $\pm\omega_c$). A little reflection shows that this convolution yields the spectrum shown in Fig. 3.1g (see example 1.12, Fig. 1.44). This result may also be obtained directly from the identity

$$f(t) \cos^2 \omega_c t = \tfrac{1}{2}f(t)[1 + \cos 2\omega_c t] = \tfrac{1}{2}[f(t) + f(t) \cos 2\omega_c t] \quad (3.2)$$

Therefore, if

$$f(t) \leftrightarrow F(\omega)$$

then

$$f(t) \cos^2 \omega_c t \leftrightarrow \tfrac{1}{2}F(\omega) + \tfrac{1}{4}[F(\omega + 2\omega_c) + F(\omega - 2\omega_c)] \quad (3.3)$$

It is evident from the spectrum in Fig. 3.1g that the original signal $f(t)$ can be recovered by using a low-pass filter which will allow $F(\omega)$ to pass and will attenuate the remaining components centered around $\pm 2\omega_c$.

A possible form of low-pass filter characteristic is shown (dotted) in Fig. 3.1g. The system required at the receiving end to recover the signal $f(t)$ from the received modulated signal $f(t)\cos \omega_c t$ is shown in Fig. 3.1f. It is interesting to observe that multiplication of $f(t)$ by $\cos \omega_c t$ translates its spectrum by $\pm \omega_c$. The new spectrum can be retranslated to its original position by another translation of $\pm \omega_c$, which is accomplished by the multiplication of the modulated signal by $\cos \omega_c t$ at the receiver. (In the process, we get an additional spectrum at $\pm 2\omega_c$, which is filtered out.) The process at the receiving end is therefore exactly the same as that required at the transmitting end. Hence this method of recovering the original signal is called synchronous detection or coherent detection (also homodyne detection).

It is obvious from this discussion that in this system it is necessary to generate the local carrier at the receiver. The frequency and the phase of the local carrier are extremely critical. Consider, for example, the local carrier with a small frequency error $\Delta \omega$. The received signal is $f(t) \cos \omega_c t$ and the local carrier is $\cos (\omega_c + \Delta \omega)t$. The product is given by

$$f(t) \cos \omega_c t \cos (\omega_c + \Delta \omega)t = \tfrac{1}{2}f(t)[\cos (\Delta \omega)t + \cos (2\omega_c + \Delta \omega)t] \quad (3.4)$$

The term $f(t) \cos (2\omega_c + \Delta \omega)t$ represents the spectrum of $f(t)$ centered at $\pm(2\omega_c + \Delta \omega)$ and can be filtered out by a low-pass filter. The output of this filter will yield the remaining term $\tfrac{1}{2}f(t) \cos (\Delta \omega)t$ in Eq. 3.4. Thus instead of recovering the original signal $f(t)$, we obtain the signal $f(t) \cos (\Delta \omega)t$. In general, $\Delta \omega \to 0$ and $f(t) \cos (\Delta \omega)t$ represents $f(t)$ multiplied by a slowly varying gain constant. This is obviously a rather undesirable kind of distortion. It is therefore of utmost importance to have identical carrier frequencies at the transmitter and the receiver. The phase of the local oscillator is also critical. We shall discuss this topic in more detail in Section 3.5.

In order to achieve the precision frequency and phase control of the local oscillator, a very expensive and elaborate circuitry is required at the receiver. In most of these systems a very small amount of free carrier (pilot carrier) is transmitted along with the modulated signal. At the receiver the pilot carrier is separated by an appropriate filter and is amplified. This weak carrier is then used to phase lock the local oscillator which generates the strong carrier of the same frequency as that of the transmitter.

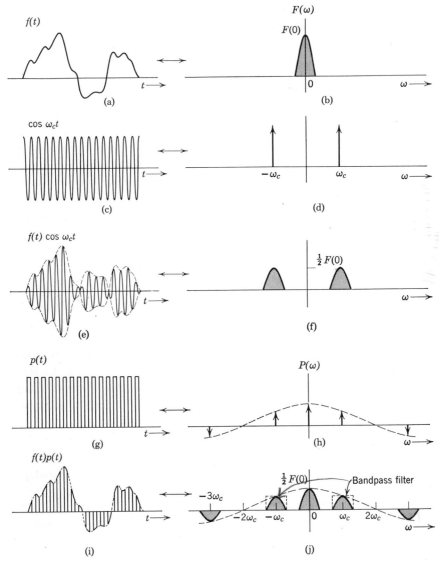

Figure 3.2

Techniques of Frequency Translation

It is obvious from the modulation theorem that a spectrum of any signal can be translated by $\pm \omega_c$ radians per second in the frequency domain by multiplying the signal with a sinusoidal signal of frequency ω_c. This, however, is not the only way to achieve this. We can easily show that a spectrum can be translated by an amount of $\pm \omega_c$ by multiplying the signal by any periodic signal of frequency ω_c, regardless of its waveform. This is obvious intuitively since any periodic waveform of frequency ω_c contains sinusoidal components of frequencies 0, ω_c, $2\omega_c$, $3\omega_c$, ..., etc. Hence the multiplication of a signal $f(t)$ by an arbitrary periodic waveform of frequency ω_c will translate the spectrum of $f(t)$ by 0, $\pm\omega_c$, $\pm 2\omega_c$, $\pm 3\omega_c$, etc. We are, however, interested only in that part of the spectrum that is centered around $\pm\omega_c$. This desired spectrum can be separated by using a bandpass filter which will allow the components of frequencies centered around $\pm\omega_c$ to pass and will attenuate all of the other frequencies.

As an example, consider a signal $f(t)$ (Fig. 3.2a) whose spectrum $F(\omega)$ is shown in Fig. 3.2b. Multiplication of this signal by a sinusoidal signal $\cos \omega_c t$ (Fig. 3.2e) shifts the spectrum by $\pm\omega_c$ (Fig. 3.2f). Now, instead of a sinusoidal signal, we shall multiply $f(t)$ by a square wave (Fig. 3.2g) of frequency ω_c. The spectrum of a periodic square wave $p(t)$ is shown in Fig. 3.2h. This spectrum $P(\omega)$ is a sequence of impulses located at $\omega = 0$, $\pm\omega_c$, $\pm 3\omega_c$, $\pm 5\omega_c$, ..., etc. (see Fig. 1.34). It is evident that the spectrum of $f(t)p(t)$ is given by $(1/2\pi)F(\omega) * P(\omega)$. The result of this convolution performed graphically is shown in Fig. 3.2j.

It is easy to see from this figure that the multiplication of $f(t)$ by $p(t)$ shifts the spectrum of $f(t)$ by $\omega = 0$, $\pm\omega_c$, $\pm 3\omega_c$, $\pm 5\omega_c$, ..., etc. This result is true for any periodic function of frequency ω_c, regardless of its waveform. In the special case of a square wave, the even harmonics $\pm 2\omega_c$, $\pm 4\omega_c$, ..., etc., are zero. But for a general periodic signal this need not be the case. We therefore conclude that multiplication of a signal $f(t)$ by any periodic signal of frequency ω_c, regardless of the waveform, shifts its spectrum by $\omega = 0$, $\pm\omega_c$, $\pm 2\omega_c$, $\pm 3\omega_c$, ..., etc. This result can be readily obtained analytically. Let $\varphi(t)$ be a periodic signal of frequency f_c Hz ($\omega_c = 2\pi f_c$). The Fourier transform of a general periodic signal was determined in Chapter 1 (Eq. 1.102).

In general, we have

$$\varphi(t) \leftrightarrow 2\pi \sum_{n=-\infty}^{\infty} \Phi_n \, \delta(\omega - n\omega_c) \qquad (3.5)$$

when Φ_n represents the coefficient of the nth harmonic in the exponential Fourier series for $\varphi(t)$. It follows from the convolution theorem that

$$f(t)\varphi(t) \leftrightarrow \frac{1}{2\pi} F(\omega) * 2\pi \sum_{n=-\infty}^{\infty} \Phi_n \, \delta(\omega - n\omega_c) \qquad (3.6)$$

$$\leftrightarrow \sum_{n=-\infty}^{\infty} \Phi_n F(\omega) * \delta(\omega - n\omega_c)$$

$$\leftrightarrow \sum_{n=-\infty}^{\infty} \Phi_n F(\omega - n\omega_c) \qquad (3.7)$$

It is evident from Eq. 3.7 that the spectrum of $f(t)\varphi(t)$ contains the spectrum $F(\omega)$ itself and $F(\omega)$ translated by $\pm\omega_c, \pm2\omega_c, \dots,$ etc. Note that the amplitudes of the successive cycles of $F(\omega)$ are multiplied by constants $\Phi_0, \Phi_1, \Phi_2, \dots,$ etc. When $\varphi(t)$ is a square wave, Φ_n can be found from Eq. 1.66a by substituting $T = 2\delta$ and $A = 1$.

$$\Phi_n = \frac{1}{2} Sa\left(\frac{n\pi}{2}\right)$$

Note that

$$Sa\left(\frac{n\pi}{2}\right) = \frac{\sin(n\pi/2)}{(n\pi/2)} = \begin{cases} (-1)^{(n-1)/2}\left(\dfrac{2}{n\pi}\right) & n \text{ odd} \\ 1 & n = 0 \\ 0 & n \text{ even} \end{cases}$$

Hence, from Eq. 3.5, we get

$$p(t) \leftrightarrow \pi\,\delta(\omega) + 2 \sum_{\substack{n=-\infty \\ (n \text{ odd}, 3, 5, \dots)}}^{\infty} \frac{(-1)^{(n-1)/2}}{n}\,\delta(\omega - n\omega_c) \qquad (3.8)$$

and

$$f(t)p(t) \leftrightarrow \frac{1}{2} F(\omega) + \frac{1}{\pi} \sum_{\substack{n=-\infty \\ (n \text{ odd}, 3, 5, \dots)}}^{\infty} \frac{(-1)^{(n-1)/2}}{n}\,F(\omega - n\omega_c) \qquad (3.9)$$

Figure 3.2j precisely represents the spectrum represented by Eq. 3.9. In amplitude modulation, however, we are interested in the frequency spectrum centered around $\pm\omega_c$ only. This can be obtained by

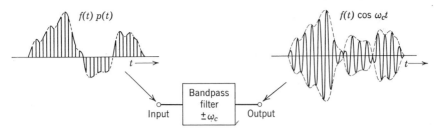

Figure 3.3   Effect of filtering on a modulated square wave.

using a bandpass filter which allows the frequency components centered at $\pm \omega_c$ to pass and attenuates the other frequency components. A simple R-L-C resonant circuit tuned at $\omega = \omega_c$ will pass a band of frequencies centered at $\pm \omega_c$, and will filter out the remaining frequency components. It is therefore evident that if we pass the signal $f(t)p(t)$ (Fig. 3.2i) through such a bandpass filter centered at $\pm \omega_c$, the resultant output will be given by $f(t) \cos \omega_c t$ as shown in Fig. 3.3.

The process of frequency translation is also called *frequency conversion* or *frequency mixing*. The systems which perform this function are called *frequency converters* or *frequency mixers*. A modulator or a demodulator both perform the operation of frequency translation, and hence they are also referred to as frequency converters or frequency mixers.

In our discussion of the techniques of frequency translation we shall often be referring to low-pass filters, high-pass filters, and bandpass filters. It is possible to design filters with magnitude characteristics (or phase characteristics) as close to the ideal characteristic as possible by using larger numbers of elements. But in numerous cases the undesired frequency components to be filtered out are separated so widely from the desired frequency components that very simple forms of filters may be used.

Modulating Systems (Frequency Converters or Frequency Mixers)

We shall now consider some simple circuits to produce modulation. The process of modulation translates the frequency spectrum. Hence the response of a modulator contains frequencies that are different from those present in the input signal. It is therefore impossible to produce modulation by using linear time-invariant systems, because the response

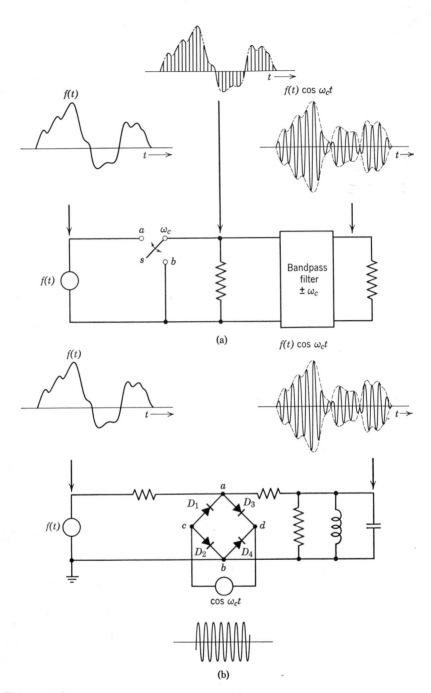

Figure 3.4 (a) Schematic diagram of a chopper-type modulator. (b) A chopper-type balanced modulator (ring modulator), using diodes as switches.

158

of such systems cannot contain frequencies other than those present in the input signal. Modulation can, however, be effected by using time-varying linear systems (such as switching or chopping circuits) or circuits that use nonlinear elements. The nonlinearity provides the actual mechanism for modulation, but often a system producing modulation can be represented as a linear time-varying system.

The schematic diagram of a chopper type of modulator is shown in Fig. 3.4a. The switch s alternates between terminals a and b with frequency ω_c. For one-half of the period the switch connects the terminal c to the signal $f(t)$, and for the remaining half-period terminal c is grounded. The output waveform at terminal c is therefore chopped at a frequency ω_c. The chopping operation may be viewed as a multiplication of $f(t)$ by a square wave $p(t)$. As discussed previously, such a chopped waveform contains the spectrum of $f(t)$ translated by $\omega = 0$, $\pm \omega_c$, $\pm 3\omega_c$, ..., etc., and the desired modulated signal $f(t) \cos \omega_c t$ may be recovered by passing this chopped signal through a bandpass filter centered a $\pm \omega_c$ (Fig. 3.3).

A practical arrangement to achieve such a circuit is shown in Fig. 3.4b. The diodes in this circuit act as the necessary switch here. When the signal $\cos \omega_c t$ is of such polarity as to make the terminal c positive with respect to the terminal d, all of the diodes conduct, assuming that the signal $\cos \omega_c t$ is much larger than the signal $f(t)$. Under these conditions the voltage across the diode D_1 is the same as that across D_2, and hence the terminal a is at the same potential as that of terminal b. Thus the output terminal a is connected to the ground. When the polarity of the signal $\cos \omega_c t$ makes the terminal d positive with respect to the terminal c, all of the diodes are reverse-biased and act as an open circuit. In this condition, the terminal a is connected to the signal $f(t)$ through a resistance R. It is obvious that the diodes switch the terminal a to the signal $f(t)$ and ground alternately at a frequency ω_c. At the output terminal, a parallel resonant circuit tuned to frequency ω_c acts as a bandpass filter. The output voltage is the desired modulated signal which is proportional to $f(t) \cos \omega_c t$. Note that the modulator circuit discussed here is a linear circuit since a multiplication of $f(t)$ by a constant will increase the output by the same constant. This circuit, however, is time-varying since its parameters change periodically. The modulator shown in Fig. 3.4b is known as a ring modulator.

A linear modulator, in general, may be described as a system whose gain (or the transfer function) can be varied with time by applying a time-varying signal at a certain point. The gain G may be varied proportional to the signal $f(t)$. Thus

$$G = Kf(t)$$

The carrier $\cos \omega_c t$ is applied at the input terminal (Fig. 3.5a). It is evident that the output will be a modulated signal $Kf(t) \cos \omega_c t$. Alternatively, a carrier may be used to vary the gain parameter (Fig. 3.5b), and $f(t)$ may be applied at the input terminals. The example of the ring modulator falls under the latter category. The ring modulator acts as a system whose gain varies between unity and zero at the carrier frequency. The variation of gain with time in this case is not sinusoidal but rectangular. This, of course, gives rise to unwanted translations at higher harmonics of ω_c, which are filtered out.

In practice, the gain parameters of active devices like vacuum tubes (μ) and transistors (β) depend upon the values of bias voltages and currents. Thus the gain of these devices can be made to vary with time by varying the bias signals, using appropriate signals. Details of such modulating (and demodulating) systems using vacuum tubes and transistors may be found in texts on electronic circuits.

As stated before, the modulation can also be achieved by using nonlinear devices. A typical nonlinear device characteristic is shown in Fig. 3.6a. A semiconductor diode is a good example of such a device.

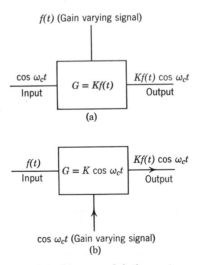

Figure 3.5   Linear modulating systems

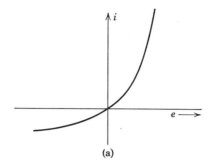

(a)

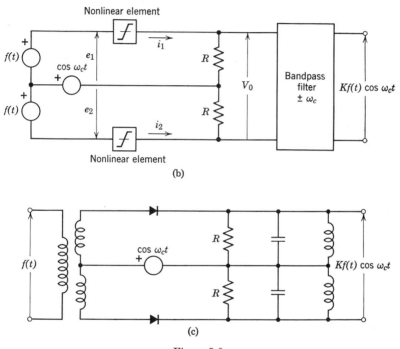

Figure 3.6

A nonlinear characteristic such as this may be approximated by a power series:

$$i = ae + be^2$$

Transistors and vacuum tubes also exhibit similar relationships between the input and the output under large signal conditions. A possible arrangement for the use of nonlinear elements for modulation is shown in Fig. 3.6c.

To analyze this circuit, we shall consider the nonlinear element in series with the resistance R as a composite nonlinear element whose terminal voltage e and the current i are related by a power series:

$$i = ae + be^2$$

The voltages e_1 and e_2 (Fig. 3.6b) are given by

$$e_1 = \cos \omega_c t + f(t)$$

and

$$e_2 = \cos \omega_c t - f(t)$$

It is obvious that the currents i_1 and i_2 are given by

$$i_1 = ae_1 + be_1{}^2$$
$$= a[\cos \omega_c t + f(t)] + b[\cos \omega_c t + f(t)]^2 \qquad (3.10a)$$

and

$$i_2 = a[\cos \omega_c t - f(t)] + b[\cos \omega_c t - f(t)]^2 \qquad (3.10b)$$

The output voltage v_o is given by

$$v_o = i_1 R - i_2 R$$

Substitution of Eq. 3.10 in this equation yields

$$v_o(t) = 2R[2bf(t) \cos \omega_c t + af(t)]$$

The signal $af(t)$ in this equation can be filtered out by using a bandpass filter tuned to ω_c at the output terminals. Semiconductor diodes can be conveniently used for the nonlinear elements in this circuit. A practical form of such a modulator is shown in Fig. 3.6c. All of the modulators discussed above generate a suppressed-carrier amplitude-modulated signal and are known as *balanced modulators*.

Demodulation (Detection) of Suppressed-Carrier Modulated Signals

At the receiver end, to recover the original signal $f(t)$ we need to demodulate the received signal $f(t) \cos \omega_c t$. As seen before, the process of demodulation is also equivalent to translation of the spectrum and can be achieved by multiplying the modulated signal $f(t) \cos \omega_c t$ by the signal $\cos \omega_c t$ (synchronous detection). Therefore the same circuits as those used for the process of modulation can be employed for the purpose of demodulation. There is, however, one difference between the modulating and demodulating circuits. The output spectrum of the modulator was centered around frequencies $\pm \omega_c$, and hence it was necessary to use a bandpass filter tuned to ω_c at the output of the modulator circuit. In the case of the demodulator, however, the output spectrum is $F(\omega)$ and is centered at $\omega = 0$. Hence we need to use a low-pass filter at the output terminals of the demodulator in order to filter out the undesired high-frequency components which are centered at $\pm\omega_c$, $\pm2\omega_c$, $\pm3\omega_c$, ..., etc. The demodulator using switching (chopper-type) and nonlinear elements is shown in Figs. 3.7a and 3.7b. Note that a low-pass filter is provided at the output terminals of each circuit by an R-C circuit.

The demodulation may be accomplished by multiplying the modulated signal $[f(t) \cos \omega_c t]$ by any periodic signal of frequency ω_c. If $\varphi(t)$ is a periodic signal of frequency ω_c, then its Fourier transform $\Phi(\omega)$ may be written as (Eq. 3.5):

$$\varphi(t) \leftrightarrow 2\pi \sum_{n=-\infty}^{\infty} \Phi_n \, \delta(\omega - n\omega_c)$$

It is obvious that if the modulated signal $f(t) \cos \omega_c t$ is multiplied by $\varphi(t)$, the resultant spectrum will be given by

$$f(t) \cos \omega_c t \varphi(t) \leftrightarrow \pi[F(\omega - \omega_c) + F(\omega + \omega_c)] * \sum_{n=-\infty}^{\infty} \Phi_n \, \delta(\omega - n\omega_c)$$

$$\leftrightarrow \pi \sum_{n=-\infty}^{\infty} \Phi_n \{F[\omega - (n + 1)\omega_c] + F[\omega - (n - 1)\omega_c]\}$$

$$(3.11)$$

It is evident that this spectrum contains a term $F(\omega)$ which can be filtered out by using a low-pass filter.

A Chopper Amplifier

The principle of frequency translation also finds a useful application in d-c and low-frequency amplifiers. Because of practical considerations of the sizes required for the coupling capacitors, it is very difficult to build amplifiers to amplify very low frequencies. Since the capacitor acts as an open circuit at lower frequencies, the sizes of the coupling capacitors required for a multistage amplifier for satisfactory gain are extremely large. Hence to amplify d-c signals and signals of very low frequencies, direct coupling is used. The direct coupling, however, introduces a serious problem of drift in the quiescent operating point of the amplifier. The drift introduced by environmental changes varies the output signal, and this variation cannot be distinguished from that

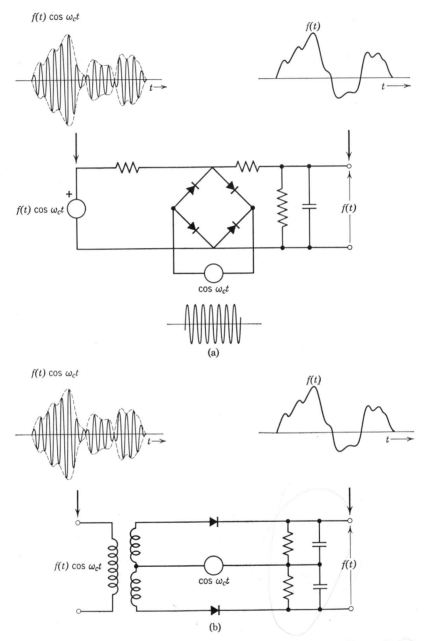

Figure 3.7 (a) A ring demodulator. (b) A demodulator that uses nonlinear elements.

introduced by the input signal itself. This problem is overcome by using a chopper amplifier which essentially shifts the spectrum of the input signal from a lower frequency range to a suitable higher range of frequencies where it can be easily amplified. The amplified signal is now demodulated to get back the amplified form of the original low-frequency signal.

Any of the circuits discussed above may be used. It is customary, however, to use a mechanical chopper for modulation and demodulation. The mechanical chopper has a switch which vibrates between two terminals and makes and breaks the contacts with these terminals periodically. Since the processes of modulation and demodulation both need the carrier of the same frequency, it is necessary to use the same chopper for modulation and demodulation, as shown in Fig. 3.8.

Consider a low frequency signal $f(t)$ and its spectrum $F(\omega)$ as shown in Figs. 3.8b and 3.8c, respectively. The signal $f(t)$ is applied at the input terminals of the chopper amplifier as shown in Fig. 3.8a. The input chopper grounds $f(t)$ for every half-cycle. The chopped signal as it appears across the terminals aa' is shown in Fig. 3.8d. This signal is equal to the signal $f(t)$ multiplied by a square wave $p(t)$. The spectrum of the chopped signal across terminals aa' is $F_{aa'}(\omega)$ as shown in Fig. 3.8e (see Fig. 3.2j).

The input capacitor blocks the spectrum around $\omega = 0$. Hence $F_{bb'}(\omega)$, the spectrum of the signal that appears across terminals bb', is identical to $F_{aa'}(\omega)$ with spectrum at $\omega = 0$ missing. The spectrum $F_{bb'}(\omega)$ is shown in Fig. 3.8g. The signal appearing across terminals bb' is $f_{bb'}(t)$ and is the inverse Fourier transform of $F_{bb'}(\omega)$. We note here that

$$F_{bb'}(\omega) = F_{aa'}(\omega) - \tfrac{1}{2}F(\omega)$$

Hence

$$f_{bb'}(t) = f_{aa'}(t) - \tfrac{1}{2}f(t)$$

Thus the signal $f_{bb'}(t)$ can be obtained by subtracting half the original signal $f(t)$ from the chopped signal $f_{aa'}(t)$. This yields the waveform $f_{bb'}(t)$, in a bipolar form as shown* in Fig. 3.8f. This result is also made

---

* Actually, the pulses in Fig. 3.8f will not follow the envelope exactly but will have some sag. This is because the capacitor cannot suppress the complete low frequency spectrum at $\omega = 0$. The capacitor blocks the signal $\omega = 0$ perfectly, but it can suppress the other components in the range $(0, \omega_m)$ only partially. However, since ω_m is very small (very low frequency signal), one is justified in assuming complete suppression of the spectrum in the range $(0, \omega_m)$ and the sag in the pulses will be negligible.

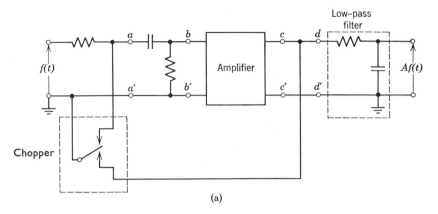

(a)

Figure 3.8 (continued)

obvious by observing that the input capacitor blocks the average component of $f_{aa'}(t)$. This converts the signal to the bipolar form $f_{bb'}(t)$. The signal $f_{bb'}(t)$ is the input signal to the amplifier. Note that $f_{bb'}(t)$ now does not contain very low frequency components and hence can be amplified easily. The output of the amplifier is shown in Fig. 3.8h. This signal is now demodulated by the same chopper. The chopper grounds the output signal every half-cycle. Observe that the grounding half-cycles at the input and at the output are complimentary.

The demodulated signal which appears across dd' is $f_{dd'}(t)$ shown in Fig. 3.8i. This signal is really an amplified, chopped version of the original signal $f(t)$. Hence one can recover $f(t)$ from the signal by passing it through a low-pass filter (R-C circuit) shown in Fig. 3.8a. The final output acquires a sign reversal. Most of the amplifiers, however, also have an additional signal reversal (180° phase shift). In such cases, the output signal is an amplified waveform $f(t)$ with no sign reversal.

3.3 AMPLITUDE MODULATION WITH LARGE CARRIER POWER (AM)

We have seen that the suppressed-carrier systems need very complex circuitry at the receiver for the purpose of generating a carrier of

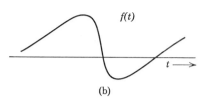

(b)

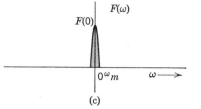

(c)

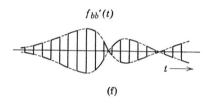

(d)

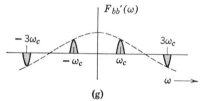

(e)

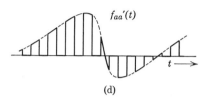

(f)

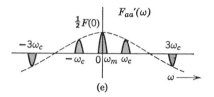

(g)

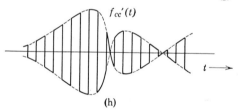

(h)

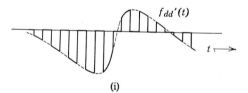

(i)

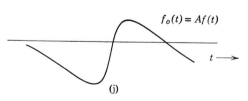

(j)

Figure 3.8 (concluded)

exactly the right frequency required for synchronous detection. But such systems are very efficient from the point of view of power requirements at the transmitter. In point-to-point communications, where there is one transmitter for each receiver, substantial complexity in the receiver system can be justified, provided that it results in a large enough saving in expensive high-power transmitting equipment. On the other hand, for a broadcast system with a multitude of receivers for each transmitter, it is more economical to have one expensive high-power transmitter and simpler, less expensive receivers. For such applications a large carrier signal is transmitted along with the suppressed-carrier modulated signal $f(t) \cos \omega_c t$, thus obviating the need to generate the carrier signal at the receiving end. Therefore the transmitted signal is now $\varphi_{AM}(t)$, given by

$$\varphi_{AM}(t) = f(t) \cos \omega_c t + A \cos \omega_c t \qquad (3.12a)$$

It is obvious that the spectrum of $\varphi_{AM}(t)$ is the same as that of $f(t) \cos \omega_c t$, except that there are two additional impulses at $\pm \omega_c$ (Fig. 3.9):

$$\varphi_{AM}(t) \leftrightarrow \tfrac{1}{2}[F(\omega + \omega_c) + F(\omega - \omega_c)]$$
$$+ \pi A[\delta(\omega + \omega_c) + \delta(\omega - \omega_c)] \quad (3.12b)$$

The modulated signal $\varphi_{AM}(t)$ is shown in Fig. 3.9. This signal (Eq. 3.12a) can be written as

$$\varphi_{AM}(t) = [A + f(t)] \cos \omega_c t \qquad (3.13)$$

It is evident that the modulated signal $\varphi_{AM}(t)$ may be viewed as a carrier signal $\cos \omega_c t$ whose amplitude is given by $[A + f(t)]$. The envelope of the modulated signal is the waveform $f(t)$ shifted by a constant A. Therefore the recovery of signal $f(t)$ in this case simply reduces to envelope detection. Note the constant A should be kept sufficiently large in order to preserve the envelope waveform exactly as $f(t)$. If A is not large enough (Fig. 3.9e), then the envelope waveform is not the same as that of $f(t)$. Under these conditions, $f(t)$ cannot be recovered by a simple process of envelope detection, but has to be detected by a method of synchronous detection (multiplying by $\cos \omega_c t$). Therefore A should be made large enough so that $[A + f(t)]$ is always positive. This is possible if

$$A > |f(t)|_{\max} \qquad (3.14)$$

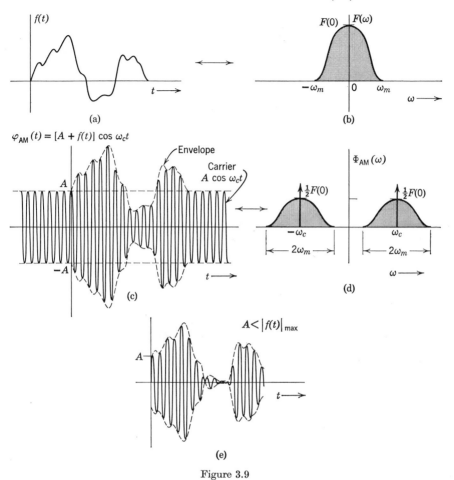

Figure 3.9

The technique of envelope detection will be described later in this section.

The modulated signals, which contain large amounts of the carrier signal so as to satisfy the condition of Eq. 3.14, are simply called amplitude-modulated signals (AM). Thus the signal $[A + f(t)] \cos \omega_c t$ (Fig. 3.9e) is referred to as an AM signal, whereas the signal $f(t) \cos \omega_c t$ (Fig. 3.1d) is known as an AM-SC signal. We shall presently see that AM signals are much easier to generate and demodulate than AM-SC signals. Some of the methods used for generation and demodulation of AM signals will now be discussed.

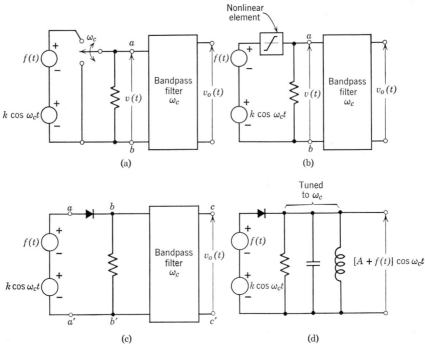

(a)

(b)

(c)

(d)

Figure 3.10 Generation of an AM signal.

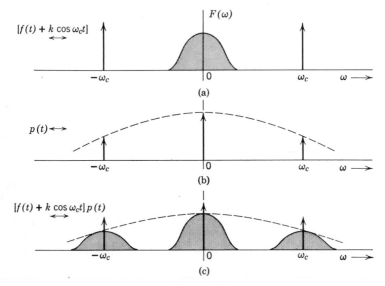

$[f(t) + k \cos \omega_c t] \longleftrightarrow$

$F(\omega)$

$-\omega_c$ 0 ω_c $\omega \longrightarrow$

(a)

$p(t) \longleftrightarrow$

$-\omega_c$ 0 ω_c $\omega \longrightarrow$

(b)

$[f(t) + k \cos \omega_c t] p(t) \longleftrightarrow$

$-\omega_c$ 0 ω_c $\omega \longrightarrow$

(c)

Figure 3.11

Generation of AM Signals

As in the case of AM-SC signals, the AM signals can also be generated by a chopper-type (switching) modulator and modulators that use nonlinear devices. In the chopper-type modulator (Fig. 3.10a), the modulating signal $f(t)$ with the carrier signal in series is connected across a chopper which vibrates at a frequency ω_c. The chopper action is equivalent to multiplication of the input signal by a square wave $p(t)$ of frequency ω_c. The spectrum of the resulting signal $v(t)$ can be obtained by convolving the spectra of $[f(t) + k \cos \omega_c t]$ with that of $p(t)$ as shown in Fig. 3.11. The convolution yields the desired spectrum centered at $\pm \omega_c$ and additional unwanted frequency components at $\omega = 0$, $\pm 3\omega_c$, $\pm 5\omega_c$, etc., which can be filtered out by bandpass filter tuned to ω_c. It is left as an exercise for the reader to derive the result analytically.

A chopper may be constructed by using a diode, as shown in Fig. 3.10c. If we assume the diode to be ideal (zero forward resistance and infinite reverse resistance), and if the carrier amplitude is much greater than the peak value of $f(t)$, then the diode merely acts as a switch which shorts out when the carrier signal is positive and opens when the carrier signal is negative. Therefore a diode chops the input signal at frequency ω_c. The spectrum of the resultant signal $v(t)$ is shown in Fig. 3.11c. When this signal is passed through a bandpass filter tuned to ω_c, the desired signal is obtained. Note that the diode here cuts out the negative part of the composite signal $[f(t) + k \cos \omega_c t]$. This is essentially a half-wave rectification of the input signal. Hence this type of modulator is also known as the rectifier type of modulator.

In a modulator system using the nonlinear device, the modulating mechanism is provided by the nonlinear device (Fig. 3.10b). If we assume that the composite element formed by the nonlinear device and the resistor R in Fig. 3.10b have a power series relationship between the voltage and current,

$$i \stackrel{\shortmid}{=} ae + be^2 \tag{3.15}$$

then it can be easily shown that the signal $v(t)$ consists of terms representing the modulated signal and unwanted terms which can be filtered out by a bandpass filter tuned to ω_c. A semiconductor diode more closely resembles the nonlinear element satisfying Eq. 3.15 than the ideal diode.

Demodulation of AM Signals

AM signals can be detected by using the synchronous detection techniques that were discussed for AM-SC signals. However, it is possible to demodulate AM signals by much simpler techniques. The detectors for AM signals may be classified as rectifier detectors and envelope detectors. The two types of detectors superficially appear to be equivalent, but they operate on entirely different principles. The rectifier detector actually operates on the principle of synchronous detection, whereas the envelope detector is a nonlinear circuit whose output tends to follow the envelope of the input signal. We shall consider each type individually.

Rectifier Detector

This circuit (Fig. 3.12) is essentially the same as a rectifier-type modulator except that the carrier signal is not required. The detector circuit merely rectifies the modulated signal. The rectified signal is the same as the original signal except that the negative cycles are cut off. This is really equivalent to multiplying the signal by unity for positive values and by zero for negative values. Hence it is evident that the rectification is really equivalent to multiplication of the modulated

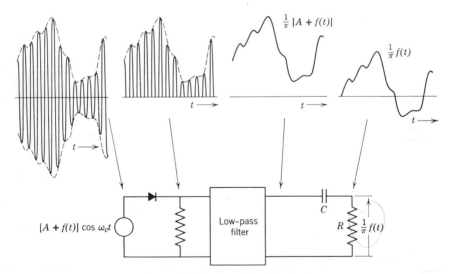

Figure 3.12　A rectifier detector.

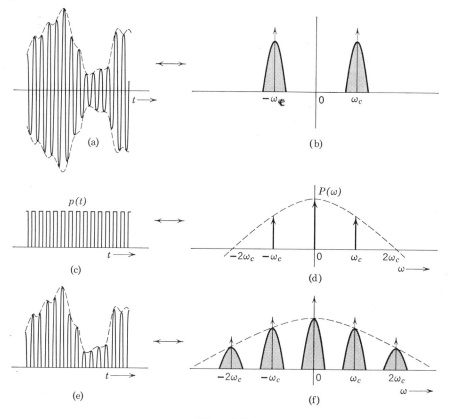

Figure 3.13

signal by a square wave $p(t)$ of frequency ω_c. The spectrum of the rectified signal is therefore obtained by convolving the spectrum of the modulated signal with that of $p(t)$. The result of the graphical convolution is shown in Fig. 3.13f. It is obvious from this figure that the signal $f(t)$ can be recovered by passing the rectified signal through a low-pass filter. The output of a low-pass filter still contains a d-c term (impulse at the origin). This can be eliminated by placing a capacitor C in the output circuit (Fig. 3.12). The convolution can be readily determined analytically by convolving the spectra of $\varphi_{\text{AM}}(t)$ (Eq. 3.12b) and $p(t)$ (Eq. 3.8). Alternatively, we may use directly the result in Eq. 3.9:

$$\varphi_{\text{AM}}(t)p(t) \leftrightarrow \frac{1}{2}\,\Phi_{\text{AM}}(\omega) + \frac{1}{\pi} \sum_{n=-\infty}^{\infty} \frac{(-1)^{(n-1)/2}}{n}\,\Phi_{\text{AM}}(\omega - n\omega_c) \quad (3.16)$$

From Eq. 3.12b, we have

$$\Phi_{AM}(\omega) = \tfrac{1}{2}[F(\omega + \omega_c) + F(\omega - \omega_c)]$$
$$+ \pi A[\delta(\omega + \omega_c) + \delta(\omega - \omega_c)] \quad (3.17)$$

We substitute Eq. 3.17 in Eq. 3.16. This gives us the entire spectrum shown in Fig. 3.13f. We are interested only in the low frequency component of this spectrum (the spectrum centered at $\omega = 0$), which is given by $n = \pm 1$ terms in the summation in Eq. 3.16. The reader can easily see that this output $e_o(t)$ is given by

$$e_o(t) \leftrightarrow \frac{1}{\pi} F(\omega) + 2A\ \delta(\omega)$$

and

$$e_o(t) = \frac{1}{\pi}[A + f(t)] \quad (3.18)$$

The output $e_o(t)$ in Eq. 3.18 can be doubled by the use of a full-wave rectifier instead of the half-wave rectifier in Fig. 3.12.

Note that the rectifier type of detection is essentially a synchronous detection since the operation of rectification is equivalent to multiplication of the modulated signal by a periodic signal (square wave) of frequency ω_c. But it is important to realize that the multiplication is performed without any carrier signal. This is the result of a high carrier content in the modulated signal itself. If there were no carrier present (as in the case of a suppressed carrier), then the rectifier operation would not be equivalent to multiplication of the input signal by $p(t)$. In general, for an AM signal, if the condition in Eq. 3.14

$$[A + f(t)] > 0 \qquad \text{for all } t$$

is not satisfied, then the rectifier type of detector cannot be used. If this condition is satisfied, the zero crossings of the received signal $[A + f(t)] \cos \omega_c t$ are located periodically at half-period points of the carrier and the process of rectification is equivalent to multiplication of the signal by $p(t)$. If, however, the condition $[A + f(t)] \gg 0$ for all t, the amplitude $A + f(t)$ changes signs from positive to negative and vice versa. In so doing it adds in the modulated signal extra zero crossings which are not necessarily periodic. It can be easily seen that the rectification under this condition is not equivalent to multiplication by $p(t)$, and the method of rectification cannot be used to demodulate these signals. In all such cases where this condition is not satisfied, the signal $f(t)$ may be recovered by synchronous detection requiring an external carrier signal for multiplication at the receiver.

The above discussion also suggests another possibility of detection of suppressed-carrier signals (in general, signals not satisfying the condition of Eq. 3.14). We may add a sufficient amount of carrier to such signals as to make

$$[A + f(t)] > 0 \qquad \text{for all } t$$

and then rectify and filter this signal to recover $f(t)$. Thus, instead of using the carrier to multiply the signal $f(t) \cos \omega_c t$, we add a sufficient amount of carrier to the modulated signal to make it possible to detect it by rectifier-detection techniques. This topic is discussed in detail in Section 3.6.

Envelope Detector

In an envelope detector the output of the detector follows the envelope of the modulated signal. The envelope detector is essentially a rectifier circuit with a capacitor across the output terminals as shown in Fig. 3.14.

On the positive cycle of the input signal, the capacitor C charges up to the peak voltage of the input signal. As the input signal falls below this peak value, the diode is cut off because the capacitor voltage (which is very nearly the peak voltage) is greater than the input signal voltage, thus causing the diode to open. The capacitor discharges through the resistor R at a slow rate. During the next positive cycle, at its peak, the input signal becomes greater than the capacitor voltage at the peak value, and the diode conducts. The capacitor again charges to the peak value of this new cycle. The capacitor discharges slowly during the cutoff period, thus changing the capacitor voltage very little.

During each positive cycle, the capacitor charges up to the peak voltage of the input signal and holds onto this voltage until the next

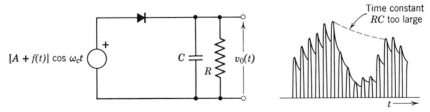

Figure 3.14 An envelope detector.

positive cycle. The time constant RC of the output circuit is adjusted so that the exponential decay of the capacitor voltage during the discharge period will approximately follow the envelope (see Problem 9). The voltage across the capacitor now has an undesired ripple of frequency ω_c, which may be filtered out by another low-pass filter.

Superficially it may appear that the envelope detector is really a rectifier detector for which a low-pass filter is provided by the R-C circuit. This is not true. The rectifier detector is a linear time varying parameter system, whereas the envelope detector is a nonlinear system. The two systems operate on entirely different principles, although the final outcome is the same. It can be easily seen why the rectifier followed by a low-pass filter yields the envelope of the modulated signal. Rectification cuts off the negative cycles. The low-pass filter yields the low frequency component which is the average of the remaining signal (positive cycles). This is obviously an envelope of the modulated signal.

It is evident from the discussion so far that the envelope detector output is π times that of the rectifier detector (Eq. 3.18). The envelope detector therefore is not only simpler than the rectifier detector but is also more efficient. Hence the envelope detector is almost universally used for the purpose of detecting AM signals. All of the commercial AM receivers have envelope detectors.

Power Content of Sidebands and Carrier in AM

In AM signals the carrier itself does not carry any information, and hence the power transmitted in the carrier signal represents a waste. It is interesting to find the relative power contents of the carrier and the sidebands (which carry the effective information). The modulated signal is given by

$$\varphi_{AM}(t) = \underset{\text{carrier}}{A \cos \omega_c t} + \underset{\text{sidebands}}{f(t) \cos \omega_c t}$$

The carrier power P_c is the mean square value of $A \cos \omega_c t$ and is obviously $A^2/2$:

$$P_c = \frac{A^2}{2}$$

The sideband power P_s is the mean square value of $f(t) \cos \omega_c t$ and is half the mean square value of $f(t)$ (see Example 2.1, Eq. 2.25):

$$P_s = \tfrac{1}{2}\overline{f^2(t)}$$

Barlo

The total power P_t is $P_c + P_s$:

$$P_t = P_c + P_s = \tfrac{1}{2}[A^2 + \overline{f^2(t)}]$$

The percentage of the total power carried by sidebands is η, given by

$$\eta = \frac{P_s}{P_t} \times 100\% = \frac{\overline{f^2(t)}}{A^2 + \overline{f^2(t)}} \times 100\% \qquad (3.19)$$

Note that for AM, $|f(t)|_{max} \leqslant A$. For a special case when $f(t)$ is a sinusoidal signal,

$$f(t) = mA \cos \omega_m t$$

m is called the modulation index, which must be less than or equal to unity ($m \leqslant 1$). Two signals with $m = 0.5$ and 1 are shown in Fig. 3.15. In this case

$$\overline{f^2(t)} = \frac{(mA)^2}{2}$$

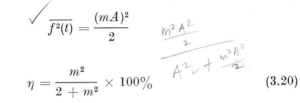

and

$$\eta = \frac{m^2}{2 + m^2} \times 100\% \qquad (3.20)$$

With the condition that $m \leqslant 1$, it can be easily seen that

$$\eta_{max} = \tfrac{1}{3} \times 100\% = 33.3\%$$

Hence for the highest modulation index ($m = 1$), the efficiency of transmission is 33%. Under these conditions 67% of the power is carried by the carrier and, as such, represents waste. For value of m less than unity, the efficiency is less than 33%. Note that for AM-SC there is no carrier and the efficiency is 100%.

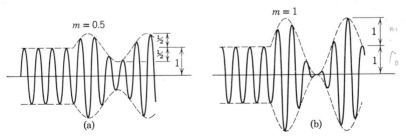

Figure 3.15

3.4 SINGLE SIDEBAND TRANSMISSION

In the process of amplitude modulation, the original spectrum $F(\omega)$ is translated by $\pm\omega_c$, as shown in Fig. 3.16b. The unmodulated signal occupies the bandwidth of ω_m (Fig. 3.16a), whereas the same signal after modulation occupies a bandwidth of $2\omega_m$. It is therefore evident that the price of the frequency translation discussed thus far is paid in terms of doubling the bandwidth. However, this need not be the case.

A glance at Fig. 3.16b shows that in transmitting the complete spectrum shown in this figure, we are transmitting redundant information. The spectrum $F(\omega)$ has been shifted at ω_c and $-\omega_c$. These two spectra are identical. Each of them contains the complete information of $F(\omega)$. So why transmit both spectra? Why not transmit only one

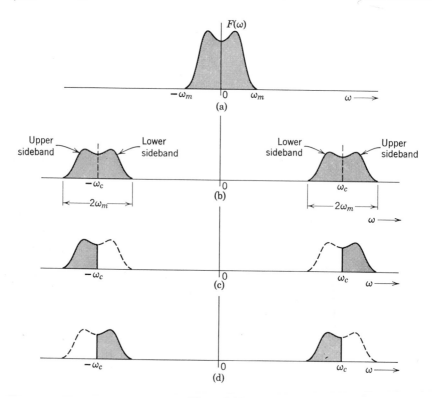

Figure 3.16

of the two? This, however, is impossible, because as we have shown in Chapter 1, for any physical signal, the spectrum is an even function of ω. A spectrum which is not symmetrical about the vertical axis passing through the origin does not represent a real signal, and hence it cannot be transmitted. But there is a way out of this.

We observe that the spectrum centered around ω_c is composed of two parts: one portion lies above ω_c and is known as the upper sideband, and the other portion lies below ω_c and is known as the lower sideband. Similarly, the spectrum centered around $-\omega_c$ has upper and lower sidebands (Fig. 3.16b). We now observe from Fig. 3.16b that the two upper sidebands (or the two lower sidebands) of the spectrum contain the complete information in $F(\omega)$. Hence, instead of transmitting the complete spectrum in Fig. 3.16b, it is sufficient to transmit only either the upper or the lower sidebands of the spectrum (shown in Figs. 3.16c and d). Note that the two upper sidebands or the two lower sidebands are each an even function of ω and hence represent a real signal. The original signal $f(t)$ can be recovered from these upper and lower sidebands by appropriate frequency translation. To transmit the sidebands we now need only one-half as much bandwidth (ω_m). This mode of transmission is referred to as single sideband transmission (SSB), in contrast to the double sideband transmission (DSB) discussed previously.

Generation of Single Sideband Signals*

To generate an SSB signal, all that is required is to filter out one of the sidebands from the modulated signals obtained from the balanced modulators discussed previously. The suppressed-carrier amplitude-modulated signal obtained from the balanced modulator is passed through an appropriate bandpass filter which will allow the desired sidebands to pass and filter out the remaining sidebands. The filter required to perform this function must have very nearly an ideal filter characteristic at frequency ω_c. In other words, the filter must have a sharp cutoff characteristic at ω_c in order to reject all of the frequencies on one side of ω_c and accept all of the frequencies on the other side of ω_c. From a practical standpoint, it is easier to design a filter with a sharp cutoff characteristic at lower frequencies. For this reason the

---

* For more information on SSB techniques, the reader is referred to a special issue of IRE on SSB transmission. Single Sideband Issue, *Proc. IRE*, Vol. 44, No. 12, December, 1956.

spectrum $F(\omega)$ thus is first translated to a lower frequency $\pm\omega_{c_1}$, where one of the sidebands is filtered out. After this filtering, the spectrum is translated to a desired higher frequency $\pm\omega_c$ from $\pm\omega_{c_1}$. Actually, the translation may be achieved successively in more than one step. The spectrum $F(\omega)$ is translated to a first lower frequency $\pm\omega_{c_1}$ where one of the sidebands is attenuated. The single sideband spectrum at $\pm\omega_{c_1}$ still contains some residual undesired sidebands due to imperfect filtering. This spectrum is then translated to the intermediate frequency ω_{c_2}, where again it is subjected to the filtering process to remove the residual of the unwanted sidebands. The spectrum is finally translated to the desired higher frequency ω_c.

The filtering problem is greatly simplified if the modulating signal does not contain significant low frequency components. In such cases the SSB filters need not have a sharp cutoff frequency since there is negligible power in the frequency components in the transition region (centered at the carrier). Voice signal provides such an example where the low frequency components have relatively small power. In the television video signals, on the other hand, this is not the case.

Phase-Shift Method

It is also possible to generate SSB signals by an indirect method of spectral phase shifting. To obtain some insight, we first consider the case of a sinusoidal signal $f(t) = \cos \omega_s t$. Here, $F(\omega)$ is represented by two impulses at $\pm\omega_s$ (Fig. 3.17a). The modulated signal with carrier $\cos \omega_c t$ is given by $\cos \omega_s t \cos \omega_c t$ and has the spectrum of $F(\omega)$ shifted by $\pm\omega_c$ (Fig. 3.17b). The SSB spectrum (lower sideband) is given by two impulses at $\pm(\omega_c - \omega_s)$, as shown in Fig. 3.17c. It is evident that the signal corresponding to this SSB spectrum (Fig. 3.17c) is given by $\cos (\omega_c - \omega_s)t$. Therefore the generation of an SSB signal for a special case of $f(t) = \cos \omega_s t$ is equivalent to generation of signal $\cos (\omega_c - \omega_s)t$.

From the trigonometric identity, we have

$$\cos (\omega_c - \omega_s)t = \cos \omega_s t \cos \omega_c t + \sin \omega_s t \sin \omega_c t$$

Thus the desired SSB signal can be produced by adding $\cos \omega_s t \cos \omega_c t$ and $\sin \omega_s t \sin \omega_c t$. The signal $\cos \omega_s t \cos \omega_c t$ can be easily produced from any balanced modulator discussed previously. The signal $\sin \omega_s t \sin \omega_c t$ can be expressed as $\cos (\omega_s t - \pi/2) \cos (\omega_c t - \pi/2)$.

Hence this signal can be produced by a balanced modulator, provided that both the signal $\cos \omega_s t$ and the carrier $\cos \omega_c t$ are shifted in phase

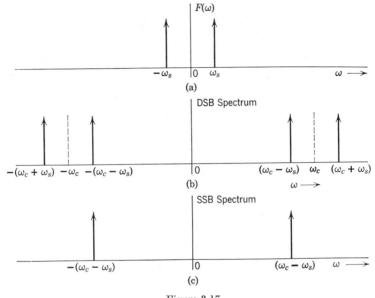

(a)

DSB Spectrum

(b)

SSB Spectrum

(c)

Figure 3.17

by $-\pi/2$ (Fig. 3.18). Although we have derived this result from a special case of $f(t) = \cos \omega_s t$, it holds true for any general waveform. This is because every waveform can be expressed as a continuous sum of sinusoidal (or exponential) signals. Hence the SSB-SC signal corresponding to $f(t)$ is given by (Fig. 3.18)

$$\varphi_{\text{SSB}}(t) = f(t) \cos \omega_c t + f_h(t) \sin \omega_c t$$

where $f_h(t)$ is the signal obtained by shifting the phase of each frequency component of $f(t)$ by $-\pi/2$. The schematic diagram of such an arrangement is shown in Fig. 3.18.

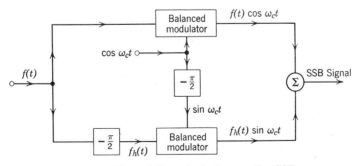

Figure 3.18   Phase-shift method of generating SSB.

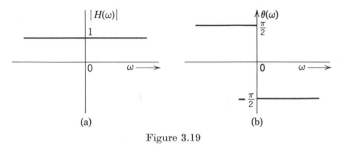

Figure 3.19

A rigorous proof for this result for any general signal $f(t)$ will now be given. A phase-shift system to shift the phase of the frequency components by $-\pi/2$ has a unit magnitude function. Thus the magnitudes of the frequency components remain unchanged, but the phase of all of the positive frequency components is shifted by $-\pi/2$. Since the phase spectrum is an odd function of ω, the phases of all of the negative frequency components are shifted by $+\pi/2$. The magnitude and phase spectrum of a phase-shift system is shown in Fig. 3.19.

$$|H(\omega)| = 1$$

$$\theta(\omega) = \underline{/H(\omega)} = \frac{\pi}{2} - \pi u(\omega)$$

Therefore the transfer function $H(\omega)$ of this phase-shift system is given by

$$H(\omega) = |H(\omega)|\, e^{j\theta(\omega)}$$
$$= e^{j[\pi/2 - \pi u(\omega)]} = je^{-j\pi u(\omega)} \tag{3.22}$$

And if

$$f(t) \longleftrightarrow F(\omega)$$

then

$$f_h(t) \longleftrightarrow jF(\omega)e^{-j\pi u(\omega)} \tag{3.23}$$

From the modulation theorem, we have

$$f(t)\cos \omega_c t \longleftrightarrow \tfrac{1}{2}[F(\omega + \omega_c) + F(\omega - \omega_c)] \tag{3.24a}$$

and from Eqs. 3.23 and 1.116b, it follows that

$$f_h(t)\sin \omega_c t \longleftrightarrow -\tfrac{1}{2}[F(\omega + \omega_c)e^{-j\pi u(\omega+\omega_c)} - F(\omega - \omega_c)e^{-j\pi u(\omega-\omega_c)}] \tag{3.24b}$$

and

$$[f(t)\cos \omega_c t + f_h(t)\sin \omega_c t] \longleftrightarrow \tfrac{1}{2}F(\omega - \omega_c)[1 + e^{-j\pi u(\omega-\omega_c)}]$$
$$+ \tfrac{1}{2}F(\omega + \omega_c)[1 - e^{-j\pi u(\omega+\omega_c)}] \tag{3.25}$$

Note that

$$u(\omega - \omega_c) = \begin{cases} 0 & \omega < \omega_c \\ 1 & \omega > \omega_c \end{cases}$$

Hence

$$1 + e^{-j\pi u(\omega - \omega_c)} = \begin{cases} 2 & \omega < \omega_c \\ 0 & \omega > \omega_c \end{cases}$$

But this is by definition $2u(\omega_c - \omega)$. Hence

$$1 + e^{-j\pi u(\omega - \omega_c)} = 2u(\omega_c - \omega) \tag{3.26a}$$

Similarly,

$$1 - e^{-j\pi u(\omega + \omega_c)} = 2u(\omega_c + \omega) \tag{3.26b}$$

Substituting Eq. 3.26 in Eq. 3.25, we obtain

$$f(t) \cos \omega_c t + f_h(t) \sin \omega_c t \leftrightarrow [F(\omega - \omega_c)u(\omega_c - \omega)$$
$$+ F(\omega + \omega_c)u(\omega + \omega_c)] \tag{3.27}$$

The spectrum on the right side of Eq. 3.27 expresses precisely the lower sidebands of $[F(\omega - \omega_c) + F(\omega + \omega_c)]$ The term $F(\omega - \omega_c)u(\omega_c - \omega)$ represents the lower sidebands of $F(\omega - \omega_c)$ because $u(\omega_c - \omega) = 0$ for $\omega > \omega_c$, causing the suppression of the upper sideband of $F(\omega - \omega_c)$. Similarly, $F(\omega + \omega_c)u(\omega + \omega_c)$ represents the lower sideband of $F(\omega + \omega_c)$ because $u(\omega + \omega_c) = 0$ for $\omega < -\omega_c$, causing the suppression of the upper sideband of $F(\omega + \omega_c)$. Thus the signal in Eq. 3.25 expresses the lower sideband SSB signal. The reader can show that if instead of adding we subtract $f_h(t) \sin \omega_c t$ from $f(t) \cos \omega_c t$, the resulting signal is an upper sideband SSB signal. Thus an SSB-SC signal $\varphi_{SSB}(t)$ may be expressed as

$$\varphi_{SSB}(t) = f(t) \cos \omega_c t \pm f_h(t) \sin \omega_c t \tag{3.28}$$

where the positive sign on the right-hand side yields the lower sideband SSB and the negative sign yields the upper sideband SSB. The signal $f_h(t)$ is the response of the phase shifter (Fig. 3.19) to the signal $f(t)$. We can easily express $f_h(t)$ in terms of $f(t)$ by using Eq. 3.23:

$$f_h(t) \leftrightarrow jF(\omega)e^{-j\pi u(\omega)}$$

Note that

$$e^{-j\pi u(\omega)} = \begin{cases} -1 & \omega > 0 \\ 1 & \omega < 0 \end{cases}$$

$$= -\text{sgn} \, (\omega)$$

Hence

$$f_h(t) \leftrightarrow -jF(\omega) \operatorname{sgn}(\omega) \tag{3.29}$$

From Eq. 1.113, we have

$$\frac{j}{\pi t} \leftrightarrow \operatorname{sgn}(\omega)$$

Application of the time convolution theorem to Eq. 3.29 yields*

$$f_h(t) = \frac{1}{\pi} f(t) * \frac{1}{t}$$

$$= \frac{1}{\pi} \int_{-\infty}^{\infty} \frac{f(\tau)}{t - \tau} d\tau \tag{3.30}$$

Demodulation of SSB-SC Signals

To recover $f(t)$ from the SSB signal, we have to retranslate the spectrum in Fig. 3.16c or d back to its original position ($\omega = 0$). This can be achieved easily by synchronous detection. Multiplication of the SSB signal by $\cos \omega_c t$ (synchronous detection) is equivalent to convolution of the spectrum of the SSB signal with the spectrum of $\cos \omega_c t$ (two impulses at $\pm \omega_c$). This is shown in Fig. 3.20 for upper sidebands. It is clear that the convolution yields $F(\omega)$ and an additional SSB-SC signal which has a carrier of $2\omega_c$. The latter part can be filtered out by a low-pass filter. Thus the demodulation of SSB signals can be accomplished by synchronous detection. This result can be obtained analytically as follows.

For a synchronous detection, the demodulator output $e_d(t)$ is given by (using Eq. 3.28)

$$e_d(t) = \varphi_{\text{SSB}}(t) \cos \omega_c t = f(t) \cos^2 \omega_c t \pm f_h(t) \sin \omega_c t \cos \omega_c t$$

$$= \underbrace{\tfrac{1}{2} f(t)}_{\text{message}} + \underbrace{[f(t) \cos 2\omega_c t \pm f_h(t) \sin 2\omega_c t]}_{\text{SSB with carrier } 2\omega_c} \tag{3.31}$$

The quantity inside the brackets on the right-hand side of Eq. 3.31 is identical to SSB signal in Eq. 3.28 except that its carrier frequency is $2\omega_c$. Thus the synchronous detection of a SSB signal yields the original

---

* Equation 3.30 defines the Hilbert transform of $f(t)$. The function $f_h(t)$ is the Hilbert transform of $f(t)$. The function $f_h(t)$ is also called the quadrature function of $f(t)$ because each frequency component of $f(t)$ is in phase quadrature with that of $f_h(t)$, as seen from Fig. 3.19. The integral in Eq. 3.30 is an improper integral and the divergence at $t = \tau$ is allowed by taking the Cauchy principle value of the integral.

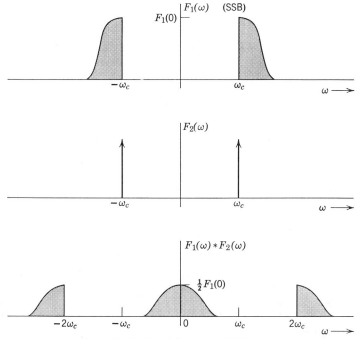

Figure 3.20   Demodulation of SSB signal.

signal $f(t)$ and another SSB signal with a carrier $2\omega_c$. This high frequency SSB component can be filtered out to recover $f(t)$.

The synchronous detection of SSB signals can be performed by either of the circuits shown in Fig. 3.7. For synchronous detection, we need to generate a local carrier of frequency ω_c and of the right phase (in phase with the carrier of the received signal). Any error in the frequency or the phase of the local carrier gives rise to distortion. The nature of this distortion is discussed in detail in Section 3.5.

Demodulation of SSB (with Large Carrier)

We now consider SSB signals with large amounts of carrier. Such signals are called SSB signals (in contrast to SSB-SC) and can be expressed as

$$\varphi(t) = A \cos \omega_c t + [f(t) \cos \omega_c t + f_h(t) \sin \omega_c t]$$

It can be easily seen that $f(t)$ can be recovered from $\varphi(t)$ by synchronous detection (multiplying $\varphi(t)$ by $\cos \omega_c t$). If, however, A, the free carrier amplitude, is large enough, $f(t)$ can be recovered from $\varphi(t)$ by the envelope or the rectifier detector. This can be readily shown by rewriting $\varphi(t)$ as

$$\varphi(t) = [A + f(t)] \cos \omega_c t + f_h(t) \sin \omega_c t$$
$$= e(t) \cos (\omega_c t + \theta)$$

where

$$e(t) = \{[A + f(t)]^2 + f_h^2(t)\}^{1/2}$$

and

$$\theta(t) = -\tan^{-1}\left[\frac{f_h(t)}{A + f(t)}\right]$$

It is evident that $e(t)$ is the envelope of SSB signal $\varphi(t)$. If $\varphi(t)$ is applied at the input of the envelope detector, the output will be $e(t)$:

$$e(t) = \{[A + f(t)]^2 + f_h^2(t)\}^{1/2}$$
$$= A\left[1 + \frac{2f(t)}{A} + \frac{f^2(t)}{A^2} + \frac{f_h^2(t)}{A^2}\right]^{1/2}$$

If $A \gg |f(t)|$, then $A \gg |f_h(t)|$ and the terms $f^2(t)/A^2$ and $f_h^2(t)/A^2$ can be ignored; then

$$e(t) \simeq A\left[1 + \frac{2f(t)}{A}\right]^{1/2}$$

Using binomial expansion and discarding higher order terms (since $f(t)/A \ll A$), we get

$$e(t) \simeq A\left[1 + \frac{f(t)}{A}\right]$$
$$= A + f(t)$$

It is evident that for large carrier, the envelope of $\varphi(t)$ has the form of $f(t)$, and the signal can be demodulated by an envelope detector. The video signal in television broadcast is transmitted by SSB with large carrier.*

3.5 EFFECTS OF FREQUENCY AND PHASE ERRORS IN SYNCHRONOUS DETECTION

The AM signals can be demodulated by the rectifier detector or an envelope detector without a local carrier at the receiver. The need to

---

* The TV system uses a slightly modified form of SSB, known as a vestigial sideband system. This is discussed fully in Section 3.8.

generate the local carrier at the receiver is obviated in AM systems because of the presence of a large carrier in the transmitted signal. However, in AM-SC, the suppressed carrier amplitude modulated systems (DSB-SC and SSB-SC), one must generate a local carrier at the receiver for the purpose of synchronous detection. Ideally, the frequency of the local carrier must be identical to that of the carrier at the transmitter. Similarly, the phase of the local carrier should be identical to that of the reference carrier in the receiver signal. Any discrepancy in the frequency and phase of the local carrier gives rise to a distortion in the detector output. We have already observed the effect of frequency error in the local carrier (Eq. 3.4). We shall now discuss this topic in more detail for DSB and SSB signals.

I. DSB-SC

Let the received signal be $f(t) \cos \omega_c t$ and the local carrier be $\cos [(\omega_o + \Delta \omega)t + \phi]$. The local carrier frequency and phase errors in this case are $\Delta \omega$ and ϕ, respectively. The synchronous detection is accomplished by multiplying the received signal by the local carrier and transmitting the product through a low-pass filter as shown in Fig. 3.1f. The product of the received signal and the local carrier is $e_d(t)$, given by

$$e_d(t) = f(t) \cos \omega_c t \cos [(\omega_c + \Delta \omega)t + \phi]$$

$$= \tfrac{1}{2} f(t)\{\cos [(\Delta \omega)t + \phi] + \cos [(2\omega_c + \Delta \omega)t + \phi]\} \quad (3.32)$$

The second term on the right-hand side represents the signal with spectrum centered at a high frequency $(2\omega_c + \Delta \omega)$ and is filtered out by a low-pass filter of cutoff frequency ω_m (Fig. 3.1f). The output of the filter is given by

$$e_o(t) = \tfrac{1}{2} f(t) \cos [(\Delta \omega)t + \phi] \qquad (3.33)$$

It is evident from this equation that the output signal is not $f(t)$ as required, but it is $f(t)$ multiplied by another time function. The output signal is therefore distorted. Note that if $\Delta \omega$ and ϕ are both zero (no phase or frequency error), then

$$e_o(t) = \tfrac{1}{2} f(t)$$

as expected.

Let us consider these two special cases:

1. $\Delta\omega = 0$ and $\phi \neq 0$ (phase error only)
2. $\Delta\omega \neq 0$ and $\phi = 0$ (frequency error only)

If $\Delta\omega = 0$, Eq. 3.33 reduces to

$$e_o(t) = \tfrac{1}{2}f(t)\cos\phi$$

This output is obviously proportional to $f(t)$ when ϕ is a constant. The output is maximum when $\phi = 0$ and is minimum (zero) when $\phi = \pm\pi/2$. Thus the phase error in the local carrier causes the attenuation of the output signal proportional to the cosine of the phase error. There is, however, no distortion in the signal waveform but only an attenuation as long as ϕ is constant. Unfortunately, however, the phase error ϕ generally varies randomly with time. This is the result of variations in the propagation path due to random variations in ionosphere. It leads to random variations in the phase of the incoming signal, which, in effect, causes the phase difference between the incoming signal and the local carrier to vary randomly with time. The gain factor $\cos\phi$ at the receiver thus varies randomly and is obviously undesirable.

Next we consider the case where $\phi = 0$ and $\Delta\omega \neq 0$. In this case Eq. 3.33 becomes

$$e_o(t) = \tfrac{1}{2}f(t)\cos(\Delta\omega)t$$

The output here is not merely an attenuated replica of the original signal but is distorted. Since $\Delta\omega$ is usually small, the output is the signal $f(t)$ multiplied by a low frequency sinusoid. The signal $f(t)$ goes through a time-varying attenuation. This is a rather serious type of distortion; hence it is important that the local oscillators be properly synchronized. This is usually accomplished by various feedback circuits. The phase and frequency errors required to actuate the feedback circuit can be obtained from a system shown in Fig. 3.21. Here the local oscillator output is split into two quadrature components (cosine and sine). The output $e_1(t)$ is the product of the incoming signal $f(t)\cos\omega_c t$ multiplied by the local carrier $\cos(\omega_c t + \phi)$ and passed through a low-pass filter which removes the components of frequencies beyond ω_m radians per second:

$$f(t)\cos\omega_c t\cos(\omega_c t + \phi) = \underbrace{\tfrac{1}{2}f(t)\cos\phi}_{e_1(t)} + \underbrace{\tfrac{1}{2}f(t)\cos(2\omega_c t + \phi)}_{\text{removed by lowpass filter}}$$

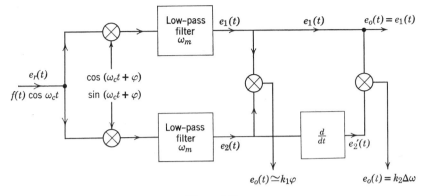

Figure 3.21

Thus

$$e_1(t) = \tfrac{1}{2}f(t) \cos \phi$$

Similarly, it can be seen that

$$e_2(t) = \tfrac{1}{2}f(t) \sin \phi$$

The product $e_1(t)e_2(t)$ is given by

$$e_1(t)e_2(t) = \tfrac{1}{4}f^2(t) \cos \phi \sin \phi$$
$$= \tfrac{1}{8}f^2(t) \sin 2\phi$$
$$\simeq \tfrac{1}{4}f^2(t)\phi \qquad \text{for} \quad \phi \ll 1$$

Thus the output is proportional to the phase error ϕ. The polarity of the output depends upon whether the phase error is positive or negative. This signal can be used as a control voltage to adjust the phase of the local oscillator. This phase control loop controls not only the phase but also the oscillator frequency over a narrow range.

If the signal $e_2(t)$ is differentiated and then multiplied by $e_1(t)$, the output $e_1(t)e_2'(t)$ can be shown to be proportional to the frequency error. This can be seen from the fact that the instantaneous frequency change of a signal is given by the rate of change of phase. Hence the instantaneous frequency error in signal $\cos(\omega_c t + \phi)$ is obviously $d\phi/dt$. Thus $\Delta\omega = d\phi/dt$. We have

$$e_2'(t) = \frac{de_2}{dt} = \tfrac{1}{2}[f(t) \cos \phi \frac{d\phi}{dt} + f'(t) \sin \phi]$$

For small values of ϕ, $\sin \phi \simeq 0$ and $\cos \phi \simeq 1$:

$$e_2'(t) \simeq \tfrac{1}{2}f(t)\frac{d\phi}{dt} = \tfrac{1}{2}(\Delta\omega)f(t)$$

and

$$e_1(t)e_2'(t) = \tfrac{1}{4}f^2(t)\cos \phi \,(\Delta\omega)$$
$$\simeq \tfrac{1}{4}f^2(t)\,(\Delta\omega)$$

Thus the product is proportional to the frequency error $\Delta\omega$.

As the loop adjusts the frequency and phase errors, eventually the proper synchronization is reached and the output $e_1(t)$ is the desired output

$$e_1(t) = \tfrac{1}{2}f(t)$$

and the output $e_2(t) = 0$.

2. SSB-SC

The incoming SSB signal at the receiver is given by* (Eq. 3.28):

$$\varphi_{\mathrm{SSB}}(t) = f(t)\cos \omega_c t + f_h(t)\sin \omega_c t$$

Let the local carrier be $\cos [\omega_c + \Delta\omega)t + \phi]$. The product of the incoming signal and the local carrier is $e_d(t)$, given by

$$e_d(t) = \varphi_{\mathrm{SSB}}(t)\cos [\omega_c + \Delta\omega)t + \phi]$$
$$= [f(t)\cos \omega_c t + f_h(t)\sin \omega_c t]\cos [(\omega_c + \Delta\omega)t + \phi]$$
$$= \tfrac{1}{2}f(t)\{\cos [(\Delta\omega)t + \phi] + \cos [(2\omega_c + \Delta\omega)t + \phi]\}$$
$$\qquad - \tfrac{1}{2}f_h(t)\{\sin [(\Delta\omega)t + \phi] - \sin [(2\omega_c + \Delta\omega)t + \phi]\}$$

The terms with frequencies $2\omega_c + \Delta\omega$ represent double carrier frequency terms and can be filtered out by a low-pass filter (Fig. 3.20a). The output of the filter is $e_o(t)$, given by

$$e_o(t) = \tfrac{1}{2}f(t)\cos [(\Delta\omega)t + \phi] - f_h(t)\sin [(\Delta\omega)t + \phi] \qquad (3.36)$$

Observe that if $\Delta\omega$ and ϕ are both zero, the output is

$$e_o(t) = \tfrac{1}{2}f(t)$$

as expected. It is interesting to compare the effect of phase and frequency errors between DSB and SSB systems. If $\Delta\omega = 0$, we observed

---

* Here we are considering lower sideband SSB. The discussion, however, applies equally well to upper sideband SSB.

that for DSB the signal remains undistorted, although it is attenuated by a factor cos ϕ. For SSB signal, however, when $\Delta\omega = 0$, the output is given by

$$e_o(t) = \tfrac{1}{2}[f(t) \cos \phi - f_h(t) \sin \phi] \qquad (3.37)$$

It is evident that the output contains an unwanted signal $f_h(t)$ which cannot be filtered out. We shall now show that this distortion is a phase distortion. If all the frequencies of $f(t)$ are shifted in phase by ϕ radians, the result will be precisely $e_o(t)$ in Eq. 3.37. This can be easily shown by taking the Fourier transform of Eq. 3.37.

$$E_o(\omega) = \tfrac{1}{2}[F(\omega) \cos \phi - F_h(\omega) \sin \phi]$$

But from Eq. 3.23, $F_h(\omega) = jF(\omega)e^{-j\pi u(\omega)}$. Hence

$$E_o(\omega) = \tfrac{1}{2}F(\omega)[\cos \phi - je^{-j\pi u(\omega)} \sin \phi]$$

$$= \begin{cases} \tfrac{1}{2}F(\omega)e^{j\phi} & \omega > 0 \\ \tfrac{1}{2}F(\omega)e^{-j\phi} & \omega > 0 \end{cases} \qquad (3.38)$$

It is evident from Eq. 3.38 that $e_o(t)$ can be obtained from $f(t)$ by shifting the phases of all frequency components by ϕ radians (note that for negative frequencies the phase shift is $-\phi$ due to the antisymmetry property of the phase spectrum). Thus the error in the phase in the local carrier gives rise to phase distortion in the detector output. The output signal is a distorted form of $f(t)$ where each frequency component of $f(t)$ undergoes a constant phase shift. The phase-shift distortion is usually not very serious with voice communication because the human ear is relatively less sensitive to phase distortions. This distortion may change the quality of speech, but the voice is still intelligible. (This distortion gives a Donald Duck voice effect.) However, in music transmission and video signals the phase distortion may be intolerable.

It can be seen by setting $\phi = 0$ in Eq. 3.36 that the effect of frequency error in SSB is similar to that observed in DSB (for small values of $\Delta\omega$).

3.6 CARRIER REINSERTION TECHNIQUES OF DETECTING SUPPRESSED CARRIER SIGNALS

A possibility of detection of suppressed carrier signals by reinserting a carrier at the receiver was mentioned earlier. This technique generally applies to DSB-SC and SSB-SC as well. After the sufficient amount

of carrier has been reinserted, one may use either rectifier detection or
envelope detection. The phase and the frequency of the reinserted
carrier should be properly synchronized with those at the transmitter
in order to avoid the distortion. We shall consider DSB-SC and
SSB-SC signals separately.

I. DSB-SC

The received signal is $f(t) \cos \omega_c t$. Let the reinserted carrier be
$A \cos (\omega_c t + \phi)$. Then the resulting signal $r(t)$ is given by

$$r(t) = f(t) \cos \omega_c t + A \cos (\omega_c t + \phi) \tag{3.39}$$

$$= [f(t) + A \cos \phi] \cos \omega_c t - [A \sin \phi] \sin \omega_c t$$

$$= \sqrt{[A + f(t)]^2 - 2Af(t)[1 - \cos \phi]} \cos (\omega_c t + \theta) \tag{3.40}$$

$$= e(t) \cos (\omega_c t + \theta) \tag{3.41}$$

where

$$e(t) = \{[A + f(t)]^2 - 2Af(t)[1 - \cos \phi]\}^{1/2} \tag{3.42}$$

$$\theta = \tan^{-1} \left[\frac{A \sin \phi}{f(t) + A \cos \phi} \right]$$

It is evident from Eq. 3.41 that $e(t)$ (Eq. 3.42) is the envelope of $r(t)$.
If $\phi = 0$, then the envelope $e(t)$ reduces to

$$e(t) = A + f(t)$$

Hence $f(t)$ can be recovered from $r(t)$ by rectifier detector or envelope
detector. Note that in this case

$$r(t) = [A + f(t)] \cos \omega_c t$$

This is, of course, the AM signal and can be detected by the rectifier
or the envelope detector provided $[A + f(t)] > 0$ for all t.

If, however, ϕ, the phase error, is not zero, then the distortion is
introduced. Rearranging Eq. 3.42, we get

$$e(t) = A \left\{ 1 + \frac{2f(t)}{A} \cos \phi + \left[\frac{f(t)}{A} \right]^2 \right\}^{1/2}$$

If $A \gg |f(t)|$, then

$$e(t) \simeq A + f(t) \cos \phi \tag{3.43}$$

The desired signal component of the output is $f(t) \cos \phi$. If ϕ is constant, the desired signal component at the output remains undistorted

but attenuated by a factor $\cos \phi$. This is exactly the same result as that obtained for synchronous detection technique.

Next we consider the case of frequency error $\phi = 0$ and $\Delta\omega \neq 0$. Here

$$r(t) = f(t) \cos \omega_c t + A \cos [\omega_c t + (\Delta\omega)t]$$

Note that this equation is identical to Eq. 3.39 except that ϕ in Eq. 3.39 has been replaced by $(\Delta\omega)t$. Hence, corresponding to Eq. 3.43, we have

$$e(t) \simeq A + f(t) \cos (\Delta\omega)t \qquad \text{for} \quad A \gg |f(t)|$$

Again this distortion is identical to that observed in the synchronous detection technique when there is a frequency error in the local oscillator carrier.

2. SSB-SC

We shall first consider the local carrier with phase error only:

$$
\begin{aligned}
r(t) &= \varphi_{\text{SSB}}(t) + A \cos (\omega_c t + \phi) \\
&= [f(t) \cos \omega_c t + f_h(t) \sin \omega_c t] + A \cos (\omega_c t + \phi) \\
&= [A \cos \phi + f(t)] \cos \omega_c t + [f_h(t) - A \sin \phi] \sin \omega_c t \\
&= e(t) \cos (\omega_c t + \theta)
\end{aligned}
$$

where

$$e(t) = [A^2 + 2Af(t) \cos \phi - 2Af_h(t) \sin \phi + f^2(t) + f_h^2(t)]^{1/2} \qquad (3.44)$$

and

$$\theta = \tan^{-1}\left[\frac{A \sin \phi - f_h(t)}{A \cos \phi + f(t)}\right]$$

If

$$A \gg |f(t)|$$

then

$$A \gg |f_h(t)|$$

and

$$e(t) \simeq A + f(t) \cos \phi - f_h(t) \sin \phi$$

The rectifier detector or envelope detector will yield $e(t)$. The constant A can be blocked out by a capacitor and the resultant output $e_o(t)$ is given by

$$e_o(t) = f(t) \cos \phi - f_h(t) \sin \phi \qquad (3.45)$$

This output is identical to that in Eq. 3.37 (except for a constant multiplier). We have already shown in Section 3.5 that $e_o(t)$ in Eq. 3.45 represents the original signal $f(t)$ with phase distortion. Thus the local carrier phase error in the carrier reinsertion technique gives rise to distortion identical to that observed in the synchronous detection technique. The reader can show along similar lines that the effect of frequency error in the local carrier in the reinsertion technique is identical to that observed in the synchronous detection technique.

Note that when $\phi = 0$, Eq. 3.45 reduces to

$$e_o(t) = A + f(t)$$

Thus if a large carrier is reinserted in an SSB-SC signal, the demodulation can be accomplished by the rectifier or envelope detector. Instead of reinserting the carrier at the receiver, we may add a large carrier at the transmitter. The received signal is then SSB with a large carrier and can be demodulated by an envelope (or rectifier) detector similar to AM signals. The SSB signal with large carrier can be obtained from an AM signal by suppressing one of its sidebands. This mode of transmission (SSB with large carrier) has the advantages of both SSB and AM systems. It needs only half the bandwidth of AM. At the same time, it has the advantages of simplicity of the envelope detector, as in AM. This mode of transmission is used with slight modification in television video signals (see Section 3.8).

From this discussion we conclude that a suppressed carrier signal can be demodulated by reinserting a sufficiently large carrier at the receiver and subsequently demodulating by a rectifier or an envelope detector. This result is expected for DSB signals, since addition of a sufficiently large carrier to a DSB-SC signal converts it into an AM signal. The result is not so obvious for SSB-SC signals. It can be explained qualitatively as follows. When a large carrier is added to an SSB-SC signal, the carrier dominates the resulting signal, and the original SSB-SC signal essentially rides upon this carrier. The resulting signal therefore has zero crossings very nearly the same as those of the carrier. Hence use of rectification in this case is largely equivalent to multiplying the signal by a square wave $p(t)$. This is obviously equivalent to synchronous detection.

We have also shown that the phase and frequency errors in the reinserted carrier give rise to distortions similar to those observed in synchronous detection techniques.

3.7 COMPARISON OF VARIOUS AM SYSTEMS

We have discussed various aspects of AM (DSB and SSB) and AM-SC (DSB-SC and SSB-SC) systems. It is interesting to compare these systems from various points of views.

The AM has advantage over AM-SC at the receiver end. The detectors required from AM are relatively simpler (rectifier or envelope detectors) than those required for suppressed carrier systems. For this reason all public communication systems use AM. In addition, AM signals are easier to generate at high power levels compared to suppressed carrier signal. Balanced modulators required in the latter are somewhat difficult to design.

The suppressed carrier systems have the advantage over AM in that they require much less power to transmit the same information. This makes the transmitter less expensive than those required for AM. But the receivers in these systems are more complex since they must be capable of generating a local carrier of the right phase and frequency. For a point-to-point communication systems where there are only a few receivers per transmitter, this complexity in the receiver may be justified.

The effect of selective fading (due to multipath propagation), however, is much more disastrous on AM than AM-SC. The fading occurs because of the arrival of the signal at the receiver over more than one path of propagation, each of a different length. This causes the phases of signals arriving by different paths to differ. The resultant signal is the sum of all the signals arriving by various paths. The signal waves along their path are reflected by the ionosphere and the ground. The ionospheric conditions change randomly with time, thus causing the path length to vary randomly. The phases of signals arriving by various paths at the receiver therefore change randomly. Hence the strength of the resultant signal varies randomly. This phenomenon is called *fading*. The fading is also sensitive to frequencies, and because of this its effects become even more serious. Thus the carrier and each of the sidebands undergo different amounts of fading. This effect is called *selective fading*. The selective fading disturbs different sidebands by different amounts and causes distortion. It also disturbs the relationship between magnitudes of the carrier and the sidebands. It may cause more severe fading of the carrier than the sidebands to the point where the condition (3.14) is no longer fulfilled. Such a wave-form, when detected by an envelope detector (or a rectifier detector), will be heavily distorted. The effect of selective fading becomes severe at high frequencies. Therefore at high frequencies, suppressed carrier systems are preferred.

Next we compare DSB-SC with SSB-SC. Here we find that SSB is almost always preferable. The following are the advantages of SSB over DSB:

1. The bandwidth required for SSB signals is half that required for DSB signals.*

2. The selective fading effect discussed earlier disturbs the phase relationships of two sidebands in DSB. This gives rise to the distortion in the demodulator output. In SSB since there is only one sideband, this possibility does not exist. Under long-range propagation conditions, the effects of selective fading prove more deleterious to DSB than to SSB.

For these reasons SSB-SC is preferred to DSB-SC; SSB-SC is used in long-range high frequency communications, particularly in audio range where phase distortions are relatively unimportant. Amateur-operated radios use SSB signals.

SSB, however, has a disadvantage in comparison to DSB in one respect. The generation of high level SSB signals is much more difficult than that of DSB signals (see Section 3.4). This disadvantage is over-come in what is called the vestigial sideband transmission. This mode of transmission is, in reality, a compromise between SSB and DSB; it combines the advantages of both systems and eliminates their dis-advantages. This mode of transmission will now be discussed.

3.8 VESTIGIAL SIDEBAND TRANSMISSION

It was mentioned in Section 3.3 that SSB signals are relatively difficult to generate. If one uses filtering techniques to generate SSB by filtering out one sideband of a DSB signal, the required filter must have a very sharp cutoff characteristic. This is rather difficult. To

---

* Use of quadrature multiplexing (see Problem 3 at the end of this chapter) can balance out this difference. This method, however, introduces more cross talk (co-channel interference) due to nonidealities in the bandpass channel in DSB than that attainable by frequency multiplying in SSB.

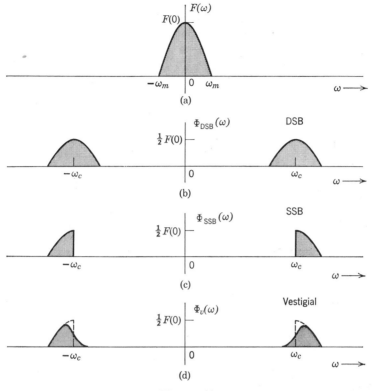

Figure 3.22

overcome this problem, a compromise between SSB and DSB is sought in what is known as the vestigial sideband transmission. In this mode, instead of rejecting one sideband completely (as in SSB), a gradual cut-off of one sideband, as shown in Fig. 3.22d, is accepted. The cutoff characteristic is such that the partial suppression of the transmitted sideband (upper sideband in Fig. 3.22d) in the neighborhood of the carrier is exactly compensated by the partial transmission of the corresponding part of the suppressed sideband (lower sideband in Fig. 3.22d). Because of this arrangement, the desired signal can be recovered exactly by an appropriate detector. If a large free carrier is transmitted along with the vestigial sidebands, the signal can be recovered at the receiver by an envelope (or rectifier) detector. For the case of suppressed carrier vestigial sidebands, the signal can be recovered by synchronous detector (or by reinsertion of the carrier and the subsequent envelope detection).

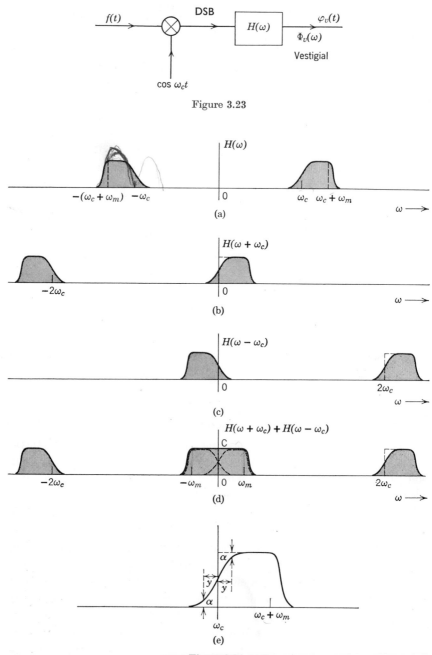

Figure 3.23

Figure 3.24

We shall now find the appropriate filter characteristics required to produce the vestigial sideband signal from a DSB signal.

Let $H(\omega)$ be the transfer function of the required filter (Fig. 3.23). If $f(t)$ is the message signal and $F(\omega)$ is its Fourier transform, then $\Phi_v(\omega)$, the spectrum of the vestigial signal, is given by

$$\Phi_v(\omega) = \tfrac{1}{2}[F(\omega + \omega_c) + F(\omega - \omega_c)]H(\omega) \qquad (3.46)$$

where $\varphi_v(t)$ is the vestigial sideband signal and $\Phi_v(\omega)$ is its transform (Fig. 3.22d). The original signal $f(t)$ can be recovered from $\varphi_v(t)$ by synchronous detection. The incoming vestigial signal $\varphi_v(t)$ is first multiplied by $\cos \omega_c t$. The product $e_d(t)$ is given by

$$e_d(t) = \varphi_v(t) \cos \omega_c t \leftrightarrow \tfrac{1}{2}[\Phi_v(\omega + \omega_c) + \Phi_v(\omega - \omega_c)]$$

Substitution of Eq. 3.46 in this equation yields

$$e_d(t) \leftrightarrow \tfrac{1}{4}\{[F(\omega + 2\omega_c) + F(\omega)]H(\omega + \omega_c)$$
$$+ [F(\omega) + F(\omega - \omega_c)]H(\omega - \omega_c)\}$$

The terms $F(\omega + 2\omega_c)$ and $F(\omega - 2\omega_c)$ represent $F(\omega)$ shifted to $\pm 2\omega_c$ and are filtered out by a low-pass filter. The resulting output $e_o(t)$ is given by

$$e_o(t) \leftrightarrow \tfrac{1}{4}F(\omega)[H(\omega + \omega_c) + H(\omega - \omega_c)] \qquad (3.47)$$

For a distortionless reception, we need

$$e_o(t) \leftrightarrow kF(\omega)$$

Hence the transfer function $H(\omega)$ must satisfy

$$H(\omega + \omega_c) + H(\omega - \omega_c) = C \qquad \text{(constant)} \qquad (3.48)$$

From Eq. 3.47, we observe that since $F(\omega) = 0$ for $|\omega| > \omega_m$, Eq. 3.48 need be satisfied only for $|\omega| < \omega_m$. Hence

$$H(\omega + \omega_c) + H(\omega - \omega_c) = C \qquad |\omega| < \omega_m \qquad (3.49)$$

The term $H(\omega + \omega_c)$ and $H(\omega - \omega_c)$ represent $H(\omega)$ shifted by $-\omega_c$ and ω_c, respectively, from the origin. This is illustrated in Figs. 3.24b and c. The sum of these two terms should be constant over $|\omega| < \omega_m$. It can be easily seen from Fig. 3.24 that this is possible only if the filter cutoff characteristic around the carrier has a complementary symmetry (shown in Fig. 3.24e).

We have shown here that the signal can be recovered from a suppressed carrier vestigial sideband signal by synchronous detection. It can be shown that if a large amount of carrier is added to vestigial sideband signal, the detection can be accomplished by an envelope (or rectifier) detector. This can be easily shown qualitatively. If the carrier amplitude is very large compared to $\varphi_v(t)$, the combined signal will have zero crossings at about the same points as those of the carrier, and hence rectification of such a signal will amount to multiplication by a square wave $p(t)$. This is obviously equivalent to synchronous detection.

Vestigial sideband combines the advantages of SSB and DSB with none of their disadvantages. It requires practically the same bandwidth as that of SSB (half that of DSB) and can be obtained from DSB signals by using relatively simpler filters with gradual cutoff characteristic. It is relatively immune to selective fading. If a large carrier is added to the vestigial signal it can be demodulated by envelope detector. In this mode the vestigial sidebands combine all the advantages of AM, SSB, and DSB. In the public television systems the video signals are transmitted by vestigial sidebands. This cuts down the bandwidth of 8 mHz (for DSB or AM) to 5 mHz. A sizable amount of carrier is also transmitted along with the vestigial sidebands. This makes it possible to demodulate the signal by envelope detector at the receiver.

3.9 FREQUENCY DIVISION MULTIPLEXING

From all the discussion on modulation thus far, it is quite easy to appreciate the utility of modulation in transmitting several signals simultaneously. Suppose we wish to transmit simultaneously on a channel n signals, each of which is bandlimited to ω_m radians per second. For illustration we shall consider the use of AM, but the discussion can be equally well applied to other modes of transmission. These n signals are now modulated with carriers $\omega_1, \omega_2, \ldots, \omega_n$, so that each carrier is separated from the adjacent carrier by at least $2\omega_m$ radians per second. Each of the modulated signals has a bandwidth of $2\omega_m$ and is centered at frequencies $\omega_1, \omega_2, \ldots, \omega_n$. (There is a similar spectrum for negative frequencies.) This is shown in Figs. 3.25a and b. Figure 3.25a shows the spectra of individual signals and Fig. 3.25b shows the combined spectra of all the modulated signals at the transmitter. At the receiver, various spectra are separated by using proper bandpass filters (Fig. 3.25c).

After filtering, the signals are demodulated to obtain the original signals. Radio and television broadcasting and receiving provide

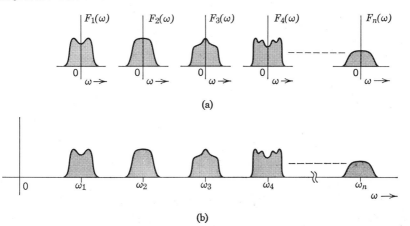

(a)

(b)

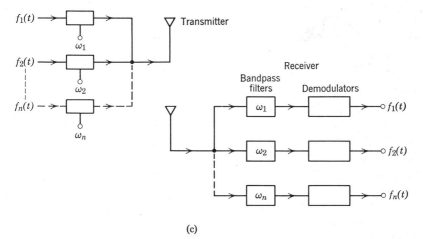

(c)

Figure 3.25

familiar examples of frequency division multiplexing. Each transmitter sends a modulated signal with a carrier which is separated from the carriers of other transmitting stations by at least $2\omega_m$. In radio broadcasting this is about 10 kHz. The broadcast receiver can pick up any desired signal by proper tuning, which allows the desired band to pass and attenuates other frequencies. This signal is now demodulated to obtain the desired signal. However, in almost all commercial AM receivers, the demodulation is not effected directly. The modulated signal received is first translated to a fixed lower frequency known as an intermediate frequency of 455 kHz. The translation of the signal

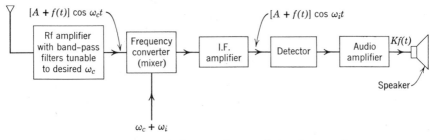

Figure 3.26   A block diagram of a superheterodyne receiver.

to an intermediate frequency is accomplished by modulating the incoming signal by a locally generated signal which differs with the incoming carrier by 455 kKz. The signal, which is now translated to a fixed intermediate frequency of 455 kHz, is amplified and demodulated to obtain the desired signal.

The advantage of conversion to an intermediate frequency is that to receive different stations it is necessary to tune only the first stage (and the local oscillator). All of the amplification is achieved at a constant intermediate frequency and needs no tuning. The process of frequency translation is also known as heterodyning. In order to translate the spectrum to a fixed intermediate frequency, the local oscillator must have a frequency above or below the incoming carrier frequency by the intermediate frequency (455 kHz). Generally, the local oscillator signal frequency is chosen higher than the incoming carrier by the intermediate frequency. For this choice, such receivers are called *superheterodyne* receivers. The block diagram of such a receiver is shown in Fig. 3.26.

PROBLEMS

1. A DSB-SC signal is given by

$$\varphi(t) = f(t) \cos \omega_1 t$$

where ω_1 is the carrier frequency. It is decided to change the carrier frequency from ω_1 to ω_2 (this is known as the frequency conversion). Show that the balanced modulator circuits shown in Figs. 3.4, 3.5, and 3.6 can be used to accomplish this by feeding $f(t) \cos \omega_1 t$ and $A \cos \omega_2 t$ at appropriate locations and using suitable filters. Assume $A \gg |f(t)|$.]

2. A balanced modulator circuit can also be used as a synchronous detector. Assuming a piecewise linear model for the diode (Fig. P-3.2a), find the

output voltage $e_0(t)$ in the circuit shown in Fig. P-3.2b if

$$\varphi_1(t) = f(t) \cos \omega_c t$$
$$\varphi_2(t) = A \cos \omega_c t$$

Assume $A \gg |f(t)|$ for all t. How can $f(t)$ be recovered from $e_0(t)$? This circuit is also used as a phase discriminator which measures the phase of

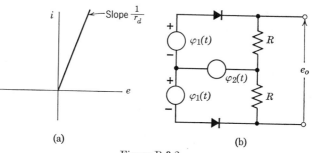

(a)   (b)

Figure P-3.2

one sinusoidal waveform with respect to the other. Show that if

$$\varphi_1(t) = \cos (\omega t + \theta)$$

and

$$\varphi_2(t) = A \cos \omega t$$

then the output voltage $e_0(t)$ contains a d-c term proportional to $\cos \theta$.

3. It is possible to transmit two different signals simultaneously on the same carrier. The two signals are modulated by carriers of the same frequency but in phase quadrature as shown in Fig. P-3.3. Show that the signals can be recovered by synchronously detecting the received signal by carriers of the same frequency but in phase quadrature (Fig. P-3.3). This scheme is known as *quadrature multiplexing*.

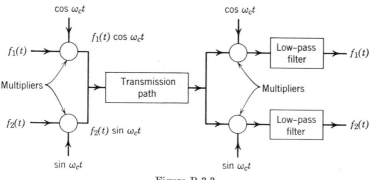

Figure P-3.3

4.(*a*) Explain qualitatively what will happen if an envelope detector (or a rectifier detector) is used to demodulate a DSB-SC signal.

(*b*) Figure P-3.4 shows a signal $f(t)$ which is amplitude-modulated DSB-SC. The modulated signal is now fed to an envelope detector. Find the output of the detector.

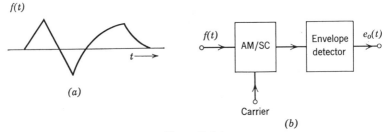

Figure P-3.4

5. The modulating signal $f(t)$ is a single sinusoid given by

$$f(t) = A \cos (2000\pi t)$$

Sketch the corresponding DSB-SC and SSB-SC signals for the carrier frequency 10 kHz. Sketch the AM signal for the modulation index 0.75.

6. The bridge circuit is commonly employed in measurements. A resistance bridge is often used to measure quantities which can vary a resistance linearly. A strain gage, for example, is a device in which a strain-sensitive

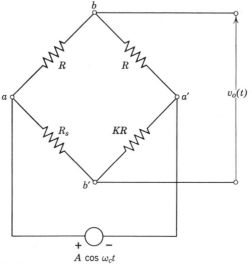

Figure P-3.6

element is attached to the body under strain. The resistance of the element then varies proportional to the strain. Similarly, some elements are temperature sensitive and their resistance varies linearly with the temperature (a thermistor, for example). A resistance bridge shown in Fig. P-3.6 has three fixed resistors, as shown. The fourth resistor is the variable resistor which varies in proportion to the quantity to be measured. Let this resistor have a quiescent value of KR. The value of this resistor R_s is given by

$$R_s = KR[1 + \alpha f(t)]$$

where $f(t)$ is the quantity to be measured (for example, strain or temperature) and α is the constant of proportionality. A sinusoidal source $A \cos \omega_c t$ is applied at the terminals aa'. Find the output at terminals bb'. What kind of modulation is this? What type of demodulation would you propose for this output? Assume $\alpha f(t) \ll 1$.

7. Show that the system shown in Fig. P-3.7 can demodulate AM signals. This is the full-wave square law rectifier arrangement. Show that the low-pass filter in this arrangement must have a cutoff frequency $2\omega_m$, where ω_m

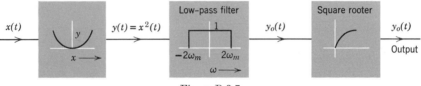

Figure P-3.7

is the highest frequency in the message signal $f(t)$. In general, show that the arrangement in Fig. P-3.7 acts as an envelope detector. Hence it cannot demodulate suppressed carrier signals.

8. Show that in Fig. P-3.7 if the first stage is a full-wave linear rectifier with characteristics as shown in Fig. P-3.8, then the AM wave can be demodulated without the final square rooter.

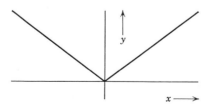

Figure P-3.8

9. The capacitor C of an envelope detector should be large enough to filter out the ripple at the carrier frequency present in the demodulated signal. However, if C is made too large, the time constant RC of the R-C circuit becomes too large and may be unable to follow the envelope of the modulated signal. Discuss carefully the effect of too low and too high values of the time constant.

(a) Determine the largest value of the time constant which will enable the detector to follow the envelope of a modulated signal shown in Fig. P-3.9. Assume the period of the modulating signal to be 10^{-3} second and

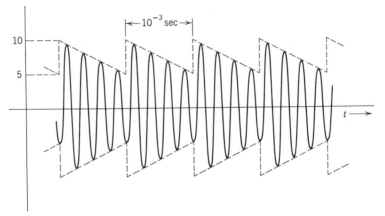

Figure P-3.9

the period of the carrier to be much smaller than 10^{-3} second. (*Hint:* Approximate the exponential decay of the R-C circuit by the first two terms of the Taylor series and equate the rate of discharge of the R-C circuit to the rate of decay of the envelope.)

(b) If the modulating signal were a sinusoidal signal of frequency ω_s, how would you determine the largest value of the time constant that can follow the envelope?

10. Sketch the waveforms of aa', bb', and cc' of a chopper-type modulator shown in Fig. 3.10c. What conditions must be satisfied by the amplitude k of the sinusoidal signal and the modulating signal $f(t)$ so that the output at terminals cc' is an AM signal? Explain. Assume that the input impedance of the bandpass filter is infinity.

11. Assume a signal $f(t)$ to be bandlimited to ω_m radians per second. Is the signal $f^2(t)$ also bandlimited? If so, find its highest frequency. What can you say about the spectrum of $f^n(t)$ in general? (*Hint:* Use frequency convolution theorem.) The spectra of two bandlimited signals are shown in Fig. P-3.11. For each case, find and sketch the spectrum of $f^2(t)$.

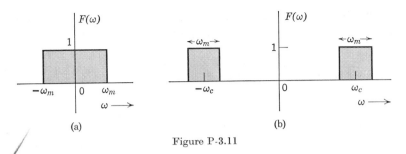

Figure P-3.11

√ 12. A periodic modulating signal $f(t)$ is shown in Fig. P-3.12. For amplitude modulation (AM), what must be the minimum carrier amplitude? For
this carrier amplitude, sketch the AM wave when the carrier frequency is

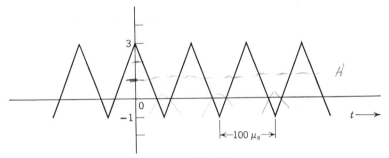

Figure P-3.12

100 kHz. Find the frequency spectrum of the AM wave. Sketch the corresponding DSB-SC wave and its spectrum.

13. Synchronous detection of an AM-SC signal may be carried out by
multiplying the received modulated signal by a pulse train $p(t)$ of frequency
ω_c as shown in Fig. 3.7. The result of this multiplication is a signal $f(t) \cos
\omega_c t \, p(t)$. Derive the analytical expression for the spectral density function
of this signal and show that the signal $f(t)$ can be recovered from the output
by using a low-pass filter. (*Hint:* Use Eqs. 3.8 and 3.11.)

14. A certain station uses DSB-SC signal of power (mean square value) P
watts to cater to a certain area. If it is decided to use SSB-SC instead, what
must be the power required to cater to the same area with the same strength?
Assume the synchronous demodulation with a locally generated carrier of
the same strength in both cases.

15. Figure 3.12 shows the half-wave diode rectifier detector. Show the
corresponding arrangment for the full-wave diode rectifier detector. Sketch
the incoming AM signal, the rectified signal, and the final output of the
low-pass filter in this arrangement. Sketch the frequency spectra of the

above signals. (*Hint:* Full-wave rectification of AM is equivalent to multi-plying it by a square wave of zero mean value. This is a bipolar square wave.)

16. A distorted form of a sinusoidal wave $\cos^3 \omega_c t$ is available. To obtain DSB-SC signal, a modulating signal $f(t)$ is multiplied by this distorted carrier waveform. Find and sketch the spectrum of the product $f(t) \cos^3 \omega_c t$. How can the desired modulated signal $f(t) \cos \omega_c t$ be obtained from this product?

17. A modulating signal $f(t)$ bandlimited to 5 kHz is multiplied by a periodic triangular signal shown in Fig. P-3.17a. Find and sketch the spectrum of the product. To obtain the modulated signal, the product is

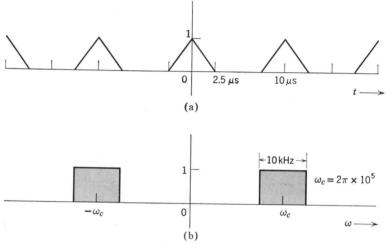

Figure P-3.17

passed through an ideal bandpass filter with a center frequency 100 kHz (Fig. P-3.17b). The output of this filter is $Af(t) \cos \omega_c t$. Determine the constant A.

18. Consider an AM signal

$$\varphi_{am}(t) = (1 + A \cos \omega_m t) \cos \omega_c t$$

where the message signal frequency $\omega_m = 5$ kHz and the carrier frequency $\omega_c = 100$ kHz. The constant $A = 15$. Can this signal be demodulated by an envelope detector? What will be the output of the envelope detector? Find the frequency spectrum of the envelope detector output.

19. Verify that the output $v_o(t)$ in Fig. 3.10c is indeed an AM signal if the composite element formed by the diode and the resistor R in series has a v-i relationship

$$i = ae + be^2$$

where i is the diode current and e is the corresponding voltage across the composite diode-resistor element. Assume the input impedance of the bandpass filter to be infinite.

20. The modulating signal $f(t)$ is given by

$$f(t) = \cos (2000\pi t) + \cos (4000\pi t)$$

Find the expression for the corresponding SSB-SC signal when the carrier frequency is 10 kHz.

21. Find the output of a phase shifter in Fig. 3.19 if the input signal is $A \cos (\omega_c t + \theta)$.

22. The modulating signal $f(t)$ is a periodic signal shown in Fig. P-3.12. Find the expression for the corresponding SSB-SC signal for this $f(t)$ when the carrier frequency is 100 kHz.

23. The vestigial sideband signal is generated by transmitting an AM wave through a vestigial sideband filter. If the transfer function of this

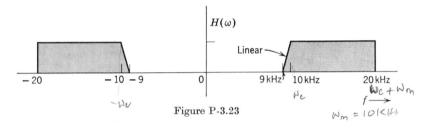

Figure P-3.23

filter is as shown in Fig. P-3.23, find the expression for the resulting vestigial sideband signal when the modulating signal $f(t)$ is given by

$$
\begin{aligned}
f(t) = (a) \quad & A \sin (100\pi t) \\
(b) \quad & A \left[\sin (100\pi t) + \cos (200\pi t) \right] \\
(c) \quad & A \sin (100\pi t) \cos (200\pi t)
\end{aligned}
$$

The carrier frequency is 10 kHz and the carrier amplitude is $4A$: Sketch the spectrum of the resulting vestigial sideband signal in each case.

24. In Problem 3.23 if the modulating signal $f(t)$ is given by

$$
\begin{aligned}
f(t) = (a) \quad & Sa\,(200\pi t) \\
(b) \quad & Sa^2\,(200\pi t)
\end{aligned}
$$

Sketch the spectrum of the vestigial sideband signal in each case (assume the carrier to be suppressed).

chapter 4

Communication Systems:
Angle Modulation

In AM signals the amplitude of the carrier is modulated by a signal $f(t)$, and hence the information content of $f(t)$ is carried by the amplitude variations of the carrier. Since a sinusoidal signal is described by three variables—amplitude, frequency, and phase—there exists a possibility of carrying the same information by varying either the frequency or the phase of the carrier. However, by definition, a sinusoidal signal represents an infinite wave train of constant amplitude, frequency, and phase, and hence the variation of any of these three variables appears to be contradictory to the definition of a sinusoidal signal. We now have to extend the concept of a sinusoidal signal to a generalized function whose amplitude, frequency, and phase may vary with time. We have already acquainted ourselves with the concept of variable amplitude in connection with AM signals. The variations in frequency and phase will now be considered.

To appreciate the concept of frequency variation it is necessary to define the instantaneous frequency. Figure 4.1a represents a sinusoidal signal $\varphi(t)$ which has a constant frequency ω_0 for $t < T$. At $t = T$, the frequency suddenly changes to $2\omega_0$ and remains constant at this value until $t = 2T$ where it changes to ω_0 again. The changes in the frequency here are abrupt, as shown in Fig. 4.1b, and present no difficulty to an understanding of the concept of variable frequency. The function $\varphi(t)$ is a sinusoidal signal which has a frequency ω_0 in intervals

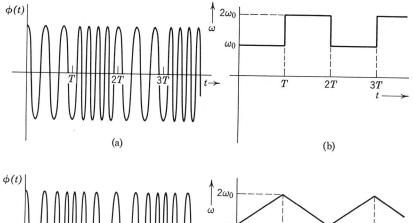

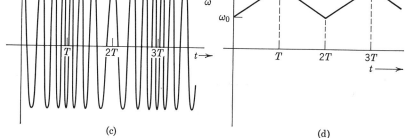

Figure 4.1   Concept of instantaneous frequency.

$2nT < t < (2n + 1)T$ and a frequency $2\omega_0$ in the intervals $(2n + 1)T < t < (2n + 2)T$ (n integral).

Next we ask: What will happen if, instead of abrupt variations, we allow a gradual variation in frequency as shown in Fig. 4.1d? Here the frequency of the signal is changing continuously at a uniform rate from ω_0 to $2\omega_0$ within a time interval T. Hence the frequency of the signal is different at every point. Now, strictly speaking, the signal $f(t)$ in Fig. 4.1c cannot be expressed by an ordinary sinusoidal expression. How can there be a continuous change in frequency in a sinusoidal signal? For this reason we have to define a generalized sinusoidal signal

$$f(t) = A \cos \theta(t) \tag{4.1}$$

Here θ is the angle of the sinusoidal signal and is a function of t. For an ordinary fixed frequency sinusoidal signal,

$$f(t) = A \cos (\omega_c t + \theta_0)$$

Hence

$$\theta(t) = \omega_c t + \theta_0$$

and

$$\omega_c = \frac{d\theta}{dt} \qquad (4.2)$$

The radian frequency ω_c here is a constant, given by the derivative of the angle $\theta(t)$. In general, this derivative may not be constant. We now define $d\theta/dt$ as the instantaneous frequency ω_i which may vary with time. We thus have a relationship between the angle $\theta(t)$ and instantaneous frequency ω_i:

$$\omega_i = \frac{d\theta}{dt}$$

$$\theta = \int \omega_i \, dt \qquad (4.3)$$

It is now easy to appreciate the possibility of transmission of information in $f(t)$ by varying the angle θ of a carrier. Such techniques of modulation, where the angle of the carrier is varied in some manner with a modulating signal $f(t)$, are known as *angle modulation*. Two methods are commonly used in angle modulation: *phase modulation* (PM) and *frequency modulation* (FM). If the angle $\theta(t)$ is varied linearly with $f(t)$, then

$$\theta(t) = \omega_c t + \theta_0 + k_p f(t) \qquad PM. \qquad (4.4)$$

Where k_p is a constant, and the resulting form is called phase modulation. Thus a signal $A \cos [\omega_c t + \theta_0 + k_p f(t)]$ represents a phase-modulated carrier. Note that the instantaneous frequency ω_i for a phase-modulated carrier is given by

$$\omega_i = \frac{d\theta}{dt} = \omega_c + k_p \frac{df}{dt} \qquad (4.5)$$

Hence in phase modulation the instantaneous frequency varies linearly with the derivative of the modulating signal. If, however, we vary the instantaneous frequency directly with the modulating signal, we have a frequency modulation. Thus for a frequency-modulated carrier the instantaneous frequency ω_i is given by

$$\omega_i = \omega_c + k_f f(t) \qquad (4.6)$$

and

$$\theta(t) = \int \omega_i \, dt$$

$$\theta(t) = \omega_c t + k_f \int f(t) \, dt + \theta_0 \qquad FM \qquad (4.7)$$

and the signal $A \cos [\omega_c t + \theta_0 + k_f \int f(t)\,dt]$ represents a frequency-modulated carrier.

It is easy to see from this discussion that although PM and FM are different forms of angle modulation, they are not essentially dissimilar. In PM the angle is varied linearly with the modulating signal, whereas in FM the angle varies linearly with the integral of the modulating signal. Indeed, if we integrate the modulating signal $f(t)$ first and then allow it to phase-modulate the carrier, we obtain a frequency-modulated wave. Similarly, if we differentiate $f(t)$ first and use it to frequency-modulate a carrier, the result is a phase-modulated wave. Actually, one of the methods of generating FM signals (the Armstrong indirect FM system) does integrate $f(t)$ and uses it to phase-modulate the carrier. In fact, PM and FM are inseparable since any variation in the phase of a carrier results in a variation in the frequency and vice versa. It is therefore unnecessary to discuss both forms of angle modulation. We shall deal here in some detail with frequency modulation alone, but our discussion is equally valid for phase modulation. Note that for angle modulation the amplitude is always constant. Figures 4.1a and c are examples of FM carriers.

If the PM and FM signals are denoted by $\varphi_{\text{PM}}(t)$ and $\varphi_{\text{FM}}(t)$, respectively, then

$$\varphi_{\text{PM}}(t) = A \cos [\omega_c t + k_p f(t)]$$

$$\varphi_{\text{FM}}(t) = A \cos \left[\omega_c t + k_f \int f(t)\,dt \right] \tag{4.8}$$

where $f(t)$ is the message signal. In these equations the initial phase θ_0 is assumed to be zero without loss of generality.

The advantages of exponential representation of sinusoidal signals are well known. For this reason we shall express the sinusoidal signal of the form (4.1) by its exponential counterpart:

$$A \cos \theta(t) \sim A e^{j\theta(t)}$$

where it is understood that we always imply the real part of the exponential signals thus represented.

$$A \cos \theta(t) = \text{Re}\,[A e^{j\theta(t)}]$$

We shall use the notation $\hat{\varphi}(t)$ for the exponential representation of $\varphi(t)$. Thus if

$$\varphi(t) = A \cos \theta(t)$$

then

$$\hat{\varphi}(t) = A e^{j\theta(t)}$$

and

$$\varphi(t) = \text{Re } \hat{\varphi}(t)$$

According to this notation, the PM and FM carriers in Eq. 4.8 can be expressed as

$$\hat{\varphi}_{\text{PM}}(t) = A e^{j[\omega_c t + k_p f(t)]}$$

$$\hat{\varphi}_{\text{FM}}(t) = A e^{j[\omega_c t + k_f \int f(t) \, dt]} \qquad (4.9)$$

For convenience let

$$\int f(t) \, dt = g(t)$$

This gives

$$\hat{\varphi}_{\text{FM}}(t) = A e^{j[\omega_c t + k_f g(t)]} \qquad (4.10)$$

4.1 NARROWBAND FM

The general expression for the FM carrier is given by Eq. 4.10. The instantaneous frequency ω_i is given by

$$\omega_i = \frac{d\theta}{dt} = \omega_c + k_f \frac{dg}{dt} = \omega_c + k_f f(t)$$

which is obviously proportional to message signal $f(t)$. The term $k_f f(t)$ represents the deviation of the carrier frequency from its quiescent value ω_c. The constant k_f therefore controls the deviation of the carrier frequency. For small values of k_f the frequency deviation is small, and we expect the spectrum of the FM signal to have a narrow band. If k_f is large the bandwidth is expected to be larger correspondingly. This is indeed true. Let us first consider the narrowband case. If k_f is so small that $k_f g(t) \ll 1$ for all t, then

$$e^{j k_f g(t)} \simeq 1 + j k_f g(t)$$

and

$$\hat{\varphi}_{\text{FM}}(t) \simeq A[1 + j k_f g(t)] e^{j\omega_c t}$$

and

$$\varphi_{\text{FM}}(t) = \text{Re } [\hat{\varphi}_{\text{FM}}(t)] = A \underbrace{\cos \omega_c t}_{\text{carrier}} - \underbrace{A k_f g(t) \sin \omega_c t}_{\text{sideband}} \qquad (4.11)$$

It is interesting to observe that the AM carrier is expressed as (Eq. 3.12a):

$$\varphi_{\text{AM}}(t) = A \cos \omega_c t + f(t) \cos \omega_c t \qquad (4.12)$$

whereas the narrowband FM carrier is given by

$$\varphi_{\text{FM}}(t) = A \cos \omega_c t - A k_f g(t) \sin \omega_c t \qquad (4.13)$$

Similarly, the narrowband PM carrier is given by

$$\varphi_{\text{PM}}(t) = A \cos \omega_c t - A k_p f(t) \sin \omega_c t \qquad (4.14)$$

Each signal has a carrier term and the sidebands which are centered at $\pm \omega_c$. If

$$f(t) \leftrightarrow F(\omega)$$

and

$$g(t) \leftrightarrow G(\omega)$$

then since

$$g(t) = \int f(t)\, dt$$

it follows from time integration property (Eq. 1.119b) that

$$g(t) \leftrightarrow \frac{1}{j\omega} F(\omega) \qquad (4.15)$$

Hence

$$G(\omega) = \frac{1}{j\omega} F(\omega) \qquad (4.16)$$

Thus, if $F(\omega)$ is bandlimited to ω_m, then $G(\omega)$ is also bandlimited to ω_m. The frequency spectrum of $\varphi_{\text{FM}}(t)$ of the FM carrier in Eq. 4.11 can be found by using Eq. 1.116b. Thus, if

$$\varphi_{\text{FM}}(t) \leftrightarrow \Phi_{\text{FM}}(\omega)$$

then

$$\Phi_{\text{FM}}(\omega) = \pi A [\delta(\omega - \omega_c) + \delta(\omega + \omega_c)]$$
$$+ \frac{jAk_f}{2} [G(\omega - \omega_c) - G(\omega + \omega_c)] \qquad (4.17)$$

Comparison of the spectrum of FM (Eq. 4.17) with that of AM (Eq. 3.12b) brings out clearly the similarities and differences between the two types of modulation. In both types there is a carrier term and sideband component centered at $\pm \omega_c$. The sideband spectrum for FM, however, has a phase shift of $\pi/2$ with respect to the carrier, whereas that of AM is in phase with the carrier. The spectrum $G(\omega) = (1/j\omega)F(\omega)$ and hence if $F(\omega)$ is bandlimited to ω_m, $G(\omega)$ is also bandlimited to ω_m. Thus a narrowband FM (and a narrowband PM) signal

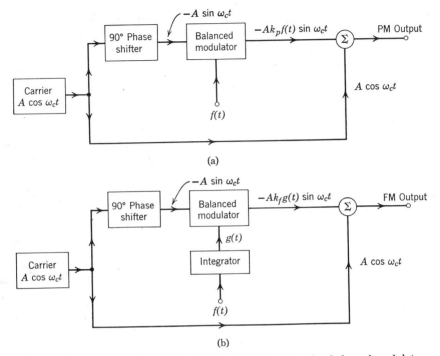

Figure 4.2   Generation of narrowband PM and FM signals, using balanced modulators.

occupies the same bandwidth ($2\omega_m$) as that occupied by an AM signal. It must be remembered, however, that despite apparent similarities, the AM and FM signals have very different waveforms. <u>In an AM signal, the frequency is constant and the amplitude varies with time,</u> whereas in an FM signal the amplitude is constant and the frequency varies with time.

Equations 4.13 and 4.14 suggest a possible method of generating narrowband FM and PM signals by using balanced modulators. The block diagram representation of such systems is shown in Fig. 4.2.

4.2 WIDEBAND FM

If the deviations in the carrier frequency are made large enough, that is, if the constant k_f is chosen large enough so that the condition $k_f g(t) \ll 1$ is not satisfied, the analysis of FM signals becomes very involved for a general modulating signal $f(t)$. Derivation of a precise expression for the bandwidth of an FM carrier for a general modulating signal $f(t)$ is not possible because FM is a nonlinear modulation (see Section 4.3 and 4.5). We shall first obtain the expression for the bandwidth of an FM carrier on a heuristic basis and later verify this result for some signals. We shall now show that W, the bandwidth of an FM carrier, is approximately given by

$$W = 2[k_f |f(t)|_{\max} + 2\omega_m] \qquad \text{(in rps)}$$

where ω_m is the bandwidth of $f(t)$. Note that

$$\omega_i = \omega_c + k_f f(t)$$

Hence $k_f |f(t)|_{\max}$ represents the maximum deviation in the carrier frequency. If we denote the maximum deviation in carrier frequency by $\Delta\omega$, the bandwidth W is given by

$$W = 2[\Delta\omega + 2\omega_m] \qquad \text{(in rps)} \qquad (4.18)$$

To derive this result, we approximate the message signal $f(t)$ by a staircase signal as shown in Fig. 4.3a. The signal $f(t)$ is bandlimited to f_m Hz. It is therefore reasonable to assume the signal to be constant over a Nyquist sampling period $1/2f_m$ seconds. The appropriate staircase approximation is shown in Fig. 4.3a. The FM carrier for a staircase approximated signal will consist of sinusoidal pulses of constant frequency and duration of $1/2f_m$ seconds.

Note that there is an abrupt change of frequency at every sampling instant. One such pulse is shown in Fig. 4.3b. The spectrum of each pulse can be obtained by using pair 13 (Table 1.1B) and the modulation theorem (also see Fig. 1.32). The spectrum of a typical pulse in Fig. 4.3b is shown in Fig. 4.3c. It can be seen from the figure that the spectrum of this pulse occupies the band $\omega_i - 2\omega_m$ to $\omega_i + 2\omega_m$ where ω_i is the frequency of the sinusoidal pulse. In this case $\omega_i = \omega_c + k_f f(t_k)$ where t_k is the kth sampling instant. Hence the spectrum lies in the region $\omega_c + k_f f(t_k) - 2\omega_m$ to $\omega_c + k_f f(t_k) + 2\omega_m$. It is evident that the spectrum of the entire FM will lie in the frequency range $\omega_c - k_f |f(t)|_{\max} - 2\omega_m$ to $\omega_c + k_f |f(t)|_{\max} + 2\omega_m$, and the bandwidth W is given by

$$W \simeq 2k_f |f(t)|_{\max} + 4\omega_m$$
$$= 2(\Delta\omega + 2\omega_m) \qquad \text{(in rps)}$$

For wideband FM $\Delta\omega \gg \omega_m$ and $W \simeq 2\,\Delta\omega$. We therefore conclude that the bandwidth of an FM carrier is approximately twice the carrier frequency deviation $\Delta\omega$. This result is quite reasonable. If the carrier

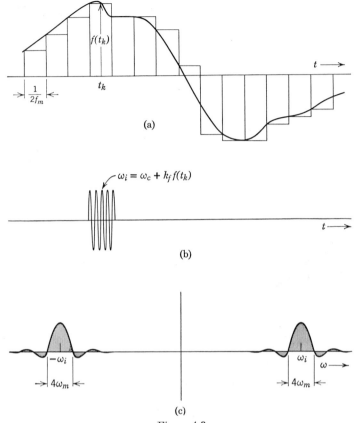

Figure 4.3

frequency ω_c has a maximum deviation $\Delta\omega$, then the FM signal frequency varies from $\omega_c - \Delta\omega$ to $\omega_c + \Delta\omega$. So, as a first guess, one would estimate that such a signal will contain frequencies in this range and thus will have a bandwidth $2\,\Delta\omega$.

We shall now verify this conclusion when the modulating signal $f(t)$ is a sinusoidal signal

$$f(t) = a \cos \omega_m t$$

We may assume that the signal $f(t)$ is switched on at $t = 0$ and hence

$$g(t) = \int f(t)\, dt = a \int_0 \cos \omega_m t\, dt$$

$$= \frac{a}{\omega_m} \sin \omega_m t \qquad (4.19)$$

Since the instantaneous frequency ω_i is given by (Eq. 4.6)

$$\omega_i = \omega_c + k_f f(t) = \omega_c + ak_f \cos \omega_m t \qquad (4.20)$$

it is evident from Eq. 4.20 that the maximum deviation in carrier frequency is ak_f radians per second:

$$\Delta\omega = ak_f \qquad (4.21)$$

Substituting Eq. 4.19 in Eq. 4.10 and using Eq. 4.21, we obtain

$$\hat{\varphi}_{\mathrm{FM}}(t) = Ae^{j[\omega_c + (\Delta\omega/\omega_m)\sin \omega_m t]}$$

The quantity $\Delta\omega/\omega_m$ is the ratio of the maximum deviation of the carrier frequency to the signal frequency ω_m and is called the *modulation index* m_f. Thus

$$\frac{ak_f}{\omega_m} = \frac{\Delta\omega}{\omega_m} = m_f \qquad (4.22)$$

and

$$\hat{\varphi}_{\mathrm{FM}}(t) = Ae^{j(\omega_c t + m_f \sin \omega_m t)}$$

$$= Ae^{jm_f \sin \omega_m t}e^{j\omega_c t} \qquad (4.23)$$

The first exponential in Eq. 4.23 is obviously a periodic function of period $2\pi/\omega_m$ and can be expanded by a Fourier series:

$$e^{jm_f \sin \omega_m t} = \sum_{n=-\infty}^{\infty} C_n e^{jn\omega_m t}$$

where

$$C_n = \frac{\omega_m}{2\pi} \int_{-\pi/\omega_m}^{\pi/\omega_m} e^{jm_f \sin \omega_m t}e^{-jn\omega_m t} \, dt$$

Letting $\omega_m t = x$, we get

$$C_n = \frac{1}{2\pi} \int_{-\pi}^{\pi} e^{j(m_f \sin x - nx)} \, dx = J_n(m_f)$$

The integral on the right-hand side cannot be evaluated in a closed form but must be integrated by expanding the integrand in infinite series. This integral has been extensively tabulated and is denoted by $J_n(m_f)$, the Bessel function of the first kind and nth order.* These functions

* E. Jahnke and F. Emde, *Tables of Functions*, Dover Publications, New York, 1945.

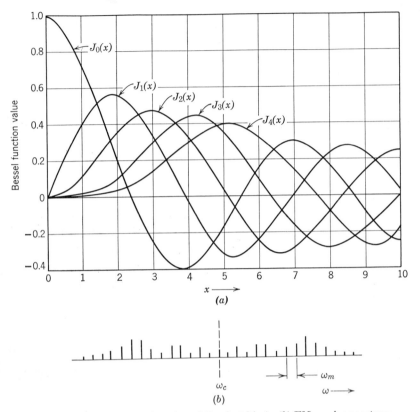

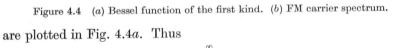

Figure 4.4   (a) Bessel function of the first kind.   (b) FM carrier spectrum.

are plotted in Fig. 4.4a.   Thus

$$e^{jm_f \sin \omega_m t} = \sum_{n=-\infty}^{\infty} J_n(m_f) e^{jn\omega_m t} \tag{4.24}$$

Furthermore it can be shown that

$$J_n(m_f) = J_{-n}(m_f) \qquad n \text{ even}$$
$$J_n(m_f) = -J_{-n}(m_f) \qquad n \text{ odd} \tag{4.25}$$

Substituting Eq. 4.24 in Eq. 4.23, we get

$$\hat{\varphi}_{\text{FM}}(t) = A e^{j\omega_c t} \sum_{n=-\infty}^{\infty} J_n(m_f) e^{jn\omega_m t}$$

and

$$\varphi_{\text{FM}}(t) = A \sum_{n=-\infty}^{\infty} J_n(m_f) \cos (\omega_c + n\omega_m)t \tag{4.26a}$$

By taking advantage of properties in Eq. 4.25, we can express Eq. 4.26a as

$$\varphi_{\text{FM}}(t) = A\{J_0(m_f)\cos\omega_c t + J_1(m_f)[\cos(\omega_c + \omega_m)t - \cos(\omega_c - \omega_m)t]$$
$$+ J_2(m_f)[\cos(\omega_c + 2\omega_m)t + \cos(\omega_c - 2\omega_m)t]$$
$$+ J_3(m_f)[\cos(\omega_c + 3\omega_m)t - \cos(\omega_c - 3\omega_m)t]$$
$$+ \cdots + \cdots \tag{4.26b}$$

Although the form in Eq. 4.26a is commonly used in the literature, we prefer Eq. 4.26b for its compactness.

It is evident from Eq. 4.26 that the modulating signal $f(t)$ of frequency ω_m gives rise to sideband frequencies $(\omega_c \pm \omega_m)$, $(\omega_c \pm 2\omega_m)$, $(\omega_c \pm 3\omega_m)$, ..., etc., as shown in Fig. 4.4b. It therefore appears that an FM carrier contains components of infinite frequencies and has an infinite bandwidth. In practice, however, the amplitudes of the spectral components of higher frequencies become negligible and hence almost all of the energy of the FM carrier is contained in the spectral components lying within a finite bandwidth. This can be easily seen from Fig. 4.4a. For $m_f \ll 1$, only $J_0(m_f)$ and $J_1(m_f)$ have any significant magnitudes. All of the higher functions, $J_2(m_f)$, $J_3(m_f)$, ..., etc., are negligible. For this case, only the carrier and the first-order sidebands are of significance. This is, of course, the narrowband FM discussed in the previous section.

For $m_f = 2$, the functions $J_5(2)$, $J_6(2)$, ..., etc., have negligible amplitudes. Hence the significant spectral components of an FM carrier for $m_f = 2$ are ω_c, $\omega_c \pm \omega_m$, $\omega_c \pm 2\omega_m$, $\omega_c \pm 3\omega_m$, and $\omega_c \pm 4\omega_m$. The bandwidth of the significant sidebands in this case is $8\omega_m$. As m_f is increased further, higher-order sidebands become significant. If we consider the significant sidebands to be those which have an amplitude of at least one per cent of that of the unmodulated carrier, then for all significant sidebands $J_n(m_f) > 0.01$. The number of significant sidebands for different values of m_f can be found from the plot of the Bessel functions. It can be seen from these plots that $J_n(m_f)$ diminishes rapidly for $n > m_f$. In general, $J_n(m_f)$ is negligible for $n > m_f$. This is particularly true for values of $m_f \gg 1$. Thus for a wideband FM where the number of significant sidebands may be considered to be the integer closest to m_f, $m_f = n$. The total bandwidth W of the FM carrier is evidently given by

$$W \simeq 2n\omega_m \simeq 2m_f\omega_m \qquad \text{(in rps)}$$

But

$$m_f = \frac{ak_f}{\omega_m} = \frac{\Delta\omega}{\omega_m}.$$

Hence

$$W \simeq 2m_f\omega_m = 2\,\Delta\omega \qquad \text{(in rps)} \qquad (4.27)$$

Actually, $W = 2(\Delta\omega + 2\omega_m)$ is a more accurate expression than $2\,\Delta\omega$. Hence we have

$$W \simeq 2(\Delta\omega + 2\omega_m) \qquad (4.28)$$

This rule gives a fairly good estimate for wideband FM. For narrowband FM where $\Delta\omega \ll \omega_m$, the bandwidth is approximately $2\omega_m$ (the same as AM as shown in Section 4.2). Hence the bandwidth required for the transmission of an FM signal is approximately twice the maximum frequency deviation of the carrier. The exact relationship between the ratio of W and $\Delta\omega$, the frequency deviation of the carrier as a function of m_f, is shown in Fig. 4.5. It is evident from this figure that for $m_f \gg 1$, the bandwidth W is approximately twice the frequency deviation of the carrier.

The Federal Communications Commission (FCC) has fixed the maximum value of Δf (frequency deviation) at 75 kHz for commercial FM broadcasting stations. The approximate bandwidth required for FM radio is therefore approximately 150 kHz.

It is evident from Fig. 4.5 that for wideband FM $(m_f \gg 1)$, the bandwidth required for transmission approaches $2\,\Delta\omega$ radians per second or $2\,\Delta f$ Hz. Since

$$\Delta\omega = ak_f = m_f\omega_m$$

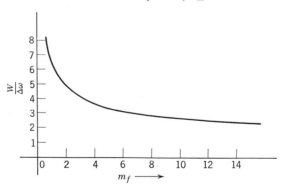

Figure 4.5 Bandwidth of FM signal as a function of the modulation index. (From *Information Transmission Modulation, and Noise,* by M. Schwartz, McGraw-Hill, New York, 1960.)

increasing m_f also increases $\Delta\omega$ proportionately for a constant value of ω_m. This is illustrated in Fig. 4.6a. Here ω_m is held constant ($f_m = 5$ kHz). The spectra of an FM carrier for $m_f = 1$, 2, 5, and 10 are shown. The frequency deviation (Δf) is 5, 10, 25, and 50 kHz, respectively. Note that the bandwidth is approximately $2\,\Delta f$ for the higher values of m_f.

Figure 4.6b shows the case where the frequency deviation $\Delta\omega$ is held constant, and $m_f = \Delta\omega/\omega_m$ is varied by varying ω_m. Here $\Delta f = 75$ kHz and m_f is varied from 10 to 5 by varying f_m from 7.5 to 15 kHz. In either case, the bandwidth B is approximately

$$B \simeq 2\,\Delta f = 150 \text{ kHz}$$

4.3 MULTIPLE FREQUENCY MODULATION

In Section 4.2 we discussed a specific case of a single frequency-modulating signal. We shall now extend these results to the case of multiple frequencies. First only two frequencies will be considered. It can then be generalized to any number of frequencies. Consider

$$f(t) = a_1 \cos \omega_1 t + a_2 \cos \omega_2 t$$
$$\omega_i = \omega_c + k_f f(t)$$
$$= \omega_c + k_f(a_1 \cos \omega_1 t + a_2 \cos \omega_2 t)$$

The maximum frequency deviation is obviously

$$\Delta\omega = (a_1 + a_2)k_f$$

and

$$\theta(t) = \int \omega_i \, dt = \omega_c t + \frac{a_1 k_f}{\omega_1} \sin \omega_1 t + \frac{a_2 k_f}{\omega_2} \sin \omega_2 t$$
$$= \omega_c t + m_1 \sin \omega_1 t + m_2 \sin \omega_2 t$$

where

$$m_1 = \frac{a_1 k_f}{\omega_1} \quad \text{and} \quad m_2 = \frac{a_2 k_f}{\omega_2}$$

Also

$$\hat{\varphi}_{\text{FM}}(t) = Ae^{j\theta(t)} = Ae^{j(\omega_c t + m_1 \sin \omega_1 t + m_2 \sin \omega_2 t)}$$
$$= Ae^{j\omega_c t}(e^{jm_1 \sin \omega_1 t})(e^{jm_2 \sin \omega_2 t}) \tag{4.29}$$

The exponentials in the parentheses are obviously periodic functions with periods $2\pi/\omega_1$ and $2\pi/\omega_2$, respectively. These exponentials can

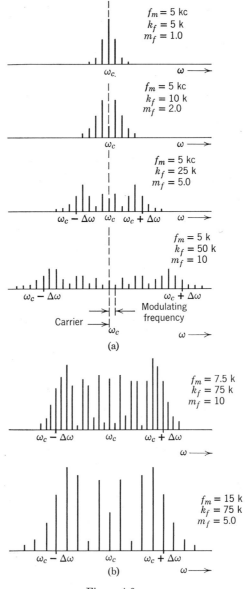

$f_m = 5$ kc
$k_f = 5$ k
$m_f = 1.0$

ω_c
$\omega \longrightarrow$

$f_m = 5$ kc
$k_f = 10$ k
$m_f = 2.0$

ω_c
$\omega \longrightarrow$

$f_m = 5$ kc
$k_f = 25$ k
$m_f = 5.0$

$\omega_c - \Delta\omega \quad \omega_c \quad \omega_c + \Delta\omega \quad \omega \longrightarrow$

$f_m = 5$ k
$k_f = 50$ k
$m_f = 10$

$\omega_c - \Delta\omega \qquad\qquad \omega_c + \Delta\omega$

Modulating frequency

Carrier $\longrightarrow$ ω_c

$\omega \longrightarrow$

(a)

$f_m = 7.5$ k
$k_f = 75$ k
$m_f = 10$

$\omega_c - \Delta\omega \qquad \omega_c \qquad \omega_c + \Delta\omega$

$\omega \longrightarrow$

$f_m = 15$ k
$k_f = 75$ k
$m_f = 5.0$

$\omega_c - \Delta\omega \qquad \omega_c \qquad \omega_c + \Delta\omega$

(b)

$\omega \longrightarrow$

Figure 4.6

224

be represented by Fourier series using Bessel functions (see Eq. 4.24). Hence

$$\hat{\varphi}_{\mathrm{FM}}(t) = Ae^{j\omega_c t}\left[\sum_{n=-\infty}^{\infty} J_n(m_1)e^{jn\omega_1 t}\right]\left[\sum_{k=-\infty}^{\infty} J_k(m_2)e^{jk\omega_2 t}\right]$$

$$= A\sum_{n=-\infty}^{\infty}\sum_{k=-\infty}^{\infty} J_n(m_1)J_k(m_2)e^{j(\omega_c+n\omega_1+k\omega_2)t} \qquad (4.30)$$

and

$$\varphi_{\mathrm{FM}}(t) = A\sum_{n=-\infty}^{\infty}\sum_{k=-\infty}^{\infty} J_n(m_1)\,J_k(m_2)\cos\left[(\omega_c + n\omega_1 + k\omega_2)t\right] \qquad (4.31)$$

It is evident from this result that when $f(t)$ is composed of two frequencies ω_1 and ω_2, the spectrum of the FM signal contains sidebands $(\omega_c \pm n\omega_1)$ and $(\omega_c \pm k\omega_2)$, corresponding to frequencies ω_1 and ω_2. In addition, there are cross modulation terms $(\omega_c \pm n\omega_1 \pm k\omega_2)$. Note that this behavior contrasts to that observed in AM. In AM, each new frequency added to the modulating signal gives rise to its own sidebands only. There are no cross modulation terms. For this reason AM is called linear modulation, whereas FM is a nonlinear type of modulation. This topic will be discussed in more detail in Section 4.5.

The results derived in this section for two frequencies can easily be extended to any number of frequencies.

4.4 SQUARE WAVE MODULATION

We shall here consider one more special case of FM modulation where the modulating signal $f(t)$ is a square wave as shown in Fig. 4.7a. The method discussed here is very general and is applicable to any periodic modulating signal. For a square wave (Fig. 4.7a), the instantaneous frequency of the FM carrier is given by

$$\omega_i = \omega_c + k_f f(t)$$

and the phase $\theta(t)$ is given by

$$\theta(t) = \int \omega_i\, dt$$

$$= \omega_c t + k_f\int f(t)\, dt$$

$$= \omega_c t + \psi(t)$$

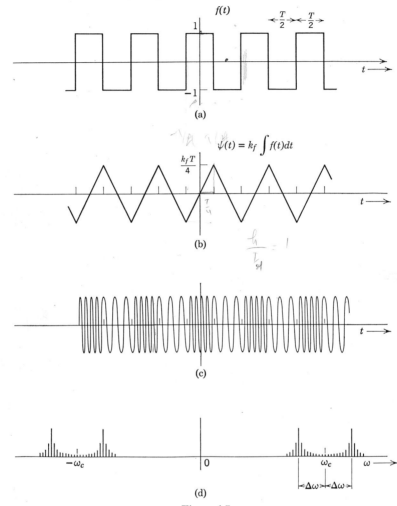

Figure 4.7

where

$$\psi(t) = k_f \int f(t)\,dt$$

is shown in Fig. 4.7b. This is a triangular periodic function with period T.

Note that the maximum carrier frequency deviation $\Delta\omega$ in this case is k_f since $|f(t)|_{\max} = 1$. Hence

$$k_f = \Delta\omega$$

and

$$\psi(t) = \begin{cases} (\Delta\omega)t & -\dfrac{T}{4} < t < \dfrac{T}{4} \\ (\Delta\omega)\left[\dfrac{T}{2} - t\right] & \dfrac{T}{4} < t < \dfrac{3T}{4} \end{cases}$$

(4.32)

and

$$\psi(t) = \psi(t \pm nT) \quad periodic$$

Also

$$\hat{\varphi}_{\text{FM}}(t) = Ae^{j\theta(t)}$$
$$= Ae^{j\psi(t)}e^{j\omega_c t}$$

The function $e^{j\psi(t)}$ is itself a periodic function of period T, and it can be expressed by Fourier series as

$$e^{j\psi(t)} = \sum_{n=-\infty}^{\infty} \alpha e^{jn\omega_s t} \qquad \omega_s = \frac{2\pi}{T}$$

where

$$\alpha_n = \frac{1}{T} \int_{-T/4}^{3T/4} e^{j\psi(t)} e^{-jn\omega_s t} \, dt$$

Substitution of Eq. 4.32 in this equation and the subsequent integration yields

$$\alpha_n = \frac{1}{2}\left\{ Sa\left[\frac{\pi}{2}(\beta - n)\right] + (-1)^n Sa\left[\frac{\pi}{2}(\beta + n)\right] \right\}$$

(4.33)

where

$$\beta = \frac{\Delta\omega}{\omega_s}$$

Hence

$$\hat{\varphi}_{\text{FM}}(t) = Ae^{j\psi(t)}e^{j\omega_c t}$$
$$= A \sum_{n=-\infty}^{\infty} \alpha_n e^{j(\omega_c + n\omega_s)t}$$

and

$$\varphi_{\text{FM}}(t) = A \sum_{n=-\infty}^{\infty} \alpha_n \cos(\omega_c + n\omega_s)t$$

The frequency spectrum of $\varphi_{\text{FM}}(t)$ is shown in Fig. 4.7d. The method discussed here is general and can be applied to any modulating signal $f(t)$ that is periodic and has a zero mean.

4.5 LINEAR AND NONLINEAR MODULATION

In the case of AM signals, the sidebands obey the principle of super-position. Thus if signals $f_1(t)$ and $f_2(t)$ give rise to sidebands φ_1 and φ_2, respectively, the sidebands created by the composite signal $f_1(t) + f_2(t)$ will be $\varphi_1 + \varphi_2$. There is no intermodulation or cross-product sidebands as we observed in FM. For this reason AM is called the linear type of modulation.*

The linear modulation lends itself easily to mathematical manipulations and generalizations. The spectrum of a modulated signal due to sum of two modulating signals can be found by calculating the spectrum due to each individual signal and then adding them together. This proves very useful in noise calculations in communication systems. For linear modulation systems, the effect of additive noise over the channel can be calculated by assuming the signal to be zero. This is not true of a nonlinear modulation where cross-modulation terms arise. For these reasons, we are interested in approximating a nonlinear modulation by a linear model. This is analogous to the case of system analysis where one can approximate a nonlinear system by a linear one over a limited range of signal amplitudes. We shall now show that FM closely approximates a linear behavior for a small modulation index.

Linearization of FM Signal for Small Modulation Index

For a small modulating index, FM closely exhibits linear behavior. Consider again the case of modulating signal $f(t)$ containing two frequencies ω_1 and ω_2. The FM signal in this case is given by Eq. 4.29. If m_1 and $m_2 \ll 1$, then for $f(t)$,

$$f(t) = a_1 \cos \omega_1 t + a_2 \cos \omega_2 t$$
$$\hat{\varphi}_{\mathrm{FM}}(t) \simeq A(1 + jm_1 \sin \omega_1 t)(1 + jm_2 \sin \omega_2 t)e^{j\omega_c}$$
$$\simeq A(1 + jm_1 \sin \omega_1 t + jm_2 \sin \omega_2 t)e^{j\omega_c t} \qquad (4.34)$$

Note that if

$$f_1(t) = a_1 \cos \omega_1 t \qquad \text{and} \quad m_1 \ll 1$$

* A general definition of linear modulation is as follows: The modulated signal is a function of the modulating signal $f(t)$. Let $\varphi[f(t)]$ be the modulated signal. Then the modulation is linear if $\{d/d[f(t)]\}\{\varphi[f(t)]\}$ is independent of $f(t)$. Otherwise it is a nonlinear modulation. The reader can easily verify according to this definition that AM is linear, whereas FM is not.

then

$$\hat{\varphi}_{\mathrm{FM}}(t) = A e^{j m_1 \sin \omega_1 t} e^{j \omega_c t}$$

$$\simeq A[1 + j m_1 \sin \omega_1 t] e^{j \omega_c t}$$

If

$$f_2(t) = a_2 \cos \omega_2 t \qquad \text{and} \quad m_2 \ll 1$$

then

$$\hat{\varphi}_{\mathrm{FM}}(t) \simeq A(1 + j m_2 \sin \omega_2 t) e^{j \omega_c t}$$

It is easy to see that under the conditions m_1, $m_2 \ll 1$, the sidebands due to modulating signal $f_1(t) + f_2(t)$ are the sum of sidebands due to $f_1(t)$ and $f_2(t)$ individually. Hence FM can be assumed to be a linear modulation for a small modulation index. The cross-modulation terms under this assumption can be ignored.

4.6 SOME REMARKS ON PHASE MODULATION

We have observed that in angle-modulated carriers the bandwidth of the resultant signal is approximately twice the maximum deviation of the carrier frequency. Thus, if the deviation in carrier frequency is kept constant, the spectrum of the modulated signal has a constant bandwidth.

For FM signals, the instantaneous frequency ω_i is given by Eq. 4.20:

$$\omega_i = \omega_c + a k_f \cos \omega_m t$$

The deviation of the carrier frequency is $\Delta \omega = a k_f$ and is independent of the frequency ω_m of the modulating signal. Hence for FM signals, the bandwidth is approximately $2 \Delta \omega = 2 a k_f$, regardless of the frequency of the modulating signal. For phase modulation, on the other hand,

$$\theta(t) = \omega_c t + a k_p \cos \omega_m t$$

and

$$\omega_i = \frac{d\theta}{dt} = \omega_c - a k_p \omega_m \sin \omega_m t \qquad (4.35)$$

It is evident from Eq. 4.35 that the maximum deviation of the carrier frequency in PM is not a constant but is given by $a k_p \omega_m$ and varies linearly with ω_m, the frequency of the modulating signal. Hence the bandwidth required for transmission of a PM carrier is not constant but depends strongly upon the waveform of the modulating signal.

In practice, however, it is easier to generate a PM signal than an FM signal. It was shown earlier that if we integrate the modulating signal first and then allow it to phase-modulate a carrier, we obtain an FM carrier. Hence in many systems the FM signals are generated by using PM generators which use the integrated signal $[\int f(t)\, dt]$ for phase modulation.

4.7 POWER CONTENTS OF THE CARRIER AND THE SIDEBANDS IN ANGLE-MODULATED CARRIERS

It was shown that the total power carried by an AM carrier was a function of the modulation index m. For an angle-modulated carrier, the amplitude of the carrier is always constant regardless of the modulation index m_f. Hence it is reasonable to expect that the power carried by angle-modulated carrier is constant regardless of the extent of the modulation. This is indeed the case. Consider Eq. 4.26a:

$$\varphi_{\mathrm{FM}}(t) = A \sum_{n=-\infty}^{\infty} J_n(m_f) \cos{(\omega_c + n\omega_m)t}$$

The FM carrier in this case is a periodic function and is thus expressed as a discrete sum of sinusoidal components. The power of $\varphi_{\mathrm{FM}}(t)$ is equal to the sum of the powers of individual components.* Hence

$$\overline{\varphi_{\mathrm{FM}}{}^2(t)} = \frac{A^2}{2} \sum_{n=-\infty}^{\infty} J_n{}^2(m_f)$$

It can be shown† that the summation on the right-hand side is 1 for all values of m_f. Hence

$$\overline{\varphi_{\mathrm{FM}}{}^2(t)} = \frac{A^2}{2}$$

The power of the unmodulated carrier $\cos \omega_c t$ is also $A^2/2$. Hence the power of an FM carrier is the same as that of the unmodulated carrier. The modulated signal, however, has a carrier component and sideband components as expressed in Eq. 4.26b. In this equation, $AJ_0(m_f)$ represents the amplitude of the carrier component and $AJ_n(m_f)$ represents the amplitude of the nth-order sideband. It is possible to make $J_0(m_f)$ as small as possible by a proper choice of m_f. In fact,

---

* This follows from Parsevals theorem (Eq. 1.34).

† G. N. Watson, *Treatise on Theory of Bessel Functions*, p. 31. Cambridge, 1922.

$J_0(m_f) = 0$ for $m_f = 2.405$, 5.52, and so on (see Fig. 4.3). Hence the power carried by the carrier component can be made as small as desired. In such a situation, most of the power is carried by the sideband components. Hence, by a proper choice of m_f, the efficiency of transmission can be made as close to 100% as desired. Note that as m_f increases the number of sidebands increases and $J_0(m_f)$ decreases, thus increasing the efficiency of transmission.

4.8 NOISE-REDUCTION CHARACTERISTICS OF ANGLE MODULATION

It is obvious from the preceding discussion that for a given modulating signal the bandwidth required for the transmission of an angle-modulated wave is much larger than that required for the AM wave. For example, if f_m is 10 kHz, then the bandwidth of the AM wave is 20 kHz, whereas the bandwidth of the corresponding FM wave may be about 150 kHz for $\Delta f = 75$ kHz. For narrowband FM the bandwidth is about 20 kHz. Under no condition is the bandwidth required for FM less than 20 kHz. This is definitely a great disadvantage of the FM system. For this reason, frequency modulation was considered wasteful and of little practical utility at the time of its discovery (about 1930).[*] This was subsequently proven false by Major Edwin H. Armstrong, a brilliant engineer, whose contributions to the field of radio systems are comparable with those of Hertz and Marconi.[†]

We now know that FM provides better discrimination against noise and interfering signals. It will be shown in Chapter 7 that the signal-to-noise ratio improves 6 db for each two-to-one increase in bandwidth occupancy. It is not difficult to obtain the reduction in noise interference by about 30:1 (about 1000:1 in power) using wideband FM. The property of noise reduction in FM also follows directly from the theory of communication. We shall show in Chapter 8 that, in general, increasing the bandwidth increases the signal-to-noise ratio of a given signal. On the other hand, if we are willing to tolerate lower signal-to-noise in a signal, then the signal can be transmitted over a lower bandwidth system.

[*] J. Carson, "Notes on the Theory of Modulation," *Proc. I.R.E.*, **16**, 966–975 (July 1928).

[†] E. H. Armstrong, "A Method of Reducing Disturbances in Radio Signalling by a System of Frequency Modulation," *Proc. I.R.E.*, **24** (May 1936).

It should be remembered that the property of discriminating noise in angle modulation becomes significant only when the bandwidth required for the transmission is larger, that is, when k_f is large. This means that m_f, the modulation index, should be large. In narrowband FM, the bandwidth required is the same as that of AM and the improvement in signal-to-noise ratio is not significant.

4.9 GENERATION OF FM SIGNALS

FM signals may be generated directly by frequency-modulating the carrier (direct FM) or by integrating the modulating signal first and then allowing it to phase-modulate the carrier (indirect FM).

Indirect FM

Because of the ease of generating a PM carrier, it is often used to generate FM signals indirectly. It can be easily seen from Eq. 4.8 that if we feed signal $\int f(t)\,dt$ at the input of a PM generator, the output will be an FM signal. This principle was used in generating a narrowband FM earlier (Fig. 4.2). The FM generator in Fig. 4.2b is essentially a narrowband PM generator to which the signal $\int f(t)\,dt$ is applied at the input. This system, however, yields a narrowband FM. We can now convert the narrowband FM into wideband FM by using a frequency multiplier. The frequency multiplier is a nonlinear device which multiplies the frequency of the input. A simple square law device, for example, can multiply the frequency by a factor 2. For a square law device the input $e_i(t)$ and the output $e_o(t)$ are related by

$$e_o(t) = [e_i(t)]^2$$

If

$$e_i(t) = \varphi_{\mathrm{FM}}(t) = \cos\left(\omega_c t + k_f \int f(t)\,dt\right)$$

then

$$e_o(t) = \cos^2\left(\omega_c t + k_f \int f(t)\,dt\right)$$
$$= \frac{1}{2}\left[1 + \cos\left(2\omega_c t + 2k_f \int f(t)\,dt\right)\right]$$

It is obvious that both the carrier frequency and the modulation index in the output signal are twice the corresponding values in the input

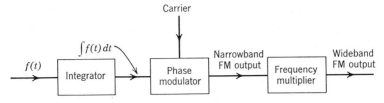

Figure 4.8   A block diagram representation of indirect FM.

signal. If one uses the nth law device in general, the carrier frequency and the modulation index can be multiplied by n. This method of obtaining a wideband FM from a narrowband PM generator is called the Armstrong indirect FM system and is shown in Fig. 4.8.

Direct FM

In direct FM, the modulating signal varies the carrier frequency directly. In general, electronic oscillators are used and one of the reactive elements (L or C) of the frequency-determining tuned circuit is varied in proportion to $f(t)$. If L and C are the inductance and the capacitance of the tuned circuit of the oscillator, the frequency of oscillation ω_i is given by

$$\omega_i = \frac{1}{\sqrt{LC}}$$

If we vary one of the elements L or C linearly with $f(t)$, we can show that ω_i, the instantaneous oscillator frequency, is also linear proportional to $f(t)$ for small variations. Suppose that the capacitor C is varied with $f(t)$ as

$$C = C_0 + af(t) = C_0\left[1 + \frac{a}{C_0}f(t)\right]$$

and

$$\omega_i = \frac{1}{\sqrt{LC_0}\left[1 + \dfrac{a}{C_0}f(t)\right]^{1/2}}$$

If $(a/C_0)f(t) \ll 1$, then

$$\omega_i \simeq \frac{1}{\sqrt{LC_0}}\left[1 - \frac{a}{2C_0}f(t)\right]$$

$$= \omega_c + k_f f(t)$$

where

$$\frac{1}{\sqrt{LC_0}} = \omega_c \qquad \text{and} \qquad \frac{-a\omega_c}{2C_0} = k_f$$

Obviously, the oscillator output is the desired FM signal. Similarly, we can show that by holding C constant and varying L with $f(t)$ an FM signal can be obtained.

Diode Reactance Modulator

The variable reactance can be obtained from a subcircuit whose terminal impedance is a reactance proportional to the modulating signal. For example, a capacitance of a reverse biased diode is a function of the voltage across its terminal. Thus by applying the modulating signal voltage $f(t)$ across a reverse biased diode, one can vary the junction capacitance in a desired way.

Saturable Reactor Modulator

In a saturable reactor modulator, the inductor is varied with the modulating signal $f(t)$. The permeability of the ferrite core of such a reactor is a function of the external magnetic field, which can be created by passing a current through another coil wound on the core. The current, proportional to the modulating signal, is passed through this coil. This causes the variation in the reactance of the main coil in proportion to the modulating signal. This method of generating FM is extremely attractive because of its simplicity and the fact that for FM broadcast, the required frequency deviation of ± 100 kHz is directly obtainable from this device without need for further frequency multiplication.

For discussion on other techniques, the reader is referred to Panter.*

The Reactance-Tube Modulator

The terminal impedance of the so-called reactance tube circuit (Fig. 4.9a) appears as a variable reactance (inductive or capacitive) proportional to the modulating signal $f(t)$. The schematic diagram of the reactance tube circuit is shown in Fig. 4.9a. The equivalent circuit

* P. F. Panter, *Modulation Noise and Spectral Analysis*, McGraw-Hill, New York, 1965.

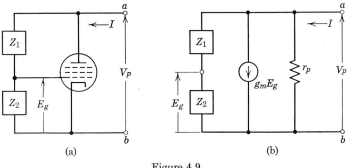

Figure 4.9

of this arrangement is shown in Fig. 4.9b. Since r_p is usually very high, it may be ignored. From Fig. 4.9b, we have

$$I = g_m E_g + \frac{V_p}{Z_1 + Z_2} \qquad (4.36)$$

and

$$E_g = \frac{Z_2}{Z_1 + Z_2} V_p \qquad (4.37)$$

Substituting Eq. 4.37 in Eq. 4.36,

$$I = \left(\frac{g_m Z_2 + 1}{Z_1 + Z_2}\right) V_p$$

Hence the admittance Y_{ab}, seen across terminals ab, is

$$Y_{ab} = \frac{I}{V_p} = \frac{1 + g_m Z_2}{Z_1 + Z_2}$$

Usually, $|g_m Z_2| \gg 1$, and if we let $Z_1 \gg Z_2$, then

$$Y_{ab} \simeq \frac{g_m Z_2}{Z_1}$$

The transconductance g_m of the tube is a function of grid voltage. If we apply a signal proportional to $f(t)$ at the grid,

$$g_m = g_{m0} + bf(t)$$

and

$$Y_{ab} = \frac{[g_{m0} + bf(t)]Z_2}{Z_1}$$

It is evident from this equation that if we let $Z_1 = 1/j\omega C_1$ and $Z_2 = R$, then Y_{ab} represents the admittance of a time varying capacitance:

$$C = C_0 + \beta f(t)$$

In each of these cases, the carrier frequency and the modulation index are kept quite small. The desired value of the carrier can be obtained by proper frequency multiplication and frequency translation. The frequency multiplication also increases the modulation index m_f.

Frequency modulation may also be obtained from voltage-controlled devices such as klystrons and multivibrators. In these devices, the frequency of oscillation can be controlled by application of the voltage at appropriate points. In a reflex klystron, the oscillation frequency is a function of the repeller voltage.* Hence the frequency can be modulated by applying $f(t)$ to a repeller. In multivibrators the oscillation frequency can be controlled by controlling the voltage at the control terminal (the control grid in vacuum tubes or the base terminals in a transistor).

4.10 DEMODULATION OF FM SIGNALS

To recover the modulating signal $f(t)$ from the FM carrier, we must provide a circuit whose output varies linearly with the frequency of the input signal. FM detectors are therefore frequency-sensitive devices, and are also called *frequency discriminators*. In general, the frequency discriminator consists of a circuit whose gain varies linearly with frequency. Thus the FM signal is converted into an AM signal by this frequency-sensitive subcircuit. The resultant AM signal is consequently demodulated by an envelope detector using a diode and R-C circuit.

Figure 4.10 shows the simple type of frequency discriminators. In Fig. 4.10a, the first R-L circuit converts the FM signal into an AM signal, which is then detected by a diode and the second R-C circuit (envelope detector). In Fig. 4.10c, the FM signal is converted into an AM signal by a tuned circuit which is slightly off-tuned at ω_c as shown in Fig. 4.10d. The voltage across the tuned circuit roughly varies linearly with the frequency as seen from Fig. 4.10d. The resulting AM signal is then detected by an envelope detector (Fig. 4.10c). The R-L discriminator in Fig. 4.10a has a poor sensitivity. The single tuned circuit discriminator in Fig. 4.10c has a better sensitivity but has a rather nonlinear

* H. A. Atwater, *Introduction to Microwave Theory*, McGraw-Hill, New York, 1962.

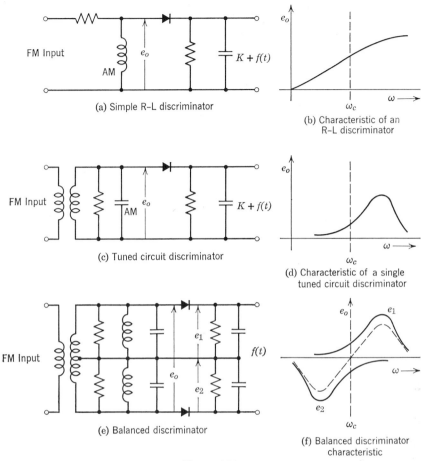

(a) Simple R-L discriminator

(b) Characteristic of an
R-L discriminator

(c) Tuned circuit discriminator

(d) Characteristic of a single
tuned circuit discriminator

(e) Balanced discriminator

(f) Balanced discriminator
characteristic

Figure 4.10

characteristic with respect to frequency. The balanced discriminator shown in Fig. 4.10c has a high sensitivity as well as a better linearity. The upper and the lower tuned circuits are tuned above and below the carrier frequency, respectively.

The voltages e_1 and e_2 are shown as a function of frequency in Fig. 4.10f. The resultant output $e_o(t)$ as a function of frequency is shown dotted. The balanced discriminator provides excellent linearity compared to the single tuned circuit type of discriminator, because the distortion caused by even harmonics is balanced out in such arrangements. In addition, any distortion caused by the residual amplitude modulation present in the input FM carrier is also balanced out.

PROBLEMS

1. In angle modulation (FM or PM), if the modulating signal is a single frequency sinusoid ($A \cos \omega_m t$), show that there is absolutely no way of ascertaining from the received carrier whether it is FM or PM.

2. A carrier 10 mHz is phase-modulated by a sinusoidal signal of frequency 10 kHz and unit amplitude. The maximum phase deviation is 10 radians for unit amplitude of the modulating signal. Calculate the approximate bandwidth of the PM carrier. If the frequency of the modulating signal is now changed to 5 kHz, find the new bandwidth of the PM carrier. If the frequency is the same as before (10 kHz) but the amplitude is doubled, find the bandwidth of the PM carrier.

3. A 100-mHz carrier is frequency-modulated by a sinusoidal signal of 10 kHz so that the maximum frequency deviation is 1 mHz. Determine the approximate bandwidth of the FM carrier. Now find the bandwidth of the FM carrier if the modulating signal amplitude is doubled. Determine the bandwidth of the FM carrier if the frequency of the modulating signal is also doubled.

4. A 100-mHz carrier is phase-modulated by a sinusoidal signal of 10 kHz with a modulator having $k_p = 100$. Determine the approximate bandwidth of the PM carrier if the modulating signal has unit amplitude. What is the approximate bandwidth if the modulating signal amplitude is doubled? What is the bandwidth if the modulating signal frequency is also doubled?

5. A high frequency carrier is phase-modulated by a modulating signal $a \cos \omega_m t$. The phase deviation constant is k_p (Eq. 4.8). Find all possible relationships between the maximum frequency deviation $\Delta\omega$, the maximum phase deviation $\Delta\varphi$, the phase deviation constant k_p, the modulating signal frequency ω_m, and the amplitude a.

6. A carrier of 10 mHz is phase-modulated by a sinusoidal signal of 10 kHz and unit amplitude, and the maximum phase deviation is 2 radians. Calculate the bandwidth of the PM carrier. (Note that the frequency deviation in this case is rather small and hence the chart in Fig. 4.5 should be used to calculate the bandwidth.)

7. An angle modulated wave is described by an equation

$$\varphi(t) = 10 \cos (2 \times 10^6 \pi t + 10 \cos 2000\pi t)$$

Find:

(a) The power of the modulated signal.
(b) The maximum frequency deviation.
(c) The maximum phase deviation.
(d) The bandwidth of the signal.

Can you determine whether this is a frequency modulated carrier or a phase modulated carrier?

8. A carrier is frequency modulated by a sinusoidal signal $f(t)$. The modulating constant $k_f = 30,000$. Find the power carried by the carrier and the total power carried by all the sidebands for each of the following cases:

(a) $f(t) = \cos 5000t$ (b) $f(t) = 2 \sin 2500t$
(c) $f(t) = \frac{1}{2} \cos 2500t$ (d) $f(t) = 2.405 \cos 30,000t$
(e) $f(t) = 10 \sin (1000t + \theta)$ (f) $f(t) = 5.52 \sin (30,000t + \varphi)$

9. Sketch the FM and PM carriers when the modulating signal $f(t)$ is as shown in Fig. P-4.9.

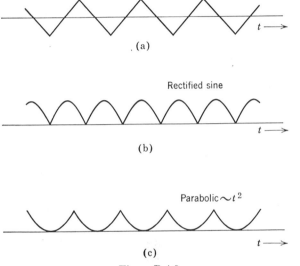

. (a)

Rectified sine

(b)

Parabolic $\sim t^2$

(c)

Figure P-4.9

10. For each of the periodic modulating signals in Fig. P-4.9, find the spectrum of the FM carrier. Find the amount of power carried by the carrier and the power carried by the sidebands.

11. Corresponding to each of the periodic modulating signals in Fig. P-4.9, find the spectrum of the PM carrier.

12. A carrier of 100 mHz is phase-modulated by a signal $f(t)$ shown in Fig. P-4.12. The modulating system constant k_p is 10^6. (a) Sketch the

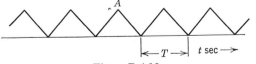

Figure P-4.12

modulated carrier. (*b*) Find and sketch the spectrum of the modulated carrier if

(1) $A = 2 \times 10^{-6}$ $T = 2 \times 10^{-6}$

(2) $A = 10^{-6}$ $T = 2 \times 10^{-6}$

(3) $A = 2 \times 10^{-6}$ $T = 10^{-6}$

(4) $A = 10^{-6}$ $T = 10^{-6}$

chapter 5

Communication Systems: Pulse Modulation

The sampling theorem discussed in Section 1.16 provides the theoretical basis for pulse-modulation techniques. It was shown that a band-limited signal, which has no spectral components above the frequency f_m Hz, is completely specified by its values at intervals spaced uniformly at $1/2f_m$ seconds (or less) apart. Instead of transmitting the complete signal continuously, we need to transmit the signal only at a finite number of instants ($2f_m$ per second). The information in the sample can be transmitted by pulse modulation. We shall now discuss various forms of pulse modulation.

5.1 PULSE-AMPLITUDE MODULATION

In this mode the sample values are transmitted by pulses whose amplitudes vary in proportion to sample values. For convenience, we shall first consider the transmission of ideal samples (impulses) as shown in Fig. 5.1. The samples are spaced $1/2f_m$ seconds apart. As observed in Section 1.16, the spectrum of the sampled signal $f_s(t)$ is given by the periodic repetition of $F(\omega)$, the spectrum of $f(t)$ (Fig. 5.1e). It is evident from Fig. 5.1e that $f(t)$ can be recovered from the sampled signal $f_s(t)$ by letting $f_s(t)$ pass through a low-pass filter with cutoff

241

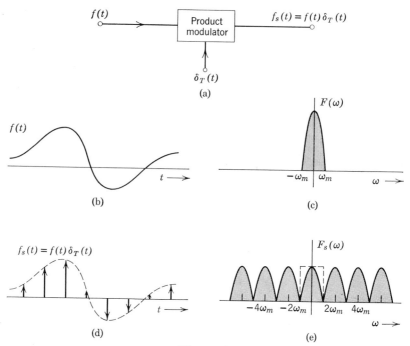

Figure 5.1

frequency f_m. The spectrum of $f_s(t)$ is $F_s(\omega)$ given by Eq. 1.136:

$$F_s(\omega) = \frac{1}{T} \sum_{n=-\infty}^{\infty} F(\omega - n\omega_0) \qquad \omega_0 = \frac{2\pi}{T}$$

$$T \leqslant \frac{1}{2f_m}$$

Figure 5.1 illustrates the case of $T = 1/2f_m$ (Nyquist interval). For this case $\omega_0 = 2\omega_m$ and

$$F_s(\omega) = \frac{1}{T} \sum_{n=-\infty}^{\infty} F(\omega - 2n\omega_m) \qquad T = \frac{\pi}{\omega_m} \qquad (5.1)$$

Natural Sampling

The process just discussed samples the function $f(t)$ at certain instants by impulses (instantaneous sampling). It is evident from Fig. 5.1e that the spectrum of such an ideal sampled signal occupies the entire bandwidth ($-\infty$ to ∞); that is, it contains components of all frequencies.

In practice, however, such an ideal sampling (instantaneous sampling) cannot be achieved, since it is impossible to generate true impulses. Usually the sampling is performed by very narrow pulses of finite width. Hence the sampling by such pulses is not instantaneous but occurs over a finite time interval. We shall now investigate the effect of such sampling.

Assume that the sampling is performed by periodic rectangular pulses of width τ seconds, repeating every T seconds. We shall denote this pulse train by $p_\tau(t)$ (Fig. 5.1c). The sampling interval T will be taken as the Nyquist interval $1/2f_m$ seconds. $P_\tau(\omega)$, the spectrum of the pulse train (Eq. 1.103, Fig. 1.34) is shown in Fig. 5.1d. The sampled signal $f_s(t)$ is a product of $f(t)$ and $p_\tau(t)$. Hence $F_s(\omega)$, the spectrum of $f_s(t)$, is obtained by convolving $F(\omega)$ with $P_\tau(\omega)$. The convolution can be readily performed graphically, with the result as shown in Fig. 5.2f. The nonideal sampling of $f(t)$ yields the spectrum similar to that of ideal sampling but with decaying amplitude. We can derive the same result analytically. In this case, we have

$$f_s(t) = f(t)p_\tau(t)$$

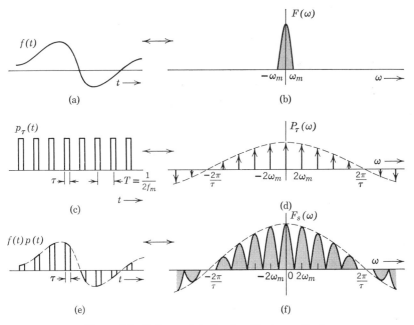

Figure 5.2   Pulse modulation (nonideal sampling).

Hence

$$F_s(\omega) = \frac{1}{2\pi} F(\omega) * P_\tau(\omega) \qquad (5.2)$$

$P_\tau(\omega)$ can be obtained from Eq. 1.103

$$T = \frac{1}{2f_m} = \frac{\pi}{\omega_m}$$

and

$$\omega_0 = \frac{2\pi}{T} = 2\omega_m$$

$$P_\tau(\omega) = 2A\tau\omega_m \sum_{n=-\infty}^{\infty} Sa(n\tau\omega_m)\,\delta(\omega - 2n\omega_m) \qquad (5.3)$$

Substituting Eq. 5.3 in Eq. 5.2, we get

$$F_s(\omega) = \frac{A\tau\omega_m}{\pi} F(\omega) * \sum_{n=-\infty}^{\infty} Sa(n\tau\omega_m)\,\delta(\omega - 2n\omega_m)$$

$$= \frac{A\tau}{T} \sum_{n=-\infty}^{\infty} Sa(n\tau\omega_m)F(\omega) * \delta(\omega - 2n\omega_m)$$

$$= \frac{A\tau}{T} \sum_{n=-\infty}^{\infty} Sa(n\tau\omega_m)F(\omega - 2n\omega_m) \qquad (5.4)$$

It is obvious that the right-hand side of Eq. 5.4 represents the spectrum $F(\omega)$ repeating itself every $2\omega_m$ radians per second, but with amplitudes varying as $Sa(n\tau\omega_m)$. This equation therefore represents the spectrum in Fig. 5.2f. Observe that the sampling pulses need not be rectangular as shown in Fig. 5.2c. We may represent all samples by any other chosen pulse form $q(t)$. This will merely cause the shape of the envelope of the spectrum $F_s(\omega)$ in Fig. 5.2f to change. [This envelope will be $Q(\omega)$, the Fourier transform of $q(t)$.]

The nonideal sampling of $f(t)$ therefore results in a repetition of the spectrum of $f(t)$ but with decaying amplitudes. The original signal $f(t)$ can be recovered from the sampled signal $f_s(t)$ by using a low-pass filter with cutoff at ω_m. The demodulation process for pulse modulation (nonideal sampling) is exactly the same as that used for impulse modulation (ideal sampling). Note that the signal can be recovered without any distortion, even with nonideal sampling. The bandwidth required for the transmission of an ideal sampled signal (impulse modulation) is infinite, whereas that required for pulse modulation will be finite since the spectrum $F_s(\omega)$ (Fig. 5.2f) decays with frequency, and it has negligible energy content at higher frequencies.

As the pulses are made wider, the spectrum decays faster, and hence a smaller bandwidth is required for the transmission. It therefore appears that the pulse modulation (nonideal sampling) is superior to the impulse modulation (ideal sampling) since it requires the smaller bandwidth for transmission. However, what is gained in the frequency domain is lost in the time domain. The pulse modulation needs a larger time interval for transmitting the sampled signal than the impulse modulation. Since the pulses have a finite width, it is possible to transmit only a finite number of signal simultaneously on a time-sharing basis (time division multiplexing) for the pulse modulation scheme, whereas it will be possible to transmit any number of signals by using impulse modulation. The pulse modulation described above is known as pulse-amplitude modulation (PAM), since the sampled signal essentially represents the periodic pulse train whose amplitude is modulated by $f(t)$.

The sampled signal just considered can be expressed as a product of $f(t)$ with a uniform pulse train

$$f_s(t) = f(t)q_\tau(t)$$

$$= f(t) \sum_{n=-\infty}^{\infty} q(t - nT)$$

where $q(t)$ is the basic sampling pulse. This type of sampling is called the natural sampling.

Instantaneous Sampling

In natural sampling each sampling pulse is multiplied by $f(t)$ over the corresponding sampling interval. As a result, each pulse in the sampled signal $f_s(t)$ has a different waveform. This is quite obvious from Fig. 5.2e where each pulse top assumes the shape of the modulating signal $f(t)$ over the corresponding interval. In contrast to this, we now consider instantaneous sampling, in which all the pulses in the sampled signal have the same form but their amplitudes are proportional to the corresponding sample values (Fig. 5.3). This type of sampling signal obviously carries the complete information of all the samples and hence contains all the information about $f(t)$ (provided the sampling interval is less than $1/2f_m$ seconds). Note that the natural sampling (Fig. 5.2) carries the information about $f(t)$ over the entire width of each sampling pulse. In contrast, the instantaneous sampling contains the information about $f(t)$ only at the sampling instants. This is why this type of sampling is called instantaneous sampling. It should be observed that the sampling by impulse train (Fig. 5.1) may be considered either as a natural sampling or instantaneous sampling.

Let us consider the basic pulse used in this sampling to be some arbitrary waveform $q(t)$ as shown in Fig. 5.3a. The sampled signal will be denoted by $x_s(t)$. Thus for instantaneous sampling

$$x_s(t) = \sum_{n=-\infty}^{\infty} f(nT)q(t - nT) \tag{5.5}$$

Note that we have reserved the notation $f_s(t)$ for $f(t)$ sampled by unit impulse train.

$$f_s(t) = \sum_{n=-\infty}^{\infty} f(nT)\,\delta(t - nT) \tag{5.6}$$

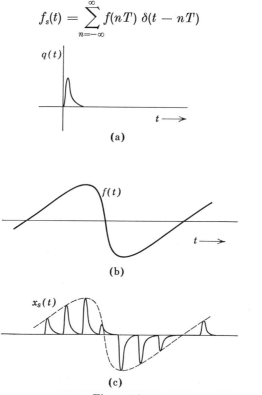

(a)

(b)

(c)

Figure 5.3

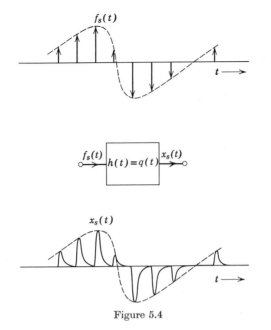

Figure 5.4

We shall now find the spectrum of $x_s(t)$ for an arbitrary pulse shape $q(t)$. This is easily obtained by observing the fact that $x_s(t)$ can be obtained as the response of a system whose unit impulse response is $q(t)$ and the input signal is $f_s(t)$ in Eq. 5.6 (or Fig. 5.1d). This is shown in Fig. 5.4. Thus, if

$$q(t) \leftrightarrow Q(\omega)$$

then

$$x_s(t) \leftrightarrow F_s(\omega)Q(\omega)$$

Using Eq. 5.1, we get

$$x_s(t) \leftrightarrow \frac{1}{T} \sum_{n=-\infty}^{\infty} Q(\omega)F(\omega - 2n\omega_m) \qquad T = \frac{\pi}{\omega_m} \qquad (5.7)$$

It is evident from Eq. 5.7 that the spectrum of the sampled function $x_s(t)$ consists of periodic repetition of $F(\omega)$, which is multiplied by $Q(\omega)$, the spectrum of $q(t)$. This is shown in Fig. 5.5 for the case of flat-topped pulses.

Note that $X_s(\omega)$ [the spectrum of $x_s(t)$] in Fig. 5.5g is not the same as that in Fig. 5.2f although it may appear so superficially. In Fig. 5.2f, the spectrum consists of $F(\omega)$ repeating periodically with decaying

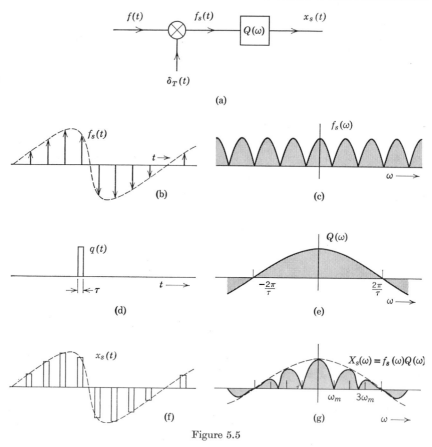

Figure 5.5

amplitudes. The shape of $F(\omega)$ itself, however, remains undistorted in each cycle. Only the amplitude changes. On the other hand, in Fig. 5.5g, $F(\omega)$ has lost its original shape. It is weighted by a multiplying factor $Q(\omega)$. At each frequency there is a different multiplying factor. Thus none of the cycles in Fig. 5.1g has a shape of $F(\omega)$. This fact can also be clearly seen from Eqs. 5.4 and 5.7. In Eq. 5.4, the multiplying factor $Sa(n\tau\omega_m)$ is a constant over any given cycle, whereas in Eq. 5.7 the multiplying factor $Q(\omega)$ is a function of frequency.

Recovering $f(t)$ from Instantaneous Sampling

The spectrum of the sampled function $x_s(t)$ (Fig. 5.5g) does not contain an undistorted form of $F(\omega)$, as was the case for natural sampling

(Fig. 5.2f). (This statement is true with one exception—impulse train sampling.) Thus it is not possible to recover $F(\omega)$ by using a low-pass filter alone. If we use a low-pass filter with cutoff frequency ω_m, the spectrum of the output will be $F(\omega)Q(\omega)$. The desired signal $f(t)$ can be recovered from this output by transmitting it through another filter whose transfer function is $1/Q(\omega)$ (Fig. 5.6a). Note that since $F(\omega)$ is bandlimited to ω_m radians per second, it is sufficient that the second filter in Fig. 5.6a have a transfer function $1/Q(\omega)$ only over the range $(0, \omega_m)$. Outside this range it may be chosen arbitrarily according to convenience of the design. We can combine both the filters into one composite filter whose frequency response is shown in Fig. 5.6b. It is evident that the transfer function $H(\omega)$ of this filter should be $1/Q(\omega)$ over the range $(0, \omega_m)$ and zero outside this range. Thus $H(\omega)$, the transfer function of the recovering filter, can be expressed as

$$H(\omega) = \begin{cases} \dfrac{1}{Q(\omega)} & |\omega| < \omega_m \\[2mm] 0 & \text{otherwise} \end{cases} \tag{5.8}$$

If the pulse $q(t)$ is extremely narrow, it tends to approach an impulse function and $Q(\omega)$ tends to be flat. In other words, $Q(\omega)$ tends to become constant over the range $(0, \omega_m)$. In such a case the composite filter in Fig. 5.6b reduces to an ideal low-pass filter with cutoff frequency ω_m and the recovery of $f(t)$ in this case is identical to that of the natural sampling. This is expected since if $q(t)$ becomes very narrow it approaches an impulse, and the sampling then approximates to the sampling by impulse train.

Alternatively, $f(t)$ may be recovered from $x_s(t)$ by multiplying $x_s(t)$ by $\delta_T(t)$, a train of impulse. This obviously yields $f_s(t)$, as shown in

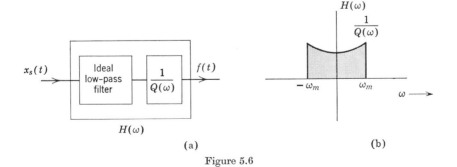

(a)   (b)

Figure 5.6

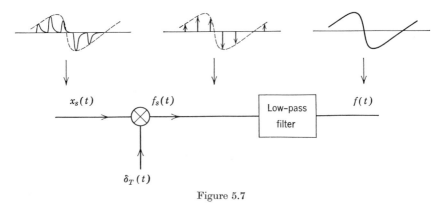

Figure 5.7

Fig. 5.7. The desired signal $f(t)$ can now be recovered from $f_s(t)$ by using an ideal low-pass filter.

Transmission of PAM Signals

The PAM signal may be directly transmitted along a pair of wires. But it cannot be easily transmitted directly by electromagnetic waves in free space since the PAM spectrum is concentrated at lower frequencies, which require impracticably large antennas. Hence the spectrum is translated to a higher frequency by the amplitude-modulation techniques that we discussed previously. At the receiver, the signal is demodulated to retranslate the spectrum to its original position. The output of this demodulator is the pulse train $f_s(t)$ from which $f(t)$ is

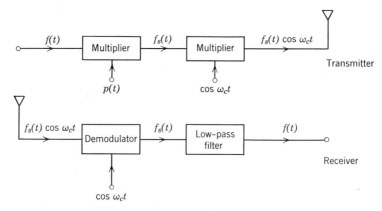

Figure 5.8   A block diagram representation of PAM/AM.

recovered by filtering through a low-pass filter. Such systems are designated as PAM/AM systems.

The spectrum of PAM signals may also be translated by techniques of angle modulation. Such systems are designated by PAM/FM or PAM/PM, as the case may be. A block diagram representation of a PAM/AM system is shown in Fig. 5.8.

5.2 OTHER FORMS OF PULSE MODULATION

To transmit a signal $f(t)$, it is necessary to transmit the information about its values at intervals $1/2f_m$ seconds apart. The significant point here is that it is not necessary to transmit the bandlimited signal continuously, but complete information about such signals can be transmitted as discrete values. Such a discrete form of information can be transmitted in a number of ways. The pulse-amplitude modulation (PAM), discussed in Section 5.1, is but one example.

In PAM signals this information is carried by the amplitudes of the pulses. We could have kept the amplitudes of all of these pulses constant but varied their width in proportion to the values of $f(t)$ at the corresponding instants. Such systems are designated as pulse-width modulation (PWM) or pulse-duration modulation (PDM) systems. Alternatively, the information may be transmitted by keeping both the amplitude and the width of the pulses constant but changing the positions of the pulses in proportion to the sampled values of $f(t)$ at the corresponding instants. These systems are known as pulse-position modulation (PPM) systems. The PAM, PWM, and PPM signals for a signal $f(t)$ are shown in Fig. 5.9.

Another very important mode of pulse modulation is pulse-code modulation (PCM). In this system each sample value of $f(t)$ is transmitted as a code formed by a pattern of pulses. The signal $f(t)$ to be transmitted is sampled, and each sample is approximated to the nearest allowable level as shown in Fig. 5.10a. In this case there are 16 allowable levels spaced 0.1 volt apart. Each voltage level is now represented by a certain pattern of pulses. One such code is shown in Fig. 5.10b. Thus instead of transmitting the individual samples, the corresponding pulse pattern is transmitted.

In the example given here, we have used binary pulses (pulses which can assume two values). In general one can use s'ary pulses (pulses that

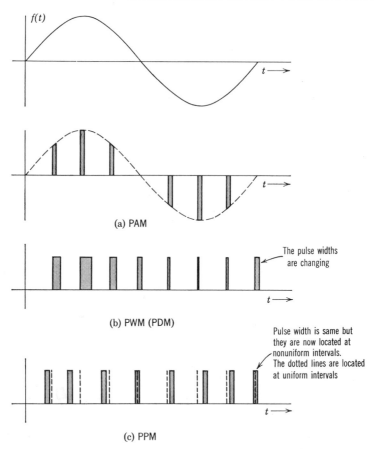

Figure 5.9   Representation of PAM, PWM, and PPM signals.

can assume s values). For larger values of s, we need less pulses per sample. Thus in PCM, instead of one pulse per sample (as in PAM), we need to transmit a group of pulses per sample. Therefore PCM requires more time interval for the transmission of the same information than other modulation systems. This disadvantage in PCM, however, is offset by its property of being more immune to noise interference. To receive a PCM signal, all that is required is to know whether the pulse is present or absent, regardless of the amplitude and shape of the pulse. Thus any external interference which may tend to introduce distortion in the height or width of the pulses has no influence whatsoever on PCM signals. (This topic is discussed at length in Chapter 7.)

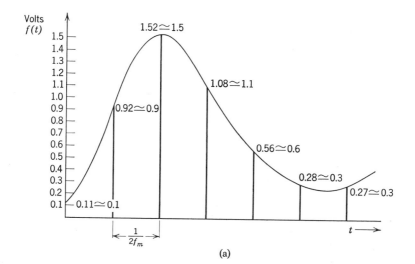

(a)

Digit	Binary equivalent	Pulse–code waveform
0	0000	
1	0001	
2	0010	
3	0011	
4	0100	
5	0101	
6	0110	
7	0111	
8	1000	
9	1001	
10	1010	
11	1011	
12	1100	
13	1101	
14	1110	
15	1111	

(b)

Figure 5.10 (b) A possible form of pulse code.

5.3 TIME DIVISION MULTIPLEXING

The sampling theorem makes it possible to transmit the complete information in a continuous bandlimited signal by transmitting mere samples of $f(t)$ at regular intervals. The transmission of these samples engages the channel only part of the time, and it is possible to transmit several signals simultaneously on the time-sharing basis. This is done by sampling all of the signals to be transmitted and interlacing these samples as shown in Fig. 5.11. In this figure the samples of two signals are shown interlaced. At the receiver the samples of each signal are separated by appropriate techniques. We shall describe briefly the outline of a time division multiplexed transmitter and receiver system.

Figure 5.12 shows a block diagram representation of a transmitter and a receiver of a time division multiplexed system. At the transmitter, the commutator is switched from channel to channel in a sequence by the timing circuit, which also generates the sampling pulses. Thus the commutator connects different channels in a sequence to the sampling circuit, which samples all of the signals in a sequence by pulses generated by the timing circuit. The commutator switching and the sampling pulses are in synchronism. The output of the sampling circuit is thus a signal which consists of samples of all of the signals interlaced. At the receiver, another timing circuit which is in synchronism with that at the transmitter is used to switch the commutator to different channels.* The samples of various signals are now properly separated. The desired signal is recovered from each channel by a low-pass filter.

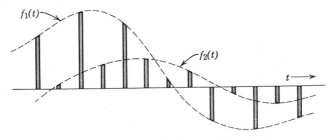

Figure 5.11   Time multiplexing of two signals.

* For details of such switching arrangement, see M. Schwartz, *Information Transmission, Modulation, and Noise,* Chapter 4. McGraw-Hill, New York, 1959.

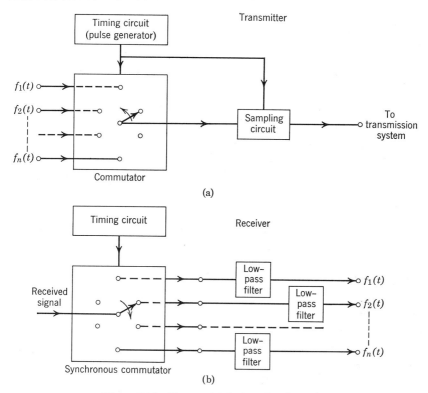

Figure 5.12   Time multiplexing of n channels.

The Sampling Rate

If the samples of a bandlimited signal are to carry complete information of the signal, then the rate of sampling must never be less than $2f_m$
samples per second. The minimum sampling rate ($2f_m$ samples per
second) is the Nyquist rate.

It is obvious that the Nyquist rate results in a repetition of the signal
spectrum without overlap and without any free interval between the
successive cycles as shown in Fig. 5.13a. Therefore, to recover the
signal $f(t)$ from such a sampled signal, it is necessary to use an ideal
low-pass filter which will allow all of the frequencies $\omega < \omega_m$ to pass
unattenuated and attenuate all of the frequencies above ω_m. A practical low-pass filter with very sharp cutoff characteristics could be built,
but a large number of elements would be needed. The stringent requirements on the filter can be relieved by sampling the signal at a rate

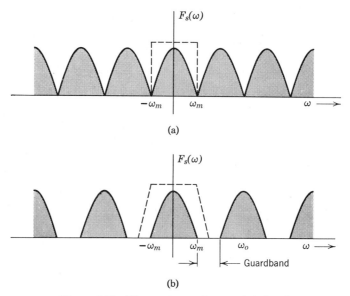

Figure 5.13   The spectrum of a sampled signal.

higher than the Nyquist rate. It can be easily seen that, under these conditions, the sampling results in repetition of the signal spectrum with a free band (*guardband*) between successive cycles as shown in Fig. 5.13*b*. It is obvious from this figure that the original signal can be recovered from the sampled signal by a low-pass filter, which need not have a sharp cutoff characteristic.

A similar situation occurs in frequency division multiplexing. The spectra of various signals are separated by a guardband for the reasons just mentioned. In time division multiplexing the samples of various signals are interlaced. An interlacing that is too close calls for stringent requirements on the receiving system which separates the samples of different signals. Hence the samples of successive signals are also separated by a small time interval, called the *guard time* interval.

5.4 BANDWIDTH REQUIRED FOR TRANSMISSION OF PAM SIGNALS

For comparisons of the PAM system with other systems, it is important to know the bandwidth required for the transmission of PAM signals. The spectrum of a PAM signal, as seen from Figs. 5.1*e*, 5.2*f*, or 5.5*f*, contains frequencies over the entire interval $(-\infty, \infty)$. Hence to transmit a PAM signal completely we need an infinite bandwidth. We shall show, however, that all the information contained in the samples can be completely transmitted by a finite bandwidth.

To time multiplex n continuous signals, each bandlimited to f_m Hz, we need to transmit $2f_m$ samples per second per signal. Hence time division multiplexing of these n signals calls for transmission of $2nf_m$ pulses per second. We shall now show that the information in these $2nf_m$ pulses can be transmitted over a bandwidth of nf_m Hz.

At first glance, it appears that we need an infinite bandwidth for transmission of PAM signals because we are transmitting rectangular pulses which have spectra occupying all of the frequency range. It must be remembered, however, that we are really not interested in the shape of the pulses. We need to know only the heights of the pulses. Hence any distortion in the pulse shape is immaterial as long as the height of the pulse is retained.

The output of the time division multiplexer consists of $2nf_m$ pulses per second, which are contributed by each of the n signals at a rate of $2f_m$ samples per second. We would now like to know the bandwidth required to transmit $2nf_m$ independent pieces of information (samples) per second. From the sampling theorem it follows that a continuous signal bandlimited to B Hz can be transmitted by $2B$ independent pieces of information (samples) per second.* Conversely, we can state that $2nf_m$ independent samples per second define a continuous signal bandlimited to nf_m Hz. We can consider these $2nf_m$ samples to be samples of some continuous signal $\varphi(t)$ bandlimited to nf_m Hz. Indeed, we can construct such a signal $\varphi(t)$ from the knowledge of these samples according to Eq. 1.140. Hence, instead of transmitting the discrete $2nf_m$ samples per second, we may transmit the corresponding continuous signal $\varphi(t)$ defined by these samples. Since this signal is bandlimited to nf_m Hz the bandwidth required is nf_m Hz.

How can we obtain $\varphi(t)$ from the $2nf_m$ samples per second? The discrete pulses are merely samples of $\varphi(t)$, and hence from the previous discussion it follows that $\varphi(t)$ can be obtained by passing these discrete samples through a low-pass filter with a cutoff frequency nf_m Hz.

* Note that the number of samples larger than $2B$ per second is not independent. A maximum of $2B$ samples are independent and the remaining samples can be expressed in terms of the $2B$ independent samples. This can be easily seen from Eq. 1.140. The complete information about the signal can be expressed in terms of a minimum $2B$ samples per second. Hence the remaining samples can always be expressed in terms of these $2B$ samples per second.

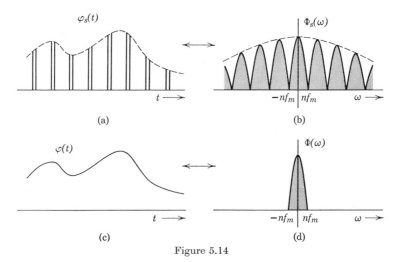

Figure 5.14

The signal represented by discrete $2nf_m$ pulses per second is $\varphi_s(t)$ and its spectrum is $\Phi_s(\omega)$. Note that $\Phi_s(\omega)$ is formed by periodic repetition of $\Phi(\omega)$. Signals $\varphi(t)$ and $\varphi_s(t)$ and their spectra are shown in Fig. 5.14. This result is very significant and will be used again in later chapters. However, it must be remembered that if it is desired to reproduce the pulse shape as well as its height, then the required bandwidth is larger than nf_m Hz.

It is interesting to note that the bandwidth required for the direct transmission of an n time division multiplexed PAM signal is nf_m. This is exactly equivalent to the bandwidth required for transmitting these n signals using AM-SSB frequency division multiplexing. However, it was pointed out that the direct transmission of PAM signals by radiation is impractical, since the energy in such a signal is concentrated at lower frequencies, necessitating unreasonable size radiating systems. In such cases the whole spectrum of PAM signals is shifted to a higher frequency by amplitude modulation. The resultant signal is called a PAM/AM signal.

The process of amplitude modulation gives rise to upper and lower sidebands, and hence the required bandwidth is doubled. Thus the bandwidth required for transmission of a PAM/AM signal of time division multiplexed n continuous signals bandlimited to f_m is $2nf_m$ Hz. It should be noted that the bandwidth required for the transmission of these n signals by AM-DSB frequency division multiplexing is also $2nf_m$ Hz. Hence we conclude that transmission of a signal by PAM needs the same bandwidth as that required for AM-SSB. Similarly, the transmission of a signal by PAM/AM requires the same bandwidth as that for AM-DSB.

5.5 COMPARISON OF FREQUENCY DIVISION MULTIPLEXED AND TIME DIVISION MULTIPLEXED SYSTEMS

We have discussed two methods of simultaneous transmission of several bandlimited signals on a channel. In frequency division multiplexed systems, all of the signals to be transmitted are continuous signals and are mixed in the time domain. The spectra of the various modulated signals, however, occupy different bands in the frequency domain and can be separated by appropriate filters. Thus the signals are all mixed in the time domain but maintain their identity in the frequency domain.

In the case of time division multiplexing, on the other hand, the samples of each signal remain distinct and can be recognized and separated in the time domain. However, the frequency spectra of the various sampled signals occupy the same frequency region and are all mixed beyond recognition. Thus the spectrum identity is maintained in frequency division multiplexed signals, whereas the waveshape identity is maintained in time division multiplexed signals. Since a signal is completely specified either by its time-domain or frequency-domain specification, the multiplexed signals can be separated at the receiver by using appropriate techniques in the respective domains.

The distinction between the two systems can be conveniently represented graphically on a communication space which is used to transmit information. The time frequency communication space is shown in Fig. 5.15 for frequency division multiplexed and time division multiplexed systems. For a frequency division multiplexed system, each signal is present on the channel all of the time and all are mixed. But each of them occupies a finite and distinct frequency interval (not occupied by any other signal). This is shown in Fig. 5.15a. On the other hand, in a time division multiplexed system, each signal occupies a distinct time interval (not occupied by any other signal). But the spectra of all of the signals have components in the same frequency interval. This is shown in Fig. 5.15b.

Quantitatively, we have already shown that the bandwidth requirements for the transmission of a given number of signals either by time

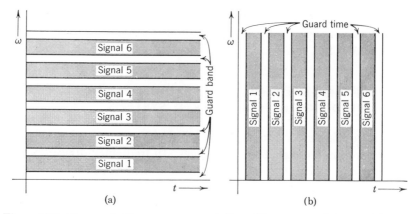

Figure 5.15   Communication space representation of frequency and time multiplexing.

division multiplexing of by frequency division multiplexing is the same (PAM and AM/SSB require nf_m Hz; PAM/AM and AM/DSB require $2nf_m$ Hz). It is therefore evident that for a given channel the number of bandlimited signals that can be simultaneously transmitted by frequency division multiplexing or by time division multiplexing is the same.

From the practical point of view, the time division multiplexed system proves superior to the frequency division multiplexed system. The first advantage is the simplified circuitry used in time division multiplexed systems compared to that used in frequency division multiplexed systems. In the latter, one needs to generate different carriers for each channel. Moreover, each channel occupies a different frequency band, and hence needs a different bandpass filter design. On the other hand, time division multiplexed systems require identical circuits for each channel, consisting of relatively simple synchronous switches or gating circuits. The only filters in the detection process are the low-pass filters which are identical for each channel. This circuitry is much simpler compared to the modulators, demodulators, carrier generators, and bandpass filters required in the frequency division multiplexed systems.

The second advantage of the time division multiplexed system is the relative immunity from interference within channels (interchannel crosstalk), which arises in frequency division multiplexed systems because of nonlinearities in the amplifiers in the path of transmission. The nonlinearities in various amplifiers produce harmonic distortion (due to frequency multiplication) and hence will introduce interference within channels (interchannel crosstalk). Hence the nonlinearity requirements in a frequency division multiplexed system are much more stringent than those for a single channel. On the other hand, for time division multiplexed systems, the signals from different channels are not applied to the system simultaneously but are allotted different time intervals. Hence the nonlinearity requirements in a time division multiplexed system are the same as that for a single channel. For these reasons the time division multiplexed systems are being used more commonly in such applications as long-distance telephone communication.

PROBLEMS

1. The signals given below are not bandlimited. However, they can be approximated as bandlimited signals. Assume a suitable criterion for such an approximation in each case and find the corresponding minimum sampling rate.

(a) $e^{-2|t|}$

(b) $e^{-2t} \cos 100t \, u(t)$

(c) $te^{-t} u(t)$

(d) $G_{20}(t)$

2. If $f(t)$ is a continuous signal bandlimited to f_m Hz and $f_s(t)$ is the sampled signal (sampled uniformly at $1/2f_m$ seconds interval), then $f(t)$ can be recovered from $f_s(t)$ by allowing it to pass through a low-pass filter. In

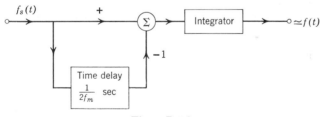

Figure P-5.2

practice, a first-order hold circuit shown in Fig. P-5.2 is commonly used to recover $f(t)$ from $f_s(t)$. The output of this circuit approximately resembles $f(t)$.

(a) Sketch the waveforms at various points of this circuit for a typical sampled signal $f_s(t)$.

(b) What is the transfer function of this arrangement? (*Hint:* Determine the unit impulse response.)

(c) Sketch the frequency response of this system. Compare it with ideal low-pass filter characteristics.

3. A signal $f(t)$ is bandlimited to a frequency f_m Hz. Figure 5.2 shows a natural sampling of $f(t)$ by a rectangular pulse. If, instead, the natural sampling is performed by a pulse of arbitrary waveform $q(t)$ with Fourier transform $Q(\omega)$, find the spectrum of the sampled signal. Assume the pulse

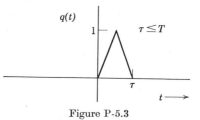

Figure P-5.3

width less than the sampling interval. In particular, if $q(t)$ is a triangular pulse as shown in Fig. P-5.3, find and sketch the sampled signal and its spectrum.

4. A signal $f(t)$ is bandlimited to f_m Hz. It is sampled by using triangular pulses shown in Fig. P-5.3. The width of the pulse is less than the sampling interval T. Find and sketch the spectrum of the sampled signal for the case of instantaneous sampling.

5. A signal $x(t)$ with spectrum $X(\omega)$ shown in Fig. P-5.5b is transmitted through a system with transfer function $H_1(\omega)$ shown in Fig. P-5.5c. The output $y(t)$ is sampled uniformly by an impulse train.

(a) What values of the sampling frequency f_s will enable the exact reproduction of $y(t)$?

(b) Choose one of these values of f_s; find and sketch the spectrum of the sampled function $y_s(t)$.

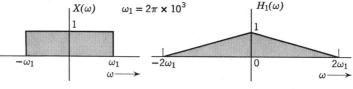

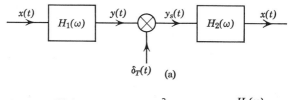

(b)

Figure P-5.5

(c) The sampled signal $y_s(t)$ is further transmitted through a system of transfer function $H_2(\omega)$. What must be $H_2(\omega)$ so that its output is exactly $x(t)$.

6. In the text, the discussion has been restricted to sampling pulses of width less than the sampling interval T. Thus all the pulses representing various sample values are nonoverlapping. This restriction, however, is not necessary. Show that one may use overlapping pulses (pulses of width greater than the sampling interval) for natural sampling and still be able to recover the original signal exactly. Show that the process of recovering the original signal from the sampled signal is identical to that used for natural sampling with pulses of width less than the sampling interval. Find and sketch the sampled signal and its spectrum when the sampling is performed by an exponential pulse

$$q(t) = e^{-t/T}$$

7. Repeat Problem 6 for the case of instantaneous sampling.

8. In instantaneous sampling, let the energy of the sampling pulse $q(t)$ be E_q. Find the power (the mean square *value*) of the sampled signal $x_s(t)$ in terms of E_q and $\overline{f^2(t)}$. Assume the pulse width less than the sampling interval T. (*Hint:* Use the results in Example 2.2, Eq. 2.31.)

9. Sketch the spectrum of $Sa(100\pi t)$. The bandwidth of this signal is 50 Hz and the minimum sampling rate is 100 samples per second. Sketch the samples of this signal taken at a minimum sampling rate starting with the first sample at $t = 0$.

Consider the case of undersampling at a rate of 50 samples per second and 25 samples per second. Sketch the samples at these rates. You will find that the samples taken in all the three cases are identical. How would you explain this? [*Hint:* The sampled signal $f(t) = f(t)\delta_T(t)$ where $T = \frac{1}{100}, \frac{1}{50}$, and $\frac{1}{25}$ in the three cases. Find the spectrum of $f_s(t)$ by using the frequency convolution theorem and show that in each case $f_s(t)$ contains the complete information of $f(t)$. This example is a pathological case.]

10. Generalize the uniform sampling theorem for signals whose spectra are bandlimited to f_m Hz but not centered at $\omega = 0$. The positive spectrum of such signals lies between f_l and f_h where $f_h - f_l = f_m$. Show that the minimum uniform sampling rate for such signals must be $2f_h/n$ samples per second where f_h is the highest frequency of the spectrum and n is the largest integer less than f_h/f_m.

chapter 6

Noise

Signals in the process of transmission always pick up some undesired signals. Indeed, any type of processing performed on a signal tends to introduce these undesired disturbances, which we shall call noise. Noise is thus an undesirable signal which is not connected with the desired signal in any way. Here, however, we shall restrict ourselves to noise signals which are random, that is, unpredictable in nature. The power-supply hum in a radio receiver, the oscillations in a feedback system, etc., are noise signals according to our definition of noise, but they are not random. They can be predicted and can be eliminated by proper design.

There are various sources of noise. We may broadly classify these sources as (a) man-made noise, (b) the erratic natural disturbances which occur irregularly, and (c) fluctuation noise which arises inside physical systems. Man-made noise arises due to the pickup of undesired signals from other sources, such as faulty contacts, electrical appliances, ignition radiation, and fluorescent lighting. Such noise can always be eliminated by removing the source of the noise. The last two types of noise sources are nonman-made types.

Erratic natural noise may arise due to lightning, electrical storms in the atmosphere, intergalactic noise, or general atmospheric disturbances. Fluctuation noise is also nonman-made, and it arises inside physical systems due to spontaneous fluctuations such as the thermal motion (Brownian movement) of free electrons inside a resistor, the (random) emission of electrons in vacuum tubes, and random generation, recombination, and diffusion of electronic carriers (holes and electrons) in semiconductors. Basically, there are two important types of fluctuation noise: shot noise and thermal noise. In this chapter we shall study noise without going into a detailed description of the statistical properties of these signals.

6.1 SHOT NOISE

Shot noise is present in both vacuum tubes and semiconductor devices. In vacuum tubes shot noise arises due to random emission of electrons from the cathode. In semiconductor devices this effect arises due to the random diffusion of minority carriers and the random generation and recombination of hole-electron pairs.

The nature of the shot noise will be illustrated by considering the electron emission from the hot cathode of a parallel plane diode shown in Fig. 6.1. At a given temperature, the average number of electrons emitted per second is constant. The process of electron emission, however, is random. This means that if we divide the time axis into a large number of small intervals of $\Delta\tau$ seconds each, the number of electrons emitted during each of these intervals is not constant but is random.

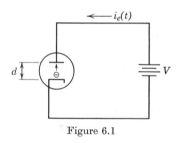

Figure 6.1

However, on the average, the rate of emission of electrons is constant, provided that it is averaged over a sufficiently long interval of time. Thus the current formed by emitted electrons is not constant but fluctuates about a mean value. If we observe this current on an oscilloscope with a slow sweep, it will appear essentially constant. However, if the current is observed with a fast sweep, where the time scale is expanded greatly, the uneven nature of the current becomes apparent. This is shown in Fig. 6.2. The diode current thus fluctuates about a certain mean value. We may consider the total current $i(t)$ as composed of a constant current I_0 and a noise current $i_n(t)$ which has a zero mean value.

$$i(t) = I_0 + i_n(t) \qquad (6.1)$$

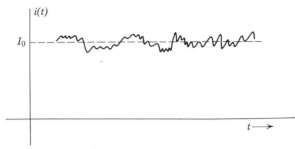

Figure 6.2

The nature of the fluctuations in $i(t)$ can be understood better by considering the process of the induction of current in the plate of the diode due to the emission of an electron. Assume that a single electron is emitted from the cathode (Fig. 6.1). This electron acquires velocity as it moves toward the plate and induces a current $i_e(t)$. If the plate and the cathode of a diode are separated by d units, then the emitted electron experiences a force of magnitude qV/d in the direction of the plate, where q is the charge of an electron and V is the applied voltage. The electron will acquire an acceleration of qV/md units, where m is the mass of an electron. The initial velocity of the emitted electron is usually much smaller than the final velocity acquired by the electron at the time it strikes the plate. Hence the initial velocity will be assumed to be zero. The velocity $v(t)$ at any time t will then be given by

$$v(t) = \frac{qV}{md} t \tag{6.2}$$

The kinetic energy (KE) acquired by an electron at any instant t is $\frac{1}{2}mv^2$ or

$$KE = \frac{q^2 V^2}{2md^2} t^2 \tag{6.3}$$

If the motion of this electron induces a charge Q on the plate, then the amount of work W that must be done to induce the charge Q on the plate with potential V is given by

$$W = QV$$

Equating this work with the kinetic energy of the electron, we obtain

$$Q = \frac{q^2 V t^2}{2md^2}$$

The current

$$i_e(t) = \frac{dQ}{dt} = \frac{q^2 V}{md^2} t \tag{6.4}$$

$$= \frac{q}{d} v(t) \tag{6.5}$$

Note that the induced current is proportional to the velocity of the electron.

The time required for the electron to reach the plate is known as the transit time τ_a and can be readily found from Eq. 6.2:

$$d = \frac{1}{2} \frac{qV}{md} \tau_a{}^2$$

and

$$\tau_a = \sqrt{\frac{2m}{qV}} d \tag{6.6}$$

Substituting Eq. 6.6 in Eq. 6.4, we get

$$i_e(t) = \begin{cases} \dfrac{2q}{\tau_a{}^2} t & (0 < t < \tau_a) \\ 0 & (t > \tau_a) \end{cases} \tag{6.7}$$

Obviously, the induced current goes to zero as soon as the electron reaches the plate at $t = \tau_a$. The current pulse induced by a single electron is shown in Fig. 6.3a. Each emitted electron induces such a pulse. Thus the total plate current is composed of a large number of these triangular pulses distributed randomly, as shown in Fig. 6.3b. The sum of all such pulses constitutes the diode current $i(t)$ shown in Fig. 6.2. Note that the area under each pulse is q units. Hence the average value of the plate current is given by

$$I_0 = \bar{n}q \tag{6.8}$$

when $\bar{n}$ is the average number of electrons emitted per second.

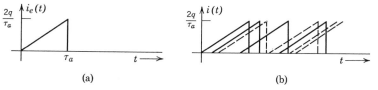

(a)   (b)

Figure 6.3

Power Density Spectrum of Shot Noise

We shall now turn our attention to the description of shot noise current $i(t)$. This current consists of two components: a constant current component I_0 and the time varying component $i_n(t)$. The component $i_n(t)$, because it is random, cannot be specified as a function of time. However, $i_n(t)$ represents a stationary random signal and can be specified by its power density spectrum. Since there are $\bar{n}$ pulses per second, it is reasonable to expect that the power density spectrum of $i_n(t)$ will be $\bar{n}$ times the energy density spectrum of $i_e(t)$. This indeed is the case.* Thus if

$$i_e(t) \leftrightarrow I_e(\omega) \tag{6.9}$$

Then $S_i(\omega)$, the power density spectrum of $i_n(t)$, is given by

$$S_i(\omega) = \bar{n} \, |I_e(\omega)|^2 \tag{6.10}$$

$I_e(\omega)$ is the Fourier transform of $i_e(t)$ and can be found as follows:

$$i_e(t) = \frac{2q}{\tau_a^2} \left[tu(t) - \tau_a u(t - \tau_a) - (t - \tau_a)u(t - \tau_a) \right] \tag{6.11}$$

Taking the Laplace transforms of both sides of Eq. 6.11 and substituting $j\omega$ for s, we get

$$i_e(t) \leftrightarrow I_e(\omega) = \frac{2q}{-\omega^2 \tau_a^2} \left[1 - e^{-j\omega\tau_a} - j\omega\tau_a e^{-j\omega\tau_a} \right] \tag{6.12}$$

Substituting Eq. 6.12 in Eq. 6.10, we get

$$S_i(\omega) = \bar{n} \, |I_e(\omega)|^2 = \frac{4I_0 q}{(\omega\tau_a)^4} \left[(\omega\tau_a)^2 + 2(1 - \cos\omega\tau_a - \omega\tau_a \sin\omega\tau_a) \right] \tag{6.13}$$

The average power density spectrum $S_i(\omega)$ can be plotted as a function of ω. As seen from Eq. 6.13, it is more convenient to plot $S_i(\omega)$ as a function of $\omega\tau_a$ (Fig. 6.4). Note that the power density spectrum is nearly flat for $\omega\tau_a < 0.5$.

The order of magnitude of τ_a can be calculated from Eq. 6.6. The ratio (q/m) for an electron is 1.76×10^{11} coulombs per kg. Hence from

---

* See, for instance, W. B. Davenport, Jr., and W. L. Root, *An Introduction to the Theory of Random Signals and Noise*, Ch. 7, McGraw-Hill, New York, 1958.

Lathi, B. P., *An Introduction to Random Signals and Communication Theory*, International Textbook Co. Scranton, Pa., 1968.

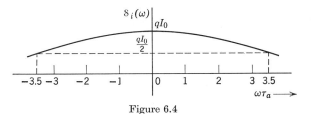

Figure 6.4

Eq. 6.6, we get

$$\tau_a = 3.36 \times 10^{-6} \frac{d}{\sqrt{V}} \text{ seconds}$$

For a diode with plate-cathode spacing $d = 1$ mm (10^{-3} meter) and with $V = 10$ volts

$$\tau_a \simeq 10^{-9} \text{ seconds}$$

The power density spectrum of the noise current component in this case will be essentially flat up to

$$\omega \simeq 0.5 \times 10^9 = 500 \times 10^6 \text{ rps}$$

This corresponds roughly to about 80 mHz. In general, the power density spectrum of a shot noise may be considered constant (qI_0) for frequencies below 100 mHz:

$$S_i(\omega) = qI_0 \tag{6.14}$$

Region of Operation in a Thermionic Diode

The operating characteristic of a diode may be divided into two regions: the temperature limited region and the space-charge limited region (Fig. 6.5). In the temperature limited region, the diode current

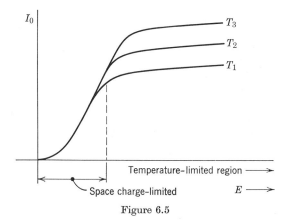

Figure 6.5

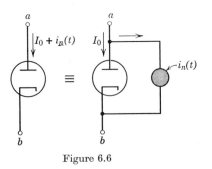

Figure 6.6

is limited by the temperature of the cathode. The electric field in this region is high enough to attract every electron emitted to the plate. The increase of the electric field therefore cannot increase the average current I_0 (Fig. 6.5). The average current can only be increased by increasing the temperature of the cathode and thereby increasing the rate of electron emission. The volt-ampere characteristic of a typical diode is shown in Fig. 6.5 for three different values of cathode temperature. At lower voltages, not all of the emitted electrons are swept to the plate, and some electrons remain inside the plate-cathode space forming what is known as the space charge. The diode operation in this region is known as the space-charge limited operation. In this region, the diode current I_0 can be increased by increasing the plate voltage (Fig. 6.5).

Note that in the temperature limited region the dynamic conductance $\partial I_0/\partial V$ of the diode is almost zero (very high dynamic or incremental resistance), whereas in the space-charge limited region, the dynamic conductance is nonzero (finite dynamic resistance).

Noise in Space-Charge Limited Region

In the discussion so far, the noise power density spectrum (Eq. 6.14) had been derived on the assumption that there is no interaction between various electrons reaching the plate. This assumption is true for a temperature limited region when there is no space charge present. Under this condition all the electrons emitted are swept away to the plate and the plate current is limited only by the number of electrons emitted per second—and thus is limited by the cathode temperature. This situation occurs when sufficiently high voltage is present at the anode to attract all electrons.

At lower voltages, an electron cloud (space charge) is present inside the anode-cathode space. The space charge depresses the potential just outside the cathode and repels some of the emitted electrons. Although the electrons are emitted independently from the cathode, there is a strong interaction among the electrons once they are emitted. Hence the arrival of an electron at the plate is dependent on the previously emitted electrons. It can be seen from a qualitative argument that the space charge has a smoothing effect upon the random fluctuation. For example, if the emission rate increases, the extra electrons increase the space charge which repels more emitted electrons back to the cathode. On the other hand, if the emission rate decreases, the space charge is reduced and more electrons reach the plate. Hence the space charge has a tendency to maintain constant current, hence the current fluctuations are smoothed out. It can be shown that the power density spectrum of the noise current $i_n(t)$ for a space-charge limited diode is given by*

$$S_i(\omega) = \alpha q I_0 \tag{6.15}$$

where

$$\alpha = 3\left(1 - \frac{\pi}{4}\right)\frac{2kT_c g_d}{q I_0} \tag{6.16}$$

Hence

$$S_i(\omega) = 1.288 kT_c g_d \tag{6.17}$$

T_c is the cathode temperature (in degrees Kelvin), k is the Boltzmann constant ($k = 1.38 \times 10^{-23}$ joules/°K), and g_d is the dynamic conductance of the diode,

$$g_d = \frac{\partial I_0}{\partial V}$$

In the space-charge region, V and I_0 are related by Child's law

$$I_0 = BV^{3\!/\!2}$$

Hence

$$g_d = \frac{\partial I_0}{\partial V} = \frac{3}{2}\frac{E}{I_0}$$

The value of the smoothing constant α varies from 0.01 to 1.

* A. J. Rack, "Effect of Space-Charge and Transit Time on the Shot Noise in Diodes," *Bell Syst. Tech. Journal* **17**, pp. 592–619, 1938. See also B. J. Thompson, D. O. North, and W. A. Harris, "Fluctuations in Space-Charge Limited Currents at Moderately High Frequencies," *RCA Review*, January 1940 et. seq.

Thus $S_i(\omega)$ can be expressed as

$$S_i(\omega) = qI_0 \qquad \text{(for temperature limited operation)}$$
$$= \alpha qI_0 \qquad \text{(for space-charge limited operation)} \qquad (6.18)$$

A diode in general can be represented by a noiseless diode in parallel with a noise current sourse $i_n(t)$ (Fig 6.6) whose power density spectrum is given by Eq. 6.18.

Shot Noise in Triodes and Multielectrode Tubes

The mechanisms of shot noise in triodes, pentodes, and other multi-electrode vacuum tubes, is basically similar to that of the space-charge limited diode. In a triode, for example, the voltage $(V_g + V_p/\mu)$ plays the same role as the plate cathode voltage in a diode. Thus the current density in a diode is given by

$$J = CV^{3/2} \qquad (6.19)$$

where V is the plate to cathode voltage. In a triode whose grid-to-cathode spacing is the same as the plate-to-cathode spacing of the diode, this relationship is given by*

$$J = C\sigma^{3/2}\left(\frac{V_g + V_p}{\mu}\right)^{3/2} \qquad (6.20)$$

where σ is a geometrical factor of the tube (usually $0.5 \leqslant \sigma \leqslant 1$), V_g and V_p are the grid and plate voltages, respectively, with respect to cathode, and μ is the amplification factor of the tube. Thus a diode with a plate cathode voltage V,

$$V = \sigma\left(V_g + \frac{V_p}{\mu}\right)$$

will have the same current density as the triode with grid and plate voltages V_g and V_p, respectively. Hence the noise power density spectrum for a triode is given by (Eq. 6.17)

$$S_i(\omega) = 1.288kT_c g_{\text{eq}}$$

where g_{eq} is the mutual conductance of the equivalent diode given by

$$g_{\text{eq}} = \frac{\partial I}{\partial[\sigma(V_g + V_p/\mu)]} = \frac{\partial I}{\partial V_g}\frac{\partial V_g}{\partial[\sigma(V_g + V_p/\mu)]}$$

$$= \frac{g_m}{\sigma}$$

* K. R. Spangenberg, *Vacuum Tubes*, McGraw-Hill, New York, 1948.

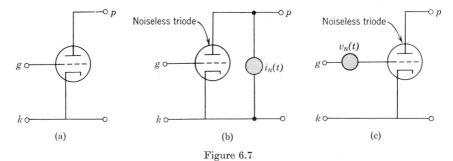

Figure 6.7

where g_m is the dynamic transconductance of the triode $\partial I/\partial V_g$. Hence

$$S_i(\omega) = \frac{1.2888kT_c g_m}{\sigma}$$

$$= 2kT \frac{0.644T_c}{\sigma T} g_m \tag{6.21}$$

Typically, T_c is $1000°$K, T, the ambient temperature, is $293°$K, and σ is 0.88. Hence

$$S_i(\omega) \simeq 2kT(2.5g_m) \tag{6.22}$$

We can therefore represent a noisy triode with a noiseless triode in parallel with a noise current $i_n(t)$ across the plate cathode terminals as shown in Fig. 6.7b. The power density spectrum of $i_n(t)$ is given in Eq. 6.22. In a triode a voltage e_g at the grid gives rise to a current $g_m e_g$ at the plate. Hence the shot noise current $i_n(t)$ may be thought of as arising due to an equivalent noise voltage $v_n(t)$,

$$v_n(t) = \frac{i_n(t)}{g_m} \tag{6.23}$$

at the grid. Hence a noisy triode may also be represented by a noiseless triode with an effective noise voltage source in the grid (Fig. 6.7c). The ratio of $i_n(t)$ to $v_n(t)$ is g_m. Since the power density is a function of the square of the signal, the ratio of the power density spectra of $i_n(t)$ and $v_n(t)$ will be $g_m{}^2$ (see the definition of the power and the power density spectrum). Hence

$$S_v(\omega) = \frac{1}{g_m{}^2} S_i(\omega) \tag{6.24}$$

where $S_v(\omega)$ is the power density spectrum of $v_n(t)$. From Eqs. 6.22 and 6.24 we have

$$S_v(\omega) = 2kT\left(\frac{2.5}{g_m}\right)$$

$$= 2kT R_{eq} \tag{6.25}$$

where

$$R_{eq} = \frac{2.5}{g_m} \tag{6.26}$$

The reason for introducing R_{eq} is that it allows us to express shot noise in a form similar to thermal noise in a resistor (to be discussed in Section 6.2).

For pentodes and other multielectrode tubes equivalent circuits similar to those for triodes hold true. In multielectrode tubes there is an additional noise component which arises due to partition of the cathode current between various electrodes. The partitioning, which is a random process, introduces an additional noise component. Both the shot noise and the partition noise can be taken into account by a single equivalent voltage source in the grid with a power density spectrum $2kT R_{eqp}$. For pentodes, the R_{eqp} is given by*

$$R_{eqp} = 1 + \frac{7.7I_s}{g_{mp}} R_{eqt} \tag{6.27}$$

where g_{mp} is the transconductance of pentode, I_s is the screen current, and $R_{eqt} = 2.5/g_{mt}$, where g_{mt} is the transconductance of the pentode operated as a triode.

In vacuum tubes (and also transistors) there is an additional source of noise, called flicker noise, which has a power density spectrum proportional to $1/f$. It is obvious that such a noise will be predominant at lower frequencies (generally below a few kilocycles). This noise arises due to slowly varying conditions at the cathode surface, and it can be reduced by a proper processing of the cathode surface in vacuum tubes and the surfaces around the junctions in transistors.

6.2 THERMAL NOISE

This type of noise arises due to the random motion of free electrons in the conducting medium such as a resistor. Each free electron inside

* *Reference Data for Radio Engineers*, 4th Edition, International Telephone and Telegraph Corporation, New York, 1956.

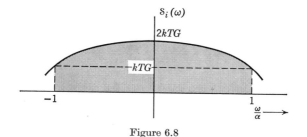

Figure 6.8

of a resistor is in motion due to its thermal energy. The path of electron motion is random and zigzag due to collisions with the lattice structure. The net effect of the motion of all electrons constitutes an electric current flowing through the resistor. The direction of current flow is random and has a zero mean value. It can be shown that the power density spectrum of the current due to free electrons is given by[*]

$$S_i(\omega) = \frac{2kTG\alpha^2}{\alpha^2 + \omega^2} \tag{6.28}$$

$$= \frac{2kTG}{1 + \left(\dfrac{\omega}{\alpha}\right)^2} \tag{6.29}$$

where k is the Boltzmann constant, T is the ambient temperature (in degrees Kelvin), G is the conductance of the resistor (in mhos), and α is the average number of collisions per second of an electron.

The power density spectrum is plotted in Fig. 6.8 as a function of ω/α. The spectrum may be assumed to be flat for $\omega/\alpha < 0.1$. The order of magnitude of α, the number of collisions per second, is of the order 10^{14}. Hence the spectrum is essentially flat up to very high frequencies. Usually, the spectrum may be considered flat up to frequencies in the range of 10^{13} Hz. Hence, for all practical purposes, the power density spectrum due to thermal noise in a resistor may be taken as

$$S_i(\omega) = 2kTG \tag{6.30}$$

The contribution in any circuit due to thermal noise therefore is limited only by the bandwidth of the circuit. Hence, the thermal noise

---

[*] Lathi, B. P., *An Introduction to Random Signals and Communication Theory*, International Textbook Co., Scranton, Pa., 1968.

J. J. Freeman, *Principles of Noise*, Ch. 4, John Wiley and Sons, New York, 1958.

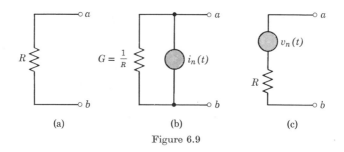

Figure 6.9

is generally considered to have a constant power density spectrum, that is, it contains all frequencies in equal amount. For this reason it is also called white noise (white implying the presence of all colors or frequencies). Thermal noise is also referred to as Johnson noise after J. B. Johnson, who investigated the noise in conductors.* Johnson found that the noise power measured in a conductor was proportional to the absolute temperature and the bandwidth of the measuring instrument. H. Nyquist of the Bell laboratories derived Eq. 6.30 based on thermodynamical reasoning.†

From this discussion it is evident that a resistor R can be represented by a noiseless conductance $G(G = 1/R)$ in parallel with noise current source (i_n) with power density spectrum $2kTG$ as shown in Fig. 6.9. The Thevenin equivalent of this arrangement (Fig. 6.9b) is shown in Fig. 6.9c. This is a voltage source equivalent circuit which has a resistance R in series with a voltage source $v_n(t)$ where $v_n(t) = Ri_n(t)$. Since the power density spectrum is a function of the square of the signal, $S_v(\omega)$, the power density spectrum of $v_n(t)$ and $S_i(\omega)$ are related by

$$S_v(\omega) = R^2 S_i(\omega)$$
$$= R^2(2kTG)$$
$$= 2kTR \qquad (6.31)$$

From the interpretation of the power density spectrum (Section 2.8), the power carried by frequency components of $v_n(t)$ in a frequency band Δf (Hz) centered at frequency ω is

$$\Delta P = 2S_v(\omega)\, \Delta f$$
$$= 4kTR\, \Delta f \qquad (6.32)$$

* J. B. Johnson, "Thermal Agitation of Electricity in Conductors," *Phys. Rev.* **32**, 97–109, July 1928.

† H. Nyquist, "Thermal Agitation of Electric Charge in Conductors," *Phys. Rev.* **32**, 110–113, July 1928.

Thus $4kTR$ is the power carried by frequency components of $v_n(t)$ in a unit bandwidth (in Hz). Note that the power of a signal by definition is its mean square value (see Section 2.8, Eq. 2.21b). If the noise voltage $v_n(t)$ is filtered by a narrowband filter of bandwidth Δf (Hz), the output voltage Δv_n of this filter will have a mean square value of $4kTR\Delta f$,

$$\overline{(\Delta v_n)^2} = 4kTR\Delta f \tag{6.33}$$

Generalization of Thermal Noise Relationship

The results in Eqs. 6.30 and 6.31 apply to a single resistive element. These relationships can be generalized to any linear bilateral passive network (networks containing R-L-C elements, for example). Such a network (shown in Fig. 6.10a) may contain several resistors. Each resistor is a source of thermal noise signal. The power density of the thermal noise voltage across the terminals aa' (Fig. 6.10a) due to all these sources can be conveniently calculated by using the generalization of Eqs. 6.30 and 6.31.

The generalized theorem (also known as generalized Nyquist theorem) states that if $Y_{ab}(\omega)$, the admittance across terminals ab of such a

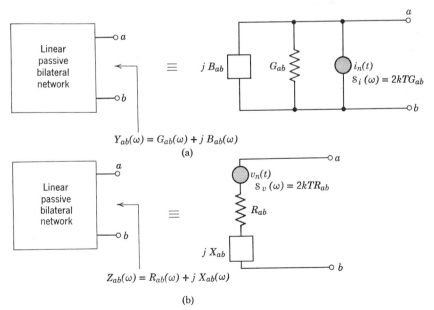

Figure 6.10

network, is given by

$$Y_{ab}(\omega) = G_{ab}(\omega) + jB_{ab}(\omega)$$

then the network can be represented by a noiseless admittance Y_{ab} in parallel with a noise current source of power density $S_i(\omega)$ given by (Fig. 6.10a)

$$S_i(\omega) = 2kTG_{ab}(\omega) \tag{6.34a}$$

Note that $G_{ab}(\omega)$, the equivalent conductance seen across terminals ab, varies with frequency in general. Hence the equivalent current noise source has a power density spectrum which is generally a function of frequency. An alternative form of this theorem (Thevenin form) states that if $Z_{ab}(\omega)$ is the impedance across terminals ab of the network where

$$Z_{ab}(\omega) = R_{ab}(\omega) + jX_{ab}(\omega)$$

then the network can be represented by a noiseless impedance Z_{ab} in series with a noise voltage source of power density spectrum $S_v(\omega)$ given by (Fig. 6.10b)

$$S_v(\omega) = 2kTR_{ab}(\omega) \tag{6.34b}$$

In general, $R_{ab}(\omega)$, the real part of $Z_{ab}(\omega)$, is a function of frequency. Hence the equivalent noise source has a power density spectrum which is a function of frequency.

It is important to remember the limitations on this theorem. It applies to any linear bilateral network in which all of the noise sources are due to resistors.* Thus it can be applied to any electrical circuit containing R-L-C elements but cannot be applied to circuits containing active elements such as vacuum tubes or transistors. The proof of this theorem is given in the appendix at the end of this chapter.

Example 6.1

Determine the power density spectrum of the thermal noise voltage across terminals ab of the circuit shown in Fig. 6.11. We have

$$Y_{ab}(\omega) = 1 + j\omega + \frac{1}{2 + 2j\omega}$$

$$= \frac{3 - 2\omega^2 + j4\omega}{2 + j2\omega}$$

---

* By suitably defining resistance, this theorem can be extended to linear dissipative systems in general, for example, to Brownian motion or gas pressure fluctuation.

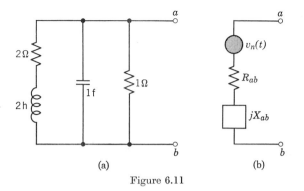

Figure 6.11

Hence

$$Z_{ab}(\omega) = \frac{2 + j2\omega}{3 - 2\omega^2 + j4\omega}$$

$$R_{ab}(\omega) = \text{Re}[Z_{ab}(\omega)] = \frac{4\omega^2 + 6}{4\omega^4 + 4\omega^2 + 9}$$

The noise voltage power density spectrum $S_v(\omega)$ is given by (Fig. 6.11b)

$$S_v(\omega) = 2kT R_{ab}(\omega) = \frac{2kT(4\omega^2 + 6)}{4 + 4\omega^2 + 9}$$

If the Norton equivalent is desired (Fig. 6.10a), then the noise current power density spectrum $S_i(\omega)$ is given by

$$S_i(\omega) = 2kT \, \text{Re}[Y_{ab}(\omega)] = 2kT \frac{3 + 2\omega^2}{2 + 2\omega^2}$$

$$= kT \frac{3 + 2\omega^2}{1 + \omega^2}$$

6.3 NOISE CALCULATIONS: SINGLE NOISE SOURCE

We are now in a position to calculate the root mean square value of noise signals in electrical systems in general. We shall deal with a single noise source first.

Consider a circuit containing only noiseless elements (Fig. 6.12). A random noise voltage source $v_{n_i}(t)$ is connected at the input terminals

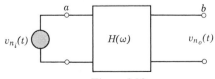

Figure 6.12

of this circuit. We shall now determine the root mean square value of the output noise signal $v_{n_o}(t)$. Let the transfer function relating the output voltage $v_{n_o}(t)$ to the input voltage $v_{n_i}(t)$ in Fig. 6.12 be $H(\omega)$. Let $S_i(\omega)$ and $S_o(\omega)$ be the power density spectra of the signals $v_{n_i}(t)$ and $v_{n_o}(t)$, respectively. Then from the discussion in Section 2.8 (Eq. 2.37) it follows that

$$S_o(\omega) = S_i(\omega) |H(\omega)|^2$$

The mean square value of a signal is given by $1/2\pi$ times the area under its power density spectrum (see Eq. 2.23b). Hence

$$\overline{v_{n_o}{}^2} = \frac{1}{2\pi} \int_{-\infty}^{\infty} S_o(\omega)\, d\omega$$

$$= \frac{1}{2\pi} \int_{-\infty}^{\infty} S_i(\omega)\, |H(\omega)|^2\, d\omega \qquad (6.35)$$

Note that the power density spectrum is always an even function of ω, and hence we have (Eq. 2.23d)

$$\overline{v_{n_o}{}^2} = \frac{1}{\pi} \int_{0}^{\infty} S_i(\omega)\, |H(\omega)|^2\, d\omega \qquad (6.36a)$$

The root mean square value of $v_{n_o}(t)$ is given by

$$\sqrt{\overline{v_{n_o}{}^2}} = \left[\frac{1}{\pi} \int_{0}^{\infty} S_i(\omega)\, |H(\omega)|^2\, d\omega \right]^{\frac{1}{2}} \qquad (6.36b)$$

Example 6.2

As an example, consider the R-C network in Fig. 6.13. We shall calculate the root mean square value of the noise voltage across the capacitor terminals aa'.

The resistor R is replaced by a noiseless resistor R and a series noise voltage source $v_{n_i}(t)$, with power density $2kTR$. The transfer function $H(\omega)$

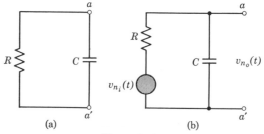

(a)   (b)

Figure 6.13

which relates the output voltage $v_{n_o}(t)$ to $v_{n_i}(t)$ is obviously

$$H(\omega) = \frac{1/j\omega C}{R + 1/j\omega RC}$$

$$= \frac{1}{j\omega RC + 1}$$

The power density spectrum of the output voltage $v_{n_o}(t)$ is given by $S_o(\omega)$, where

$$S_o(\omega) = S_i(\omega) |H(\omega)|^2$$

$$= \frac{2kTR}{1 + \omega^2 C^2 R^2}$$

The root mean square value of the output is given by (Eq. 6.36b)

$$\sqrt{\overline{v_{n_o}^2}} = \left[\frac{1}{\pi} \int \frac{2kTR}{1 + \omega^2 C^2 R^2} \, d\omega \right]^{1/2}$$

$$= \left[\frac{2kTR}{\pi CR} \tan^{-1}(\omega CR) \Big|_0^\infty \right]^{1/2}$$

$$= \sqrt{\frac{kT}{C}}$$

6.4 MULTIPLE NOISE SOURCES: SUPERPOSITION OF POWER SPECTRA

In Section 6.3 we discussed the case of a single noise source. In a given electrical system there may be a large number of noise sources such as resistors and vacuum tubes. All these sources are independent sources of random signals. We shall now show that for multiple random signals generated by independent sources, the principle of superposition applies to mean square values and power density spectra. This means whenever there are multiple sources, the mean square value of the response is equal to the sum of the mean square values of the responses computed by assuming only one source at a time. Similarly, the power density spectrum of the response is given by the sum of power spectra computed by assuming only one source at a time.

Consider a system (Fig. 6.14a) with two sources $f_1(t)$ and $f_2(t)$ at two different terminals and the response $y(t)$. Since the system is linear,

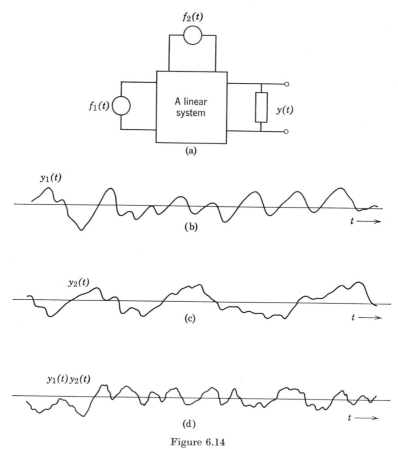

Figure 6.14

$y(t)$ is given by $y_1(t) + y_2(t)$, where $y_1(t)$ is the response due to $f_1(t)$ alone and $y_2(t)$ is the response due to $f_2(t)$ alone. We shall now show that if $f_1(t)$ and $f_2(t)$ are independent random sources with zero mean values,* then

$$\overline{y^2(t)} = \overline{y_1{}^2(t)} + \overline{y_2{}^2(t)} \qquad (6.37a)$$

and

$$S_y(\omega) = S_{y_1}(\omega) + S_{y_2}(\omega) \qquad (6.37b)$$

This can be shown as follows:

$$y(t) = y_1(t) + y_2(t)$$

* It is sufficient that either $f_1(t)$ or $f_2(t)$ has zero mean.

Therefore

$$\overline{y^2(t)} = \lim_{T \to \infty} \frac{1}{T} \int_{-T/2}^{T/2} [y_1(t) + y_2(t)]^2 \, dt$$

$$= \lim_{T \to \infty} \frac{1}{T} \int_{-T/2}^{T/2} y_1{}^2(t) \, dt + \lim_{T \to \infty} \frac{1}{T} \int_{-T/2}^{T/2} y_2{}^2(t) \, dt +$$

$$\lim_{T \to \infty} \frac{2}{T} \int_{-T/2}^{T/2} y_1(t) y_2(t) \, dt$$

$$= \overline{y_1{}^2(t)} + \overline{y_2{}^2(t)} + 2 \lim_{T \to \infty} \frac{1}{T} \int_{-T/2}^{T/2} y_1(t) y_2(t) \, dt \qquad (6.38)$$

The integral on the right-hand side is the average value of the product $y_1(t)y_2(t)$. If $y_1(t)$ and $y_2(t)$ are independent random signals with zero mean (Fig. 6.14b and c), then both $y_1(t)$ and $y_2(t)$ are equally likely to assume positive and negative values. Obviously their product $y_1(t)y_2(t)$ is also equally likely to assume positive and negative values (Fig. 6.14d). The average value of $y_1(t)y_2(t)$ must therefore be zero.* If the mean value of the input signal is zero, the mean value of the output signal is also zero for a stable system.† Therefore, if $x_1(t)$ and $x_2(t)$ have zero means, $y_1(t)$ and $y_2(t)$ also have zero mean value. The signal $y_1(t)y_2(t)$ is equally likely to be positive and negative. Hence the mean of $y_1(t)y_2(t)$ is zero, and Eq. 6.38 becomes

$$\overline{y^2(t)} = \overline{y_1{}^2(t)} + \overline{y_2{}^2(t)}$$

This proves Eq. 6.37a.

The mean square value of a signal is given by $1/2\pi$ times the area under its power density spectrum. Hence Eq. 6.38 can be expressed as

$$\frac{1}{2\pi} \int_{-\infty}^{\infty} S_y(\omega) \, d\omega = \frac{1}{2\pi} \int_{-\infty}^{\infty} S_{y_1}(\omega) \, d\omega + \frac{1}{2\pi} \int_{-\infty}^{\infty} S_{y_2}(\omega) \, d\omega$$

$$= \frac{1}{2\pi} \int_{-\infty}^{\infty} [S_{y_1}(\omega) + S_{y_2}(\omega)] \, d\omega$$

* It is easy to see that this conclusion is valid even if only one of $y_1(t)$ or $y_2(t)$ has a zero mean.

† If the system is stable, then $h(t)$ decays with time and the mean value of $h(t)$ is zero, implying that $h(t)$ has no d-c component. Hence $H(\omega)$ has no impulse at the origin, and $|H(\omega)|^2$ is finite at $\omega = 0$. If $x(t)$ is the input signal and $y(t)$ is the output signal, then

$$S_y(\omega) = |H(\omega)|^2 \, S_x(\omega)$$

If $x(t)$ has zero mean value, $S_x(\omega)$ has no impulse at $\omega = 0$. Hence $|H(\omega)|^2 \, S_x(\omega)$ is finite at $\omega = 0$ and has no impulse at $\omega = 0$. Obviously, $y(t)$ has no d-c component, implying that its mean value is zero.

This equation can be interpreted as

$$S_y(\omega) = S_{y_1}(\omega) + S_{y_2}(\omega)$$

Hence the power density spectra obey the principle of superposition. Equations 6.37a and 6.37b represent the principle of superposition of mean square values and power density spectra for two independent random signals with zero mean value.*

Thus whenever we have multiple random independent sources in a system, the power density spectrum can be obtained by assuming one source at a time and evaluating the power density of the response. The resultant power density due to all sources is the sum of all individual power densities.

Note that the principle of superposition of power density spectra applies to independent random sources.† In such cases the integral in Eq. 6.38 vanishes. The sources are then said to add incoherently. If the sources are related, then the integral in Eq. 6.38 will not be zero and Eq. 6.37 is not valid.

Whenever we have multiple independent random noise sources,‡ we can apply Eq. 6.38. Hence to calculate the power density spectrum across any terminals due to multiple noise sources, we may calculate the power density across those terminals due to each source individually, assuming the remaining sources to be zero. The sum of these individual power density spectra across those terminals will be the resultant power density spectrum due to all sources.

For a linear bilateral R-L-C network containing several resistors, the power density spectrum at the output terminals may be computed by using the generalized Nyquist theorem or the principle of superposition of power density spectra. In general, however, the Nyquist theorem proves superior. The reader may verify this fact by computing the voltage power density spectrum across terminals aa' for the R-L-C network in Fig. 6.11a by using principle of superposition of power density spectra.

---

* It can be shown that the principle of superposition of power density spectra applies under a more relaxed condition of uncorrelated signals. The independence is a much more severe constraint. Independent sources are uncorrelated, but the converse is not true. This topic cannot be fully appreciated without some background in random processes.

† Uncorrelated random sources, to be exact.

‡ Note that the mean values of all noise sources in our discussion are zero.

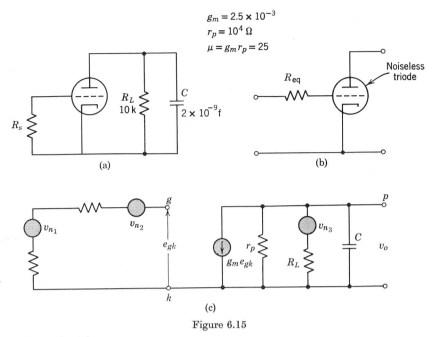

$$g_m = 2.5 \times 10^{-3}$$
$$r_p = 10^4 \,\Omega$$
$$\mu = g_m r_p = 25$$

Figure 6.15

Example 6.3

Calculate the root mean square noise voltage at the output terminals of a triode amplifier shown in Fig. 6.15a.

In order to analyze this problem, we shall represent the amplifier with noiseless elements and equivalent noise sources explicitly. A triode may be represented by a noise source at the output terminals (Fig. 6.7b) or an equivalent noise source in the input terminals (Fig. 6.7c). The second form proves to be more convenient than the first. Hence in all our future calculations we shall use the input source equivalent circuit for vacuum tubes. Note that the equivalent noise voltage in the grid has a power density spectrum of $2kTR_{\text{eq}}$ where

$$R_{\text{eq}} = \frac{2.5}{g_m}$$

This source can be taken into account by placing an equivalent resistor R_{eq} at the grid terminals, as shown in Fig. 6.15b. The thermal noise generated by R_{eq} has a power density $2kTR_{\text{eq}}$ and is equal to the equivalent shot noise voltage power density at the grid. This arrangement is satisfactory roughly below 100 mHz, where shot noise can be assumed to have a constant power density. Note that placing an equivalent noise resistor at the grid leaves the circuit unaffected because no current flows in this part of the circuit.

Three sources of noise in this amplifier are: (a) the thermal noise due to resistor R_s; (b) the shot noise of the tube accounted by the equivalent thermal noise of R_{eq} in the grid; and (c) the thermal noise due to load resistor R_L. Let $S_1(\omega)$, $S_2(\omega)$, and $S_3(\omega)$ be the noise voltage power density spectra of these three sources. Obviously,

$$S_1(\omega) = 2kTR_s = 2kT \times 10^3$$

$$S_2(\omega) = 2kTR_{eq} = 2kT \times \frac{2.5}{g_m} = 2kT \times 10^3$$

$$S_3(\omega) = 2kTR_L = 2kT \times 10^4$$

The final equivalent circuit of the amplifier where the triode is represented by its current source equivalent circuit is shown in Fig. 6.15c. The three noise source voltages are labeled v_{n_1}, v_{n_2}, and v_{n_3}. The output voltage is v_o. If $H_1(\omega)$, $H_2(\omega)$, and $H_3(\omega)$ represent the transfer functions relating v_o to v_{n_1}, v_{n_2}, and v_{n_3}, respectively, then it can be seen easily that

$$H_1(\omega) = H_2(\omega) = \frac{-g_m}{\dfrac{1}{r_p} + \dfrac{1}{R_L} + j\omega C} = \frac{-g_m r_p R_L}{R_L + r_p + j\omega r_p R_L C}$$

and

$$H_3(\omega) = \frac{r_p/1 + j\omega r_p C}{R_L + r_p/(1 + j\omega r_p C)} = \frac{r_p}{R_L + r_p + j\omega r_p R_L C}$$

If the power density spectrum of the output voltage is $S_o(\omega)$, then it follows that

$$S_o(\omega) = |H_1(\omega)|^2 S_1(\omega) + |H_2(\omega)|^2 S_2(\omega) + |H_3(\omega)|^2 S_3(\omega)$$

$$= \frac{g_m^2 r_p^2 R_L^2}{(R_L + r_p)^2 + \omega^2 r_p^2 R_L^2 C^2} (4kT \times 10^3)$$

$$+ \frac{r_p^2}{(R_L + r_p)^2 + \omega^2 r_p^2 R_L^2 C^2} (2kT \times 10^4)$$

$$= \frac{2kTr_p^2 \times 10^3}{(R_L + r_p)^2 + \omega^2 r_p^2 R_L^2 C^2} (2g_m^2 R_L^2 + 10) \qquad (6.39)$$

Note that the relative contribution to the output noise due to R_g and R_{eq} is $2g_m^2 R_L^2$ which is 1250, and that due to R_L is only 10. This observation is valid in almost all practical cases. The noise contribution due to the load resistor may be ignored since it is small in comparison to tube noise and input source noise.

Substituting the values of R_L, r_p, C, k, and T (293°K) in Eq. 6.39, we get

$$S_o(\omega) = \frac{2.52 \times 10^{-5}}{10^{10} + \omega^2} \qquad (6.40)$$

Hence

$$\overline{v_o{}^2} = \frac{1}{\pi} \int_0^\infty S_0(\omega)\, d\omega$$

$$= \frac{2.52 \times 10^{-5}}{\pi} \int_0^\infty \frac{d\omega}{10^{10} + \omega^2}$$

$$= 1.26 \times 10^{-10}$$

The rms value of $v_o(t)$ is given by

$$\sqrt{\overline{v_o{}^2}} = 11.2 \times 10^{-6} = 11.2 \; \mu v$$

6.5 EQUIVALENT NOISE BANDWIDTH

Sometimes it is convenient to define the equivalent noise bandwidth for an electrical circuit. A system with a transfer function $H(\omega)$ and the input signal power density spectrum $S_i(\omega)$ has a mean square value of the output signal

$$\overline{v_o{}^2} = \frac{1}{2\pi} \int_{-\infty}^\infty S_i(\omega)\, |H(\omega)|^2 \, d\omega$$

In most of the circuits the bandwidths are low enough to permit the assumption that shot noise and thermal noise have a constant spectral density, that is $S_i(\omega)$ is a constant, say, K. Thus

$$\overline{v_o{}^2} = \frac{K}{2\pi} \int_{-\infty}^\infty |H(\omega)|^2 \, d\omega \qquad (6.41a)$$

$$= \frac{K}{\pi} \int_0^\infty |H(\omega)|^2 \, d\omega \qquad (6.41b)$$

The integral in Eq. 6.41b is a constant for a given circuit. We define an equivalent noise bandwidth W_0 with respect to some frequency ω_0 as

$$W_0 = \frac{1}{|H(\omega_0)|^2} \int_0^\infty |H(\omega)|^2 \, d\omega \qquad (6.42)$$

Hence

$$\overline{v_o{}^2} = K \, |H(\omega_0)|^2 \, W_0 \qquad (6.43)$$

The significance of the equivalent noise bandwidth is now evident. The root mean square value of the noise signal at the output of a given system is the same as that of an ideal bandpass system of constant gain

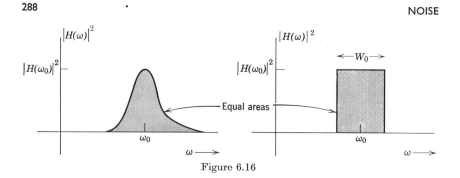

Figure 6.16

$H(\omega_0)$ and bandwidth W_0. This is clearly illustrated in Fig. 6.16. The root mean square value of the noise output is identical for both amplifiers whose gain characteristics are shown in Figs. 6.6a and b.

6.6 NOISE FIGURE OF AN AMPLIFIER

From the discussion thus far it is evident that every signal is contaminated with noise. Moreover, when a signal is being processed through any system, additional noise is being added. The ratio of signal power to noise power is a good indication of the purity of a signal (or the relative level of the signal and the noise). For simplicity, we shall call this ratio the signal-to-noise ratio.

When a signal is amplified, additional noise generated in the amplifier is being added to the original noise in the signal. This causes deterioration in the signal-to-noise ratio of the output signal compared to that of the input signal. The ratio of signal-to-noise at the input to that at the output is an indication of the noisiness of the amplifier. In any amplifier, the noise generated in the source is amplified and delivered to the load. Additional noise is generated inside the amplifier circuit and is also delivered to the load. Hence the noise in the output is contributed by the source as well as the amplifier.

We shall define the noise figure F of an amplifier as a ratio of the total noise power density in the load (or the output) to the noise power density delivered to the load (output) due solely to the source. If $S_{nto}(\omega)$ and $S_{nso}(\omega)$ represent the power density spectra of the total noise at the output and the noise at the output due solely to the source, respectively, then by definition

$$F = \frac{S_{nto}(\omega)}{S_{nso}(\omega)} \tag{6.44}$$

If $S_{nao}(\omega)$ is the power density spectrum of the noise in the load due solely to the amplifier, then

$$S_{nto}(\omega) = S_{nso}(\omega) + S_{nao}(\omega)$$

$$F = \frac{S_{nso}(\omega) + S_{nao}(\omega)}{S_{nso}(\omega)}$$

$$= 1 + \frac{S_{nao}(\omega)}{S_{nso}(\omega)} \tag{6.45}$$

Note that the load impedance also contributes to the noise at the output. This noise contribution is included in S_{nao}. However, this contribution is usually much smaller compared to the noise generated by the active device (see Example 6.3) and the source and consequently may be ignored. It is evident that the noise figure of an amplifier is a measure of the noisiness of the amplifier relative to the noisiness of the source.

The noise figure may also be expressed in an alternative form. Let $S_{si}(\omega)$ and $S_{so}(\omega)$ represent the power density spectra of the desired signal at the input and the output, respectively. Then

$$S_{so}(\omega) = S_{si}(\omega) |H(\omega)|^2 \tag{6.46}$$

If $S_{nsi}(\omega)$ represents the power density spectrum of the noise at the source (or input), then

$$S_{nso}(\omega) = S_{nsi}(\omega) |H(\omega)|^2$$

and, by definition (Eq. 6.44),

$$F = \frac{S_{nto}(\omega)}{S_{nsi}(\omega) |H(\omega)|^2}$$

The substitution of Eq. 6.46 in this equation yields

$$F = \frac{S_{si}(\omega)/S_{nsi}(\omega)}{S_{so}(\omega)/S_{nto}(\omega)} \tag{6.47}$$

This is a rather significant result. The numerator of Eq. 6.47 represents the signal-to-noise power density ratio at the input terminals, and the denominator represents the signal-to-noise power density ratio at the output terminals. Hence the noise figure measures the deterioration of the signal-to-noise power density ratio in the process of amplification. By the very definition it is evident that this ratio is always greater than unity, hence the signal-to-noise power density ratio always deteriorates in the process of amplification. The deterioration results, of course, from the noise contribution of the amplifier.

The noise figure F, defined in Eq. 6.44 (or Eq. 6.47), is a function of the frequency and hence is sometimes referred to as the spectral noise figure. In contrast, we may define an average noise figure $\bar{F}$ as the ratio of the total noise power at the output to the noise power contributed due solely to the source. Note that this definition is merely an extension of the definition in Eq. 6.44. In this definition we deal with the total noise power contributed over all of the frequencies. This therefore is an integrated or average noise figure. Thus, by definition,

$$\bar{F} = \frac{\dfrac{1}{\pi} \displaystyle\int_0^\infty S_{\mathrm{nto}}(\omega)\, d\omega}{\dfrac{1}{\pi} \displaystyle\int_0^\infty S_{\mathrm{nso}}(\omega)\, d\omega}$$

$$= 1 + \frac{\displaystyle\int_0^\infty S_{\mathrm{nao}}(\omega)\, d\omega}{\displaystyle\int_0^\infty S_{\mathrm{nso}}(\omega)\, d\omega} \tag{6.48}$$

$$= 1 + \frac{N_{\mathrm{ao}}}{N_{\mathrm{so}}} \tag{6.48b}$$

where N_{ao} is the noise power at the output due solely to the amplifier and N_{so} is the noise power at the output due solely to the source noise.

From the definition of the noise figure (spectral or average noise figure), it is evident that the noise figure of an amplifier is always greater than unity. For an ideal amplifier (noiseless amplifier), it is unity. The closeness of the noise figure to unity is the measure of superiority of the amplifier from the noise point of view. However, it should be emphasized here that the noise figure measures not the absolute but the relative quality of the amplifier. It indicates the noisiness of an amplifier relative to the noisiness of the source. It is evident from the definition that the noise figure of an amplifier can be made as close to unity as possible, merely by adding extra noise in the source. This obviously is not the proper solution for improving the performance of the amplifier, since this approach merely makes the source so noisy that the amplifier, in comparison, appears almost noise-free. The overall signal-to-noise ratio at the output, however, deteriorates badly, and consequently the output signal is much more noisy. It is therefore important not to increase the noise in the source (or to decrease signal-to-noise ratio in the input) in order to improve the noise figure. A step-up transformer in many cases solves the problem. A step-up transformer at the input increases the input noise as well as the input signal. The increased noise at the source makes the amplifier look less noisy without deteriorating the signal-to-noise ratio of the input. Hence the noise figure is reduced and the signal-to-noise ratio at the output terminals actually improves (see Example 6.5).

Example 6.4

Determine the noise figure of a triode amplifier shown in Fig. 6.17a and calculate the root mean square noise voltage at the output terminals. It is given that

$$r_p = R_L = 10 \ k\Omega, \quad g_m = 2.5 \times 10^{-3}, \quad R_g = 100 \ k\Omega,$$

$$R_s = 1 \ k\Omega \quad \text{and} \quad C = 2 \times 10^{-9} \text{ farad}$$

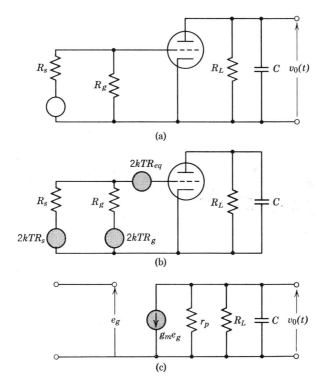

Figure 6.17

The noise contribution due to the load resistance is usually negligible compared to other components, and hence R_L may be assumed to be noiseless (see Example 6.3).

The source resistance R_s and the grid resistance R_g will be represented by noiseless resistors in series with respective equivalent noise voltage sources, and the triode will be represented by a noiseless triode and an equivalent noise voltage source in the grid (Fig. 6.17b).

The power density spectra due to R_s, R_g, and R_{eq} are given by $2kTR_s$, $2kTR_g$, and $2kTR_{eq}$, respectively, as shown in Fig. 6.17b.

Let the transfer function relating the output voltage v_o to the grid voltage e_g be represented by $H(\omega)$. It now follows that the transfer function relating the output voltage v_o to the equivalent noise voltage source due to R_{eq} is $H(\omega)$ and that relating v_o to the noise voltage source due to R_g is

$$\frac{R_s}{R_s + R_g} H(\omega)$$

Similarly, the transfer function relating v_o to the noise voltage source due to R_s is

$$\frac{R_g}{R_s + R_g} H(\omega)$$

Hence the power density spectrum of the noise voltage at the output due to R_{eq} is

$$2kTR_{eq} |H(\omega)|^2 \qquad (6.49a)$$

The power density spectrum of the noise voltage at the output due to R_g is

$$2kTR_g \left(\frac{R_s}{R_s + R_g}\right)^2 |H(\omega)|^2 \qquad (6.49b)$$

and the power density spectrum of the noise voltage at the output due to R_s is

$$2kTR_s \left(\frac{R_g}{R_s + R_g}\right)^2 |H(\omega)|^2 \qquad (6.49c)$$

Note that the grid resistance R_g is a part of the amplifier, and hence

$$S_{nao}(\omega) = 2kTR_{eq} |H(\omega)|^2 + 2kTR_g \left(\frac{R_s}{R_s + R_g}\right)^2 |H(\omega)|^2 \qquad (6.50a)$$

and

$$S_{nso}(\omega) = 2kTR_s \left(\frac{R_g}{R_s + R_g}\right)^2 |H(\omega)|^2 \qquad (6.50b)$$

Therefore the noise figure F is given by

$$F = 1 + \frac{S_{nao}}{S_{nso}}$$

$$= 1 + \frac{R_{eq} + R_g \left(\dfrac{R_s}{R_s + R_g} \right)^2}{R_s \left(\dfrac{R_g}{R_s + R_g} \right)^2}$$

$$= 1 + \frac{R_{eq}(R_s + R_g)^2 + R_g R_s^2}{R_s R_g^2} \tag{6.51}$$

$$R_{eq} = \frac{2.5}{g_m} = \frac{2.5}{2.5 \times 10^{-3}} = 1000 \text{ ohms}$$

$$R_s = 1000 \text{ ohms} \quad \text{and} \quad R_g = 10^5 \text{ ohms}$$

Hence

$$F = 2.03$$

To obtain the root mean square value of the noise voltage at the output, we find the transfer function $H(\omega)$. The equivalent circuit of the triode is shown in Fig. 6.17c. It is evident from this circuit that the transfer function which relates the output voltage v_o to the grid voltage e_g is given by

$$H(\omega) = -g_m Z$$

where

$$\frac{1}{Z} = \frac{1}{r_p} + \frac{1}{R_L} + j\omega C$$

$$= 10^{-4} + 10^{-4} + (2 \times 10^{-9})(j\omega)$$

$$= 2 \times 10^{-9} (10^5 + j\omega)$$

and

$$H(\omega) = \frac{(2.5 \times 10^{-3})}{(2 \times 10^{-9})(10^5 + j\omega)}$$

$$= \frac{1.25 \times 10^6}{(10^5 + j\omega)}$$

and

$$|H(\omega)|^2 = \frac{1.56 \times 10^{12}}{(10^{10} + \omega^2)} \tag{6.52}$$

The mean square of the noise voltage at the output is given by

$$\frac{1}{\pi} \int_0^\infty S_{nto}(\omega) \, d\omega$$

where
$$S_{nto}(\omega) = S_{nao}(\omega) + S_{nso}(\omega)$$

From Eqs. 6.50 and 6.52, we obtain

$$S_{nto}(\omega) = 2kT \, |H(\omega)|^2 \left[R_{eq} + R_g \left(\frac{R_s}{R_s + R_g} \right)^2 + R_s \left(\frac{R_g}{R_s + R_g} \right)^2 \right]$$

$$= 2 \times 1.38 \times 10^{-23} \times 290 \times \frac{1.56 \times 10^{12}}{10^{10} + \omega^2} \times 2010$$

$$= \frac{2.5 \times 10^{-5}}{10^{10} + \omega^2}$$

The mean square value of the output noise voltage is given by

$$v_o{}^2 = \frac{1}{\pi} \int_0^\infty \frac{2.5 \times 10^{-5}}{10^{10} + \omega^2} \, d\omega$$

$$= \frac{2.5 \times 10^{-5}}{\pi} \times \frac{1}{10^5} \tan^{-1} \frac{\omega}{10^5} \Big|_0^\infty$$

$$= 1.25 \times 10^{-10}$$

Hence the root mean square value of the voltage at the output terminals is given by

$$\sqrt{\overline{v_o{}^2}} = 11.2 \ \mu v$$

Example 6.5

Find the optimum value of the source resistance R_s for the triode amplifier in Example 6.4 and calculate the corresponding noise figure.

The noise figure F is given by Eq. 6.51. Generally, $R_s \ll R_g$ and Eq. 6.51 becomes

$$F \simeq 1 + \frac{R_{eq}}{R_s} + \frac{R_s}{R_g} \tag{6.53}$$

The optimum value of R_s can be obtained from $dF/dR_s = 0$. Thus

$$\frac{dF}{dR_s} = \frac{-R_{eq}}{R_s{}^2} + \frac{1}{R_g} = 0$$

or

$$(R_s)_{opt} = \sqrt{R_{eq}R_g}$$

$$= (10^3 \times 10^5)^{1/2}$$

$$= 10 \ k\Omega$$

The noise figure for $(R_s)_{\text{opt}}$ is obtained from Eq. 6.53:

$$F = 1 + 0.1 + 0.1$$
$$= 1.2$$

If the source resistance R_s is 1000 ohms, then the optimum noise figure can be attained by using a step-up transformer of the ratio $1:\sqrt{10}$. Note that it is not desirable to obtain an optimum value of the source resistance by adding an extra 9000 ohms in series with 1000 ohms. This will merely make the source noisy compared to the amplifier.

A Comment on Noise Figure

The noise figure is the ratio of the total noise power density spectrum of the output variable to the noise power density spectrum contributed only by the source to the output variable. A question may arise as to which output variable should be used, that is, whether we should use an output voltage or an output current. Actually, the noise figure is independent of the output variable. This is because the voltage and current are related by the load impedance and, since the noise figure is a function of the ratio of the signal-to-noise power density spectra, this factor (due to load impedance) cancels out.

Example 6.6

Determine the noise figure of a common base transistor amplifier.

In a transistor there are mainly three sources of noise: the shot noise, the partition noise, and the thermal noise. The shot noise can be accounted for by a current source $i_{\text{sh}}(t)$ across terminals ej (Fig. 6.18) and has a power density spectrum* (at low frequencies)

$$S_{\text{sh}}(\omega) = qI_e \tag{6.54a}$$

where I_e is the average (d-c) emitter current. The partition noise can be accounted for by a current source $i_p(t)$ across terminals jc and has a power density spectrum (at low frequencies)

$$S_p(\omega) = q\alpha_0 I_e(1 - \alpha_0) \tag{6.54b}$$

where α_0 is the d-c current gain in a common base transistor.

The thermal noise arises due to physical resistance in the base and can be accounted for by a voltage source $v_{\text{th}}(t)$ in the base lead and has a power density spectrum:

$$S_{\text{th}}(\omega) = 2kTr_b \tag{6.54c}$$

* A. Van der Ziel, "Theory of Shot Noise in Junction Diodes and Junction Transistors," *Proc. I.R.E.*, **43**, 11, 1639–1646 (November 1955).

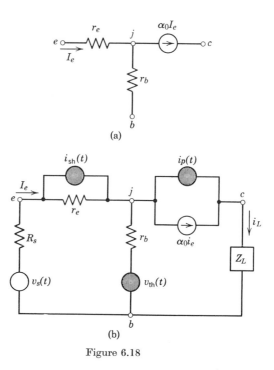

Figure 6.18

where r_b is the base spreading resistance. An equivalent circuit of a transistor is shown in Fig. 6.18a. In Fig. 6.18b, the equivalent circuit is shown with various noise generators, and source load resistors. The source resistor is represented by a noiseless resistor R_s in series with a thermal noise source $v_s(t)$ which has a power density spectrum

$$S_{R_s}(\omega) = 2kTR_s \tag{6.54d}$$

The noise generated by the transistor has three components: those due to $i_{sh}(t)$, $i_p(t)$, and $v_{th}(t)$. It is left as an exercise for the reader to show that the transfer functions relating the load current i_L to these three sources, respectively, are

$$H_{i_{sh}}(\omega) = \frac{\alpha_0 r_e}{R_s + r_b(1 - \alpha_0) + r_e} \tag{6.55a}$$

$$H_{i_p}(\omega) = \frac{r_b + r_e + R_s}{R_s + r_b(1 - \alpha_0) + r_e} \tag{6.55b}$$

$$H_{v_{th}}(\omega) = \frac{\alpha_0}{R_s + (1 - \alpha_0) + r_e} \tag{6.55c}$$

Similarly, the transfer function relating the load current i_L to the source $v_s(t)$ is given by

$$H_{v_s}(\omega) = \frac{\alpha_0}{R_s + r_b(1 - \alpha_0) + r_e} \tag{6.55d}$$

By definition,

$$F = 1 + \frac{S_{nao}(\omega)}{S_{nso}(\omega)} \tag{6.56}$$

where

$$S_{nao}(\omega) = S_{sh}(\omega) \, |H_{i_{sh}}|^2 + S_p(\omega) \, |H_{i_p}|^2 + S_{th}(\omega) \, |H_{v_{th}}|^2 \tag{6.57a}$$

and

$$S_{nso}(\omega) = S_{R_s}(\omega) \, |H_{v_s}|^2 \tag{6.57b}$$

Substituting Eqs. 6.57, 6.54, and 6.55 in Eq. 6.56, we get

$$F = 1 + \frac{r_b}{R_s} + \frac{qI_e}{2kTR_s} \left[r_e^2 + \frac{1 - \alpha_0}{\alpha_0} (r_b + r_e + R_s)^2 \right] \tag{6.58}$$

In a transistor, the dynamic emitter resistance r_e is related to the emitter current by

$$r_e = \frac{kT}{qI_e} \tag{6.59}$$

Substitution of Eq. 6.59 in Eq. 6.58 yields

$$F = 1 + \frac{r_b + r_e/2}{R_s} + \frac{(r_b + r_e + R_s)^2(1 - \alpha_0)}{2\alpha_0 r_e R_s} \tag{6.60}$$

Note that there exists an optimum value of R_s for which the noise figure is minimum.

This result holds for frequencies $\omega < \sqrt{1 - \alpha_0}\,\omega_\alpha$ where ω_α is the α cutoff frequency of the transistor. At lower frequencies (below 1 kc), the flicker noise becomes predominant. Hence the noise figure in Eqs. 6.58 and 6.60 is valid for intermediate frequencies. At high frequencies the partition noise increases. The general expression for the power density spectrum of the partition noise current generator is given by*

$$S_p(\omega) = q\alpha_0 I_e(1 - \alpha_0) \frac{1 + [\omega/\omega_\alpha \sqrt{(1 - \alpha_0)}]^2}{1 + (\omega/\omega_\alpha)^2}$$

This reduces to Eq. 6.54b at low frequencies.

* For high frequency behavior, see E. G. Nielsen, "Behavior of Noise Figure in Junction Transistor," *Proc. I.R.E.* **45**, 7, 957–963 (July 1957); G. H. Hanson and A. Vand der Ziel, "Shot Noise in Transistors," *Proc. I.R.E.* **45**, 11, 1538, 1542 (November 1957).

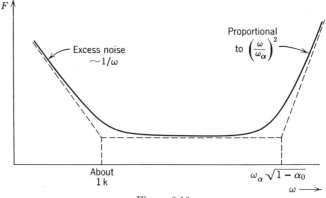

Figure 6.19

The substitution of the last equation in Eq. 6.57 yields

$$F = 1 + \frac{r_b + \dfrac{r_e}{2}}{R_e} + \frac{(r_b + r_e + R_s)^2(1 - \alpha_0)\left[1 + \dfrac{1}{1 - \alpha_0}\left(\dfrac{\omega}{\omega_\alpha}\right)^2\right]}{2\alpha_0 r_e R_s}$$

The variation of the noise figure F with frequency is shown in Fig. 6.19.

Noise Figure in a Common Emitter Amplifier

A similar procedure may be used to calculate the common emitter amplifier noise figure. It can be shown that the noise figure of a common emitter amplifier is the same as that of a common base amplifier.*

6.7 EXPERIMENTAL DETERMINATION OF A NOISE FIGURE

The noise figure of an amplifier can be conveniently determined experimentally by using the setup shown in Fig. 6.20. The amplifier has a source R_s. A diode is connected to the amplifier circuit as shown in the figure. The diode is operated in the temperature limited region so that its dynamic conductance is zero (infinite dynamic resistance). The mean square value of the output noise voltage $v_{n_o}(t)$ is measured first with the diode circuit disconnected. The reading N_{to} gives the total noise power at the output. Next the diode is connected in the circuit as shown in Fig. 6.20. The diode generates the shot noise current with power density spectrum qI_d, where I_d is the d-c current through

* E. G. Nielsen, *op. cit.*

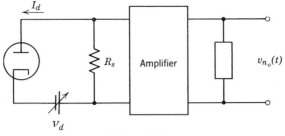

Figure 6.20

the diode. The mean square of the output noise voltage now increases. The voltage V_d is now adjusted until the mean square value of the output noise voltage is $2N_{to}$, twice the previous reading (without diode). The diode current under this condition is read to be I_d. It can be easily shown that the average noise figure $\bar{F}$ of the amplifier is given by

$$\bar{F} = \frac{qI_d R_s}{2kT} \tag{6.61}$$

This can be proved as follows. The diode has a zero dynamic conductance (being in a temperature limited region; see Fig. 6.5) and hence acts as an open circuit. It can therefore be replaced by a noise current source with power density spectrum $S_d(\omega)$ given by (see Fig. 6.6 and Eq. 6.14)

$$S_d(\omega) = qI_d$$

It is evident that addition of this noise source at the input increases the noise power at the output by N_{to}. Hence the noise power at the output due solely to the input noise source of power density qI_d is N_{to}. Now for the amplifier (without diode), the noise source at the input is a resistor R_s which may be represented by a noiseless resistor R_s in parallel with a noise current source (Fig. 6.9b) with a power density $2kTG_s$ ($G_s = 1/R_s$). If a parallel current source of power density qI_d gives rise to a noise power N_{to} at the load, a similar source of power density $2kTG_s$ will give rise to a noise power

$$2kTG_s\left(\frac{N_{to}}{qI_d}\right)$$

at the load. But by definition this is N_{so}, the noise power in the load solely due to the source noise

$$N_{so} = \frac{2kTG_s N_{to}}{qI_d}$$

The average noise figure $\bar{F}$ is by definition

$$\bar{F} = \frac{N_{to}}{N_{so}} = \frac{qI_d}{2kTG_s} = \frac{qI_dR_s}{2kT}$$

Note that this method yields the average noise figure $\bar{F}$. If it is desired to obtain the spectral noise figure F, then the measurement should be performed over the entire frequency range considering a narrowband each time. This can be done by using a tunable narrowband filter at the output. The measurement (Eq. 6.61) now represents the value of F at the center frequency of the filter. To obtain F over the entire frequency range, the narrowband filter is tuned at various frequencies until the desired frequency range is scanned.

6.8 POWER DENSITY AND AVAILABLE POWER DENSITY

In our discussion thus far the power of a signal and power density refer to a normalized load of 1-ohm resistance. The power P of a signal $f(t)$ is defined as the power dissipated by a voltage source $f(t)$ across 1-ohm resistor. If this same voltage $f(t)$ is applied across a resistance of R ohms, the power dissipated will obviously be different. To distinguish these two powers, we shall denote the normalized power (across 1 ohm) by P_n, the power dissipated across resistor R by P_R. It can be easily seen that

$$P_R = \frac{P_n}{R} \tag{6.62}$$

Thus the actual power dissipated by a signal $f(t)$ across a resistor of R ohms is $1/R$ times P_n, the power of the signal $f(t)$ (defined in Eq. 2.21a).

What is true of power is also true of a power density spectrum. The power density represents the power dissipated per unit bandwidth of frequency components of $f(t)$ across a 1-ohm resistor. Hence the power density is a normalized power density. If this same signal $f(t)$ is applied across a resistor R, the actual power density dissipated will be $1/R$ times the power density of $f(t)$

$$\mathcal{S}_R(\omega) = \frac{\mathcal{S}_n(\omega)}{R} \tag{6.63}$$

where $S_n(\omega)$ and $S_R(\omega)$ represent the power density of $f(t)$ (as defined in Eq. 2.22) and the actual power density dissipated by $f(t)$ across R ohms.

Available Power Density

Let us consider a voltage source $f(t)$ with power density $S_f(\omega)$ and an internal impedance $R_s + jX_s$. It is well known that to deliver maximum power to a load, the load impedance must be a complex conjugate of the source impedance (matched load):

$$Z_L = Z_s{}^* = R_s - jX_s$$

The matched condition is shown in Fig. 6.21. It is obvious that in matched condition, $f(t)$ sees a resistance of $2R_s$ ohms, and the actual power density dissipated by $f(t)$ is $S_f(\omega)/2R_s$. Half of this power is dissipated across the source resistance and the remaining half is delivered to the load. Hence the actual power density delivered to the load is $S_f(\omega)/4R_s$. This is the maximum power density that can be extracted from $f(t)$ by an external source. For this reason we shall call this density the available power density and denote it by $S_{av}(\omega)$

$$S_{av}(\omega) = \frac{S_f(\omega)}{4R_s} \tag{6.64}$$

To reiterate, the available power density is the actual power density that can be extracted from a given source. For a voltage source $f(t)$ with an internal impedance $R_s + jX_s$, the available power density is $1/4R_s$ times the (normalized) power density of $f(t)$. In general, R_s is a function of ω and Eq. 6.64 should be expressed as

$$S_{av}(\omega) = \frac{S_f(\omega)}{R_s(\omega)} \tag{6.65}$$

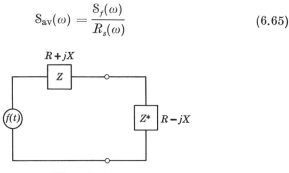

Figure 6.21

It is left as an exercise for the reader to show that for a current source $i(t)$ with power density $S_i(\omega)$ and internal admittance $Y = G_s(\omega) + jB_s(\omega)$, the available power density $S_{av}(\omega)$ is given by

$$S_{av}(\omega) = \frac{S_i(\omega)}{4G_s} \tag{6.66}$$

Available Power Density of an *R-L-C* Network

We shall now find the available power density due to thermal noise from a passive network containing *R-L-C* elements only. Such a network (Fig. 6.22) can be represented by an equivalent impedance $Z_{ab} = R_{ab}(\omega) + jX_{ab}(\omega)$. Using the generalized Nyquist theorem (Eq. 6.34b and Fig. 6.10b), the thermal noise can be represented by a voltage source of power density $2kTR_{ab}(\omega)$ as shown in Fig. 6.22b. To extract the maximum noise power out of this circuit we must use a load $R_{ab} - jX_{ab}$ across terminals ab. The available power density of this source is given by (Eq. 6.65)

$$S_{av}(\omega) = \frac{2kTR_{ab}(\omega)}{4R_{ab}(\omega)}$$

$$= \frac{kT}{2} \tag{6.67}$$

This is a rather startling result. It states that the available (thermal) noise power density from any passive *R-L-C* network is a constant given by $kT/2$. It is tacitly assumed in this discussion that all of the resistors in the network are at the same temperature T.

Note that the maximum noise power that can be extracted from an *R-L-C* network in a bandwidth Δf is given by $2\,\Delta f(kT/2) = kT\,\Delta f$.

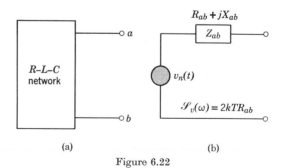

(a)   (b)

Figure 6.22

Thus kT represents the available (thermal) noise power per unit bandwidth for any two-terminal R-L-C network.

6.9 EFFECTIVE NOISE TEMPERATURE

We saw in Section 6.8 that the available power density of any R-L-C two-terminal network is $kT/2$. This result does not hold true for a two-terminal network containing sources of noise other than thermal noise (such as shot noise). Nevertheless, it is possible to extend this result to all two-terminal networks by defining the effective noise temperature T_n. If the available noise power density of any two-terminal network is $S_{av}(\omega)$, we define T_n, the effective noise temperature of the network, as

$$S_{av}(\omega) = \frac{kT_n}{2} \qquad (6.68)$$

or

$$T_n = \frac{2S_{av}(\omega)}{k} \qquad (6.69)$$

If $S_{av}(\omega)$ is constant over the frequency range of interest, T_n, the effective noise temperature, is also a constant given by Eq. 6.69. If, however, $S_{av}(\omega)$ varies with frequency, T_n is a function of frequency ω. The available noise power from any two-terminal network in a bandwidth Δf is $kT_n \, \Delta f$. Note that for a network containing only R-L-C elements, the effective noise temperature T_n is equal to the ambient temperature T of the network.

6.10 NOISE FIGURE IN TERMS OF AVAILABLE GAIN

We have seen that the noise figure can be expressed as the ratio of the signal-to-noise power density spectra at the input divided by that at the output (Eq. 6.47). It is also possible to express the noise figure in terms of available power densities and available power gain. Consider, for example, the arrangement shown in Fig. 6.23. In this figure $S_s(\omega)$ and $S_n(\omega)$ represent the power density spectra of the signal and the noise voltages, respectively. The thermal noise due to source impedance $Z_s(\omega) = R_s + jX_s$ is accounted for by $S_n(\omega)$, and hence $Z_s(\omega)$ in Fig. 6.23 is noiseless. The ratio of signal-to-noise power density

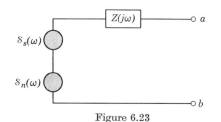

Figure 6.23

spectra across terminals ab will be denoted by (S/N)

$$(S/N) = \frac{S_s(\omega)}{S_n(\omega)} \tag{6.70}$$

Next we calculate the ratio of available power density spectra at terminals ab. Since $s(t)$ and $n(t)$ are independent, the available power density due to each source can be calculated by assuming the other to be zero. Thus the available signal power density at terminals ab is $S_s(\omega)/4R_s$. Similarly, the available noise power density is $S_n(\omega)/4R_s$. Hence $(S/N)_{\text{av}}$, the ratio of available signal-to-noise power densities at terminals ab, is given by

$$\left(\frac{S}{N}\right)_{\text{av}} = \frac{S_s(\omega)}{S_n(\omega)} \tag{6.71}$$

This is a very significant result. It states that the signal-to-noise power density ratio across any terminals is the same as the available (or maximum extractable) signal-to-noise power density ratio across the same terminals. Actually, the result is much more general. It can be easily shown that the signal-to-noise power density ratio across any terminals is equal to the actual signal-to-noise power density ratio dissipated by any load across the same terminals. Equation 6.71 is a special case of this broad result where the load impedance is matched to the source impedance.

We can take advantage of this result in finding the noise figure of an amplifier. The noise figure F of an amplifier is given by (Eq. 6.47) the ratio of signal-to-noise power density (S/N) at the input terminals divided by the same ratio at the output terminals. It is obvious from Eq. 6.71 that the noise figure F can also be expressed as the ratio of available signal-to-noise power densities at the input terminals divided by the same ratio at the output terminals. We shall designate all of the available power densities by the subscript av. Thus $(S_{si})_{\text{av}}$ represents the available signal power density at the input terminals. We now have (Eq. 6.47)

$$F = \frac{(S_{si})_{av}/(S_{nsi})_{av}}{(S_{so})_{av}/(S_{nto})_{av}}$$

$$= \frac{(S_{si})_{av}(S_{nto})_{av}}{(S_{so})_{av}(S_{nsi})_{av}} \tag{6.72}$$

For an amplifier we define an available power gain $\mathcal{G}$ as

$$\mathcal{G} = \frac{\text{Available signal power density at the output}}{\text{Available signal power density at the input}} \tag{6.73a}$$

It is thus apparent that

$$\mathcal{G} = \frac{(S_{so})_{av}}{(S_{si})_{av}} \tag{6.73b}$$

Note that the available power gain $\mathcal{G}$ is generally a function of frequency ω, and hence it may be represented as $\mathcal{G}(\omega)$. Substituting Eq. 6.73 in Eq. 6.72, we get

$$F = \frac{(S_{nto})_{av}}{\mathcal{G}(S_{nsi})_{av}} \tag{6.74}$$

where $(S_{nto})_{av}$ is the available noise power density at the output and $(S_{nsi})_{av}$ is the available noise power due to source at the input terminals. If the source impedance is a passive $R\text{-}L\text{-}C$ two-terminal network, then from Eq. 6.67 we have

$$(S_{nsi})_{av} = \frac{kT}{2} \tag{6.75}$$

If the source impedance contains noise sources other than thermal noise, then

$$(S_{nsi})_{av} = \frac{kT_n}{2} \tag{6.76}$$

If the source impedance is a passive $R\text{-}L\text{-}C$ network, then Eq. 6.74 becomes

$$F = \frac{2(S_{nto})_{av}}{\mathcal{G}kT} \tag{6.77}$$

If the source impedance contains noise sources other than thermal, the temperature T in Eq. 6.77 should be replaced by the effective noise temperature T_n.

From Eq. 6.77, we have

$$(\mathbb{S}_{nto})_{av} = \frac{F\mathcal{G}kT}{2} \tag{6.78}$$

Thus the available noise power density at the output terminals is given by $F\mathcal{G}kT/2$. Note that $(\mathbb{S}_{nto})_{av}$ is composed of two components:

1. The component $(\mathbb{S}_{nso})_{av}$ is the available noise power density at the output terminals due to noise in the source.

2. The component $(\mathbb{S}_{nao})_{av}$ is the available noise power density at the output terminals due to noise generated in the amplifier.

Hence

$$(\mathbb{S}_{nto})_{av} = (\mathbb{S}_{nso})_{av} + (\mathbb{S}_{nao})_{av} \tag{6.79}$$

Moreover, from Eq. 6.73a,

$$(\mathbb{S}_{nso})_{av} = \mathcal{G}(\mathbb{S}_{nsi})_{av} = \frac{\mathcal{G}kT}{2} \tag{6.80}$$

Hence, from Eqs. 6.78 to 6.80, we obtain

$$(\mathbb{S}_{nao})_{av} = (F - 1)\frac{\mathcal{G}kT}{2} \tag{6.81}$$

Thus the available noise power density at the output terminals due to amplifier noise is $(F - 1)(\mathcal{G}kT/2)$.

6.11 CASCADED STAGES

When the amplifier consists of more than one stage, its noise figure can be computed in terms of the noise figures of individual stages. It is intuitively obvious that the noise generated in earlier stages is amplified by later stages, and hence the noise figure of the first stage is much more significant in determining the overall noise figure of the amplifier. As an example, consider a two-stage amplifier as shown in Fig. 6.24. The two amplifiers have available power gains $\mathcal{G}_a$ and $\mathcal{G}_b$, respectively.

In order to determine F_{ab}, the overall noise figure of the cascaded amplifier, we shall first determine $(\mathbb{S}_{nto})_{av}$, the available total noise power at the output terminals. This consists of two components: $\mathbb{S}_1$, the total noise power density available at the output due to the first stage, and $\mathbb{S}_2$, the noise power available at the output due to second stage noise. The component $\mathbb{S}_1$ is obviously $\mathcal{G}_b$ times the total noise

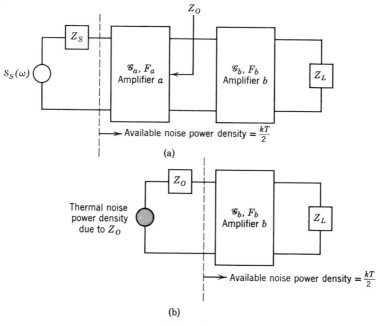

Figure 6.24

power available at the input of the second stage. But the noise power available at the input of the second stage is the total noise power available at the output of the first stage. Hence (Eq. 6.78)

$$S_1 = \mathcal{G}_b \frac{F_a \mathcal{G}_a kT}{2} \tag{6.82}$$

S_2 is also the noise power available at the output due to the second stage alone (Fig. 6.24). The source impedance for the second stage is the impedance seen at the output terminals of the first stage. Let this be Z_o. The noise component S_2 is due to noise sources inside the second amplifier alone, and it is immaterial whether Z_o, the equivalent source impedance, is thermal or not. We shall for convenience assume Z_o to be a thermal impedance and hence its available power density is $kT/2$. Note that this assumption is immaterial in the computation of S_2 as long as the same assumption is made consistently.

Let the noise figure of amplifier b in the condition shown in Fig. 6.24b be F_b. It should be remembered that we are computing F_b with the assumption that Z_o, the source impedance for the amplifier, is thermal. The available noise power density at the output due to amplifier b alone is S_2 and is given by (Eq. 6.81)

$$S_2 = \frac{(F_b - 1)\mathcal{G}_b kT}{2} \tag{6.83}$$

Hence

$$(S_{nto})_{av} = S_1 + S_2$$

$$= \frac{kT}{2} [(F_b - 1)\mathcal{G}_b + F_a \mathcal{G}_a \mathcal{G}_b] \tag{6.84}$$

But from Eq. 6.78,

$$(S_{nto})_{av} = \frac{F_{ab} \mathcal{G}_{ab} kT}{2} \tag{6.85}$$

where $\mathcal{G}_{ab}$ is the available gain of the cascades amplifier. It can be easily seen from the definition of the available gain that the gain of cascaded stages is equal to the product of the gains of individual stages,

$$\mathcal{G}_{ab} = \mathcal{G}_a \mathcal{G}_b$$

Hence

$$(S_{nto})_{av} = \frac{F_{ab} \mathcal{G}_a \mathcal{G}_b kT}{2} \tag{6.86}$$

Comparison of Eqs. 6.84 and 6.86 yields

$$F_{ab} = F_a + \frac{F_b - 1}{\mathcal{G}_a} \tag{6.87}$$

It should be remembered the F_b is the noise figure of amplifier b under the conditions that its source is a passive impedance Z_o equal to the output impedance of the amplifier a.

In general, for a multistage amplifier

$$F = F_a + \frac{F_b - 1}{\mathcal{G}_a} + \frac{F_c - 1}{\mathcal{G}_a \mathcal{G}_b} + \cdots \tag{6.88}$$

It is evident from Eq. 6.88 that the first stage is the most significant in determining the noise figure of an amplifier. Hence in low noise amplifiers the primary consideration in design of the first stage is to obtain a low noise figure even at the cost of gain.

Example 6.7

Find the noise figure of a two-stage amplifier shown in Fig. 6.25. Both tubes are identical, with parameters $g_m = 2.5 \times 10^{-3}$ mhos and $r_p = 10,000$ ohms.

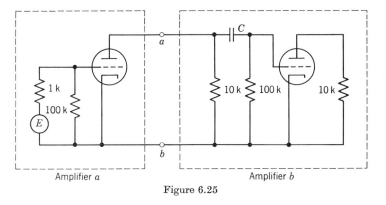

Figure 6.25

Over the frequency range of interest the coupling capacitor may be assumed to be a short circuit. For convenience, we shall arbitrarily divide the amplifier into two stages as shown by the dotted blocks in Fig. 6.25. We shall first find the noise figures F_a and F_b of the two amplifiers. This can be easily obtained from Eq. 6.51. For amplifier a,

$$R_{eq} = \frac{2.5}{2.5} \times 10^{-3} = 1000$$

$$R_s = 1000 \quad \text{and} \quad R_g = 100{,}000 \text{ ohms}$$

Substituting these values in Eq. 6.51, we get

$$F_a = 2.03$$

For amplifier b

$$R_{eq} = \frac{2.5}{2.5} \times 10^{-3} = 1000$$

R_g is the parallel combination of 10,000 ohms and 100,000 ohms. Thus $R_g = 9100$ ohms. The source impedance for amplifier b is the output impedance a and is given by the plate resistance r_p of the first tube. Thus

$$R_s = r_p = 10{,}000 \text{ ohms}$$

Substituting these values in Eq. 6.51, we get

$$F_b = 2.54$$

Next we calculate the available power gain of amplifier a. The available gain $\mathcal{G}$ of an amplifier is given by Eq. 6.73a. Note that at any one frequency the available gain $\mathcal{G}$ can also be expressed as

$$\mathcal{G} = \frac{\text{Available signal power at the load}}{\text{Available signal power at the source}} \tag{6.89}$$

Consider amplifier a, which has a source resistance of 1000 ohms. If the signal voltage is E volts, then the available power is obtained by connecting 1000 ohms across the source terminals. The maximum power available is obviously

$$\left(\frac{E^2}{2000}\right) 1000 = \frac{E^2}{4000} \text{ watts} \tag{6.90}$$

The equivalent circuit of the output terminals of amplifier a is shown in Fig. 6.26.

The equivalent current source has a magnitude $g_m e_g$ where e_g is the voltage at the grid of first tube. From Fig. 6.25 we have

$$e_g = \frac{100}{101} E$$

and

$$g_m e_g = 2.5 \times 10^{-3} \times \frac{100}{101} E \simeq 2.5 \times 10^{-3} E$$

The maximum power available at the load terminals ab is the power delivered to the matched impedance at the load. This occurs when the load at terminals ab is 10,000 ohms and the power output under this condition is

$$\left(\frac{g_m e_g}{2}\right)^2 (10,000) = 1.56 \times 10^{-2} E^2 \tag{6.91}$$

From Eqs. 6.90 and 6.91, we obtain

$$\mathcal{G}_a = \frac{1.56 \times 10^{-2}}{E^2/4000} = 62.4 \tag{6.92}$$

The overall noise figure F_{ab} is given by

$$F_{ab} = F_a + \frac{F_b - 1}{a}$$

$$= 2.03 + \frac{1.54}{62.4}$$

$$= 2.03 + 0.0247$$

$$= 2.0547$$

$$\simeq F_a$$

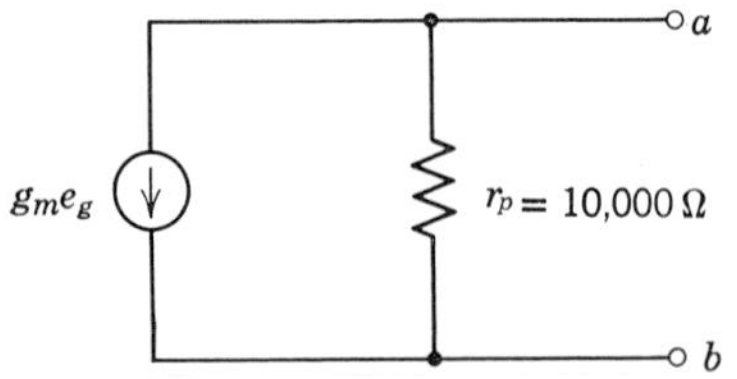

Figure 6.26

6.12 THE CASCODE AMPLIFIER

It is evident from the discussion in Section 6.11 that in a cascaded amplifier, the overall noise figure is primarily determined by the gain of the first stage. The overall noise figure is reduced by increasing the gain of the first stage. In bandpass amplifiers, increasing the voltage gain can lead to instability. Hence the gain cannot be increased beyond a certain limit without risking instability. A careful look at Eq. 6.88, however, shows that the gain $\mathcal{G}_a$ is not the voltage gain of the stage but available power gain. It is possible to design an amplifier which has a very high available power gain and at the same time has a low voltage gain. Such a design of the first stage reduces the overall noise figure without risking instability. This is the principle of the cascode amplifier. In a cascode amplifier, the first stage is designed to yield a very high available power gain, whereas the voltage gain is very small.*

APPENDIX. PROOF OF THE GENERALIZED NYQUIST THEOREM

Consider a linear bilateral passive two-terminal network (Fig. A6.1a) containing n resistors $R_1, R_2, \ldots, R_n$. Each of these resistors will be represented by a noiseless resistor in parallel with its thermal noise current source as shown in Fig. A6.1b. Let $H_k(\omega)$ be the transfer function relating $v_{ab}(t)$, the voltage across the terminals ab, to the current source $i_k(t)$. The power density spectrum of the current source $i_k(t)$ is

$$2kTG_k = \frac{2kT}{R_k}$$

Since all the current sources are independent, $S_0(\omega)$, the power density spectrum of the output voltage across terminals ab, is given by (see Section 6.4)

$$S_0(\omega) = \sum_{k=1}^{n} |H_k(\omega)|^2 \frac{2kT}{R_k}$$

$$= 2kT \sum_{k=1}^{n} \frac{|H_k(\omega)|^2}{R_k} \tag{A6.1}$$

* For an example of a cascode amplifier using vacuum tubes, refer to J. M. Pettit and M. M. McWhorter, *Electronic Amplifier Circuits*, McGraw-Hill, New York, 1961.

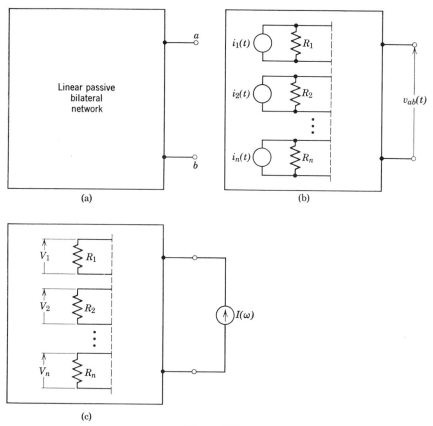

Figure A6.1

We shall now show that if $Z_{ab}(\omega)$ is the impedance seen across the terminals ab, then

$$R_{ab}(\omega) = Re\,[Z_{ab}(\omega)] = \sum_{k=1}^{n} \frac{|H_k(\omega)|^2}{R_k} \qquad (A6.2)$$

It will then follow that

$$S_o(\omega) = 2kTR_{ab}(\omega) \qquad (A6.3)$$

This is the statement of the generalized Nyquist theorem.

To prove Eq. A6.2, we consider the same network and apply a sinusoidal current of amplitude I and frequency ω across terminals ab (Fig. A6.1c). This current will be denoted by $I(\omega)$. Let the sinusoidal voltage appearing across resistor R_k pue to this current be $V_k(\omega)$ as shown in Fig. A6.1c. Since the system is linear and bilateral, the reciprocity theorem applies, and we have

$$\frac{V_k(\omega)}{I(\omega)} = H_k(\omega) \qquad (A6.4)$$

The net power dissipated in the resistor R_k is therefore

$$P_k = \frac{1}{2} \frac{|V_k(\omega)|^2}{R_k}$$

The total power P_t dissipated in the network is therefore

$$P_t = \frac{1}{2} \sum_{k=1}^{n} \frac{|V_k(\omega)|^2}{R_k}$$

Using Eq. A6.4 in this equation, we have

$$P_t = \frac{1}{2} \sum_{k=1}^{n} \frac{|H_k(\omega)|^2}{R_k} |I(\omega)|^2$$

$$= \tfrac{1}{2}|I(\omega)|^2 \sum_{k=1}^{n} \frac{|H_k(\omega)|^2}{R_k}$$

The total power dissipated in the network is equal to the total power supplied by $I(\omega)$ to the network. But the power supplied by $I(\omega)$ is given by

$$\tfrac{1}{2}|I(\omega)|^2 R_{ab}(\omega)$$

where $R_{ab}(\omega) = Re[Z_{ab}(\omega)]$. Evidently

$$\sum_{k=1}^{n} \frac{|H_k(\omega)|^2}{R_k} = R_{ab}(\omega) = Re\,[Z_{ab}(\omega)]$$

Thus we have proved Eq. A6.2 and hence the generalized Nyquist theorem.

PROBLEMS

1. Derive the root mean square value of the noise current in an R-L circuit

2. Determine the power density spectrum of the noise current flowing through a series R-L-C circuit.

3. Two resistors, each of 1000 ohms, are at temperatures 300 and 400°K, respectively. Find the voltage power density spectrum at the terminals formed by the series combination of these resistors.

4. Repeat Problem 3 if the resistors are connected in parallel.

5. Determine the root mean square noise voltage across the output terminals of a noiseless circuit (Fig. P-6.5) when 1 k resistance is connected across the input terminals and if the voltage gain transfer function of the circuit represents the following:

(a) An ideal low-pass filter with cutoff frequency f_c Hz.

(b) An ideal bandpass filter with a passband of f_c Hz centered around f_0.

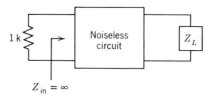

Figure P-6.5

(c) An exponential filter, $H(\omega) = Ae^{-|\omega|/\omega_c}$.

(d) A gaussian filter, $H(\omega) = Ae^{-\omega^2/\omega_c^2}$

Assume the input impedance of the circuit to be infinite. Neglect the noise contribution due to the load impedance Z_L.

6. Repeat Problem 5 if, in addition to a 1000-ohm resistor, a diode is connected across the input terminals as shown in Fig. P-6.6. Assume the

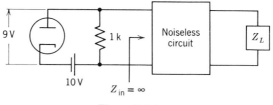

Figure P-6.6

diode to be in the space-charge limited region with a dynamic resistance of 6670 ohms and the cathode temperature to be 1000°K.

7. (a) Determine the power density spectrum of the noise voltage across terminals aa' of the resistive network in Fig. P-6.7 by the following two methods.

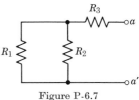

(1) By calculating the noise power density spectrum across aa' as the sum of the noise power density spectra across aa' due to each of the three resistors.

Figure P-6.7

(2) By calculating the equivalent resistance $R_{aa'}$ across aa' and finding the required power density spectrum $2kTR_{aa'}$.

(b) Determine the power density spectrum of the noise voltage across terminals aa' if the resistors R_1, R_2, and R_3 are at different temperatures T_1, T_2, and T_3, respectively.

8. Determine the noise power density spectrum and the rms noise voltage across the output terminals aa' of the low-pass filter shown in Fig. P-6.8. This filter is known as the third-order Butterworth filter.

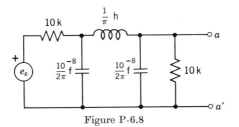

Figure P-6.8

9. Find the root mean square noise voltage across terminals bb' for each of the networks shown in Fig. P-6.9 by the following two methods:

(a) By calculating the power density spectrum at bb' as the sum of the power density spectra due to each individual resistor.

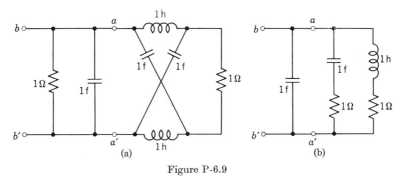

Figure P-6.9

(b) By replacing the network to the right of the terminals aa' by an equivalent immitance and using Eq. 6.34.

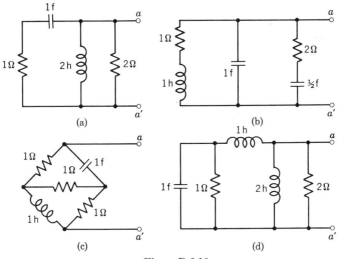

Figure P-6.10

10. Determine the power density spectrum of the noise voltage across terminals aa' of the circuits shown in Fig. P-6.10.

11. Show that the equivalent noise bandwidth of the Butterworth filter in Problem 8 referred to d-c ($f_0 = 0$) is ($10^4 \pi/3$) Hz.

12. (a) Calculate the equivalent noise bandwidth of the circuits referred to zero frequency if it is given that

$$H(\omega) = (1) \quad Ae^{-\alpha|\omega|}$$
$$= (2) \quad Ae^{-\alpha\omega^2}$$
$$= (3) \quad Sa(\omega t_0)$$
$$= (4) \quad G_{2W}(\omega)$$

(b) Calculate the equivalent noise bandwidth referred to ω_0 if

$$H(\omega) = Sa[(\omega - \omega_0)t_0] + Sa[(\omega + \omega_0)t_0]$$

13. A third-order Butterworth filter described in Problem 8 is connected at the output terminals of a triode amplifier (Fig. P-6.13). The source resistance is 1000 ohms. Determine the rms value of the noise voltage

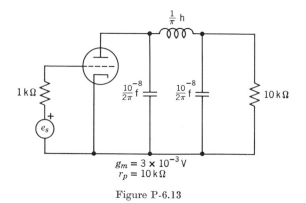

$$g_m = 3 \times 10^{-3} \, \text{V}$$
$$r_p = 10 \, \text{k}\Omega$$

Figure P-6.13

across the load. Evaluate the spectral and average noise figures of the amplifier. It is given that $g_m = 3000 \, \mu$mhos and $r_p = 10{,}000$ ohms.

14. Find the noise figure of the cascaded amplifier shown in Fig. 6.25 by a straightforward method using the definition of noise figure in Eq. 6.44.

15. Show that the noise figures of grounded-cathode, grounded-grid, and grounded-plate vacuum tube amplifiers are approximately the same.

16. Show that if the effective noise temperature of the source is constant, independent of frequency, then the average noise figure $\bar{F}$ of an amplifier

is given by

$$\bar{F} = \frac{\displaystyle\int_0^\infty F \mathcal{G}_a \, df}{\displaystyle\int_0^\infty \mathcal{G}_a \, df}$$

where $\mathcal{G}_a$ is the available power gain of the amplifier.

17. Determine the noise figure of the amplifier in Example 6.4 in the text, using Eq. 6.77 or Eq. 6.81.

chapter 7

Performance of Communication Systems

In Chapters 3 through 5 various methods of signal conditioning (modulation) were discussed. It was observed that some systems were less immune to noise interference than others. This desirable attribute required a larger bandwidth to transmit the signals. So far these conclusions have been based entirely on qualitative grounds. In this chapter, the quantitative relationship between the signal-to-noise ratio (the measure of noise immunity of the system) and the corresponding bandwidth of transmission for various systems will be derived. It will be shown that conditioning a signal so that it occupies a larger bandwidth generally makes it more immune to external noise interference.

We shall begin by discussing a method of representing noise signals.

7.1 BANDPASS NOISE REPRESENTATION

In earlier chapters, we observed that after modulation the signals have a bandpass spectrum. These signals in the process of transmission are corrupted by a wideband noise (usually white noise). At the receiver, the first obvious step is to filter the incoming signal for any noise that lies out of the useful signal band. The output of this bandpass filter is the useful signal (in a modulated form) and the bandpass noise. The output of the demodulator consists of the desired signal plus the noise

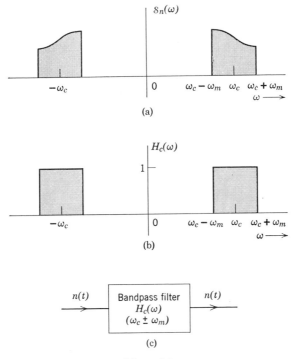

Figure 7.1

due to bandpass noise present at the demodulator input. Since, in general, we have to deal with a bandpass noise in evaluating the noise at the output, we shall develop a form of representation of a bandpass noise signal, which will be particularly convenient.

Let us consider a bandpass noise $n(t)$ with a power density spectrum $S_n(\omega)$ as shown in Fig. 7.1. We shall now show that a random bandpass noise signal $n(t)$ can be expressed as

$$n(t) = n_c(t) \cos \omega_c t + n_s(t) \sin \omega_c t \qquad (7.1)$$

where signals $n_c(t)$ and $n_s(t)$ are low frequency signals bandlimited to ω_m radians per second and the powers (mean square values) of $n(t)$, $n_c(t)$, and $n_s(t)$ are identical; that is

$$\overline{n^2(t)} = \overline{n_c{}^2(t)} = \overline{n_s{}^2(t)} \qquad (7.2)$$

To prove this result, we observe that transmission of $n(t)$ through an ideal bandpass filter shown in Fig. 7.1b would preserve the signal $n(t)$

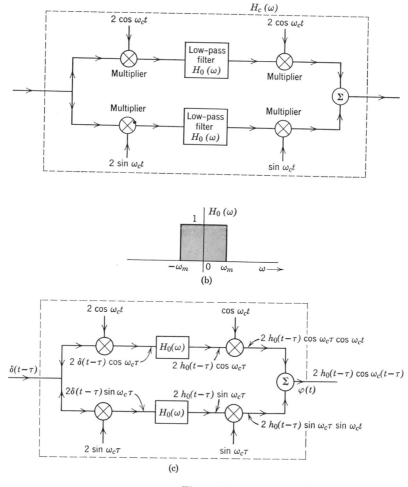

Figure 7.2

at the output. Thus the signal $n(t)$ can be transmitted through such a filter without any change as shown in Fig. 7.1c.

In the next step we show that the ideal bandpass filter $H_c(\omega)$ in Fig. 7.1b can be realized by the arrangement shown in Fig. 7.2a. The filter $H_0(\omega)$ in Fig. 7.2a is an ideal low-pass filter with cutoff frequency ω_m as shown in Fig. 7.2b. The proof is quite simple. Let us apply a delayed unit impulse $\delta(t - \tau)$ at the input of the system in Fig. 7.2a. We shall now show that the output will be given by $h_o(t - \tau) \cos \omega_c(t - \tau)$ where $h_o(t)$ is the unit impulse response of the low-pass filter in Fig. 7.2b.

The system in Fig. 7.2a is reproduced in Fig. 7.2c with $\delta(t - \tau)$ as the input signal. We observe that

$$\delta(t - \tau) \cos \omega_c t = \delta(t - \tau) \cos \omega_c \tau$$

and

$$\delta(t - \tau) \sin \omega_c t = \delta(t - \tau) \sin \omega_c \tau$$

It is therefore obvious that the inputs to the upper and lower low-pass filters in Fig. 7.2c are $2 \delta(t - \tau) \cos \omega_c \tau$ and $2 \delta(t - \tau) \sin \omega_c \tau$, respectively. If $h_o(t)$ is the impulse response of either of the low-pass filters, then $h_o(t - \tau)$ is the response of these filters to $\delta(t - \tau)$. Obviously, the outputs of these filters will be

$$2h_o(t - \tau) \cos \omega_c \tau \qquad \text{and} \quad 2h_o(t - \tau) \sin \omega_c \tau,$$

respectively. The outputs of these filters are further multiplied by $\cos \omega_c t$ and $\sin \omega_c t$, respectively, and finally summed. It is obvious that the final output signal $\varphi(t)$ is given by

$$\varphi(t) = 2h_o(t - \tau)(\cos \omega_c \tau \cos \omega_c t + \sin \omega_c \tau \sin \omega_c t)$$
$$= 2h_o(t - \tau) \cos \omega_c(t - \tau) \tag{7.3}$$

We shall now show that the response of the ideal bandpass filter in Fig. 7.1b to $\delta(t - \tau)$ is identical to $\varphi(t)$ in Eq. 7.3. It can be easily seen from the modulation theorem (Eq. 1.116a) that if the impulse response of $H_0(\omega)$ in Fig. 7.2b is $h_o(t)$, then the impulse response of $H_c(\omega)$ (Fig. 7.1b) is given by $h_c(t)$ where

$$h_c(t) = 2h_o(t) \cos \omega_c t$$

Obviously, the response of this filter to $\delta(t - \tau)$ will be* given by

$$2h_o(t - \tau) \cos \omega_c(t - \tau)$$

This is identical to the response $\varphi(t)$ in Eq. 7.3. Hence the system in Fig. 7.2a is indeed equivalent to an ideal bandpass filter.

If we apply a bandpass noise signal $n(t)$ (Fig. 7.1a) at the input of the ideal bandpass system in Fig. 7.3, the output must be $n(t)$. If we denote the outputs of the upper and the lower bandpass filters in Fig. 7.3 by $n_c(t)$ and $n_s(t)$ respectively, then it is evident that

$$n(t) = n_c(t) \cos \omega_c t + n_s(t) \sin \omega_c t \tag{7.4}$$

---

* Here we compare the response of the two systems to the delayed impulse $\delta(t - \tau)$ rather than the impulse $\delta(t)$ to ascertain the time invariance of the system in Fig. 7.2a. Since the system in Fig. 7.2a is composed of time varying elements, the comparison of impulse response can be valid only after ascertaining that the system is time invariant. The comparison of the delayed impulse (for all values of τ) assures the equivalence.

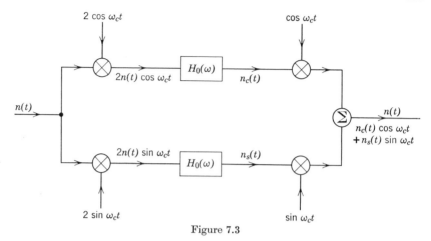

Figure 7.3

The signals $n_c(t)$ and $n_s(t)$ are the outputs of low-pass (bandlimited to ω_m) filters. Hence $n_c(t)$ and $n_s(t)$ are low frequency signals each bandlimited to ω_m radians per second. Next we shall find the power density spectra of $n_c(t)$ and $n_s(t)$. It can be seen from Fig. 7.3 that $n_c(t)$ is the response of the low-pass filter $H_0(\omega)$ to the input signal $2n(t) \cos \omega_c t$. If $S_n(\omega)$ is the power density of $n(t)$, then the power density of $n(t) \cos \omega_c t$ is given by* (Eq. 2.24b)

$$\tfrac{1}{4}[S_n(\omega + \omega_c) + S_n(\omega - \omega_c)] \tag{7.5}$$

and the power density of $2n(t) \cos \omega_c t$ is obviously

$$(4)\tfrac{1}{4}[S_n(\omega + \omega_c) + S_n(\omega - \omega_c)] = S_n(\omega + \omega_c) + S_n(\omega - \omega_c) \tag{7.6}$$

The spectra of $S_n(\omega)$, $S_n(\omega + \omega_c)$, $S_n(\omega - \omega_c)$, and $[S_n(\omega + \omega_c) + S_n(\omega - \omega_c)]$ are shown in Fig. 7.4. The spectrum of $2n(t) \cos \omega_c t$, $[S_n(\omega + \omega_c) + S_n(\omega - \omega_c)]$ (Fig. 7.4d), is passed through an ideal low-pass filter $H_0(\omega)$ (Fig. 7.3) which suppressed all frequencies beyond $|\omega| > \omega_m$. The resulting spectrum is the power density spectrum of $n_c(t)$, shown in Fig. 7.4e. It is obvious that $S_{n_c}(\omega)$, the power density spectrum of $n_c(t)$, is given by

$$S_{n_c}(\omega) = \begin{cases} S_n(\omega + \omega_c) + S_n(\omega - \omega_c) & |\omega| < \omega_m \\ 0 & |\omega| > \omega_m \end{cases} \tag{7.7}$$

Similarly, it can be seen from Eq. 2.24c and Fig. 7.3 that $S_{n_s}(\omega)$, the power density spectrum of $n_s(t)$, is identical to $S_{n_c}(\omega)$.

$$S_{n_c}(\omega) = S_{n_s}(\omega) \tag{7.8a}$$

* This result is true only if $n(t)$ is a random signal. If $n(t)$ is not a random signal, there is a possibility of additional spectrum around $\omega = 0$.

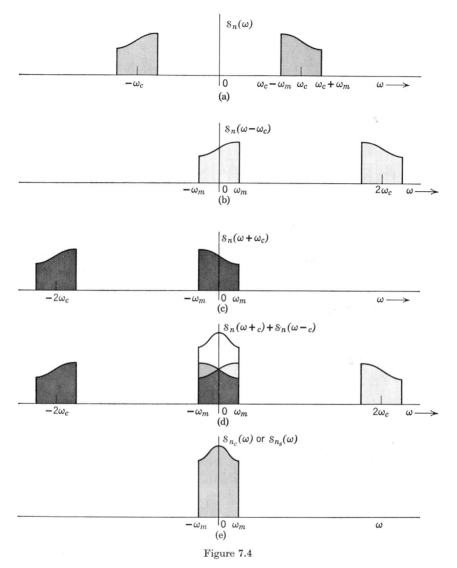

Figure 7.4

Note that when $S_n(\omega)$ is a white noise of power density $\mathcal{N}/2$,

$$S_{n_c}(\omega) = S_{n_s}(\omega) = \begin{cases} \mathcal{N} & |\omega| < \omega_m \\ 0 & \text{otherwise} \end{cases} \tag{7.8b}$$

Next we determine the powers (mean square values) of $n(t)$, $n_c(t)$, and $n_s(t)$. The power of a signal is given by $1/2\pi$ times the area under its power density spectrum. It can be seen from Fig. 7.4 that the area under $S_n(\omega)$ is equal to the area under $S_{n_c}(\omega)$ or $S_{n_s}(\omega)$. This is because $S_{n_c}(\omega)$ is formed by addition of both spectra (positive and negative) of $S_n(\omega)$. Hence it follows that

$$\overline{n^2(t)} = \overline{n_c^2(t)} = \overline{n_s^2(t)} \tag{7.9}$$

To summarize, a bandpass random noise signal $n(t)$ (bandwidth $2\omega_m$) can be expressed as

$$n(t) = n_c(t) \cos \omega_c t + n_s(t) \sin \omega_c t \tag{7.10}$$

where $n_c(t)$ and $n_s(t)$ are both low-pass noise signals bandlimited to ω_m radians per second. The mean square values of $n(t)$, $n_c(t)$, and $n_s(t)$ are identical:*

$$\overline{n^2(t)} = \overline{n_c^2(t)} = \overline{n_s^2(t)} \tag{7.11}$$

The power density spectra of $n_c(t)$ and $n_s(t)$ are identical and are related to the power density spectrum of $n(t)$ as in Eq. 7.7 (and Eq. 7.8).

Note that the power of the noise component $n_c(t) \cos \omega_c t$ is $\overline{n_c^2}/2$ and that of $n_s(t) \sin \omega_c t$ is $\overline{n_s^2}/2$ (see Eq. 2.25). Hence the power of the noise signal is equally divided between the two quadrature components.

We can express Eq. 7.10 in alternate form

$$n(t) = R(t) \cos [\omega_c t + \theta(t)] \tag{7.12}$$

where

$$R(t) = \sqrt{n_c^2(s) + n_s^2(t)} \tag{7.13a}$$

and

$$\theta(t) = -\tan^{-1} \left[\frac{n_s(t)}{n_c(t)} \right] \tag{7.13b}$$

Since both $n_c(t)$ and $n_s(t)$ are slowly varying signals, it follows from Eq. 7.13 that $R(t)$ and $\theta(t)$ are also slowly varying signals. The bandpass

---

* It should be once again stressed that these results generally apply to bandpass random noise signals. They cannot be applied indiscriminately to any bandpass signal. As an example, consider a low-pass signal $f(t)$ bandlimited to ω_m radians per second. Then the signal $\varphi(t) = f(t) \cos \omega_c t$ is a bandpass signal. If this signal is represented as in Eq. 7.10,

$$\varphi(t) = \varphi_c(t) \cos \omega_c t + \varphi_s(t) \sin \omega_c t$$

it is obvious that $\varphi_c(t) = f(t)$ and $\varphi_s(t) = 0$. Thus Eq. 7.11 does not hold. Here the signal $\varphi(t) = f(t) \cos \omega_c t$ is not a stationary random signal (even if $f(t)$ is stationary random). In such cases Eq. 7.5 is not valid. In general, any bandpass signal can be represented as in Eq. 7.10 but the relation 7.11 is valid only for stationary random bandpass signals.

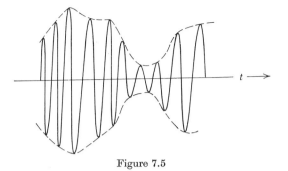

Figure 7.5

noise signal with a narrowband spectrum shown in Fig. 7.1 has the appearance of a sinusoidal signal with amplitude and phase varying slowly as shown in Fig. 7.5. It is evident from Eq. 7.12 that the envelope of this signal is given by $R(t)$ and the phase is $\theta(t)$. Note that this signal has both amplitude and angle modulation.

7.2 NOISE CALCULATIONS IN COMMUNICATION SYSTEMS

We shall now investigate the effect of different forms of modulation on the signal-to-noise ratio. It was observed earlier on a qualitative basis that the wideband communication systems usually show more immunity to noise interference. If a signal is transformed so that it occupies a larger bandwidth, it becomes more immune to noise. In other words, the signal-to-noise power ratio increases. In fact, in Chapter 8 we shall show on theoretical basis that it is possible to exchange bandwidth for signal-to-noise ratio. It will be shown that a given amount of information can be transmitted by various combinations of bandwidths and signal-to-noise ratios. If the bandwidth is reduced, we must transmit a correspondingly larger signal power (larger signal-to-noise ratio). On the other hand, if we have a larger bandwidth available for transmission, the same information can be transmitted by a smaller signal power (smaller signal-to-noise ratio). The exchange between bandwidth and signal-to-noise ratio is accomplished by various forms of modulation. We shall now study the relationship between the bandwidth of transmission and signal-to-noise ratio in various modulation systems.

If there were no noise over the channel, the modulated signal would have infinite signal-to-noise ratio as seen at the demodulator input.

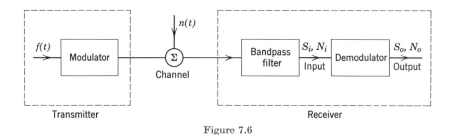

Figure 7.6

However, because of nonzero channel noise, the signal-to-noise ratio of a modulated signal is finite. The demodulator transforms the modulated signal into the modulating signal plus noise. The modulated signal at the input of the demodulator and the modulating signal at the output of the demodulator will have different bandwidths and signal-to-noise ratios. The demodulator therefore performs the task of exchanging the bandwidth for signal-to-noise ratio. We shall now study the nature of this exchange in various modulation systems.

We shall consider the case of additive noise, that is, noise assumed to interfere with the signal by simple addition. The appropriate model is shown in Fig. 7.6. Note that at the receiver the incoming signal is filtered for any noise that may lie outside the useful signal spectrum. Thus the bandwidth of the noise at the demodulator input is the same as that of the modulated signal.

For each system, we shall compare the signal-to-noise power ratio at the input of the demodulator to that at the output of the demodulator. We shall denote S_i and S_o as the power (the mean square value) of the useful signal at the input and the output, respectively, of the demodulator. Similarly N_i and N_o will represent the power (mean square value) of the interfering noise signal at the input and the output of the demodulator. Let us now examine the various forms of modulation.

7.3 NOISE IN AMPLITUDE-MODULATED SYSTEMS

I. DSB-SC

The receiver for DSB-SC is shown in Fig. 3.1f and is reproduced in Fig. 7.7.* Let $f(t)$ be the message signal bandlimited to ω_m radians per

* The use of an input bandpass filter in this case is redundant because any noise components outside the useful signal band are eventually suppressed by the final baseband filter. Hence this filter may be eliminated.

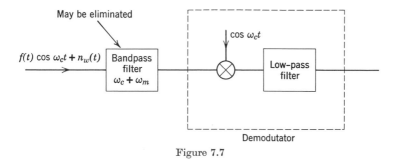

Figure 7.7

second ($\omega_m < \omega_c$, the carrier frequency). The useful signal at the demodulator input is $f(t) \cos \omega_c t$. We have shown (Eq. 2.25b) that S_i, the power (the mean square value) of $f(t) \cos \omega_c t$, is given by half the power of $f(t)$

$$S_i = \overline{[f(t) \cos \omega_c t]^2} = \tfrac{1}{2}\overline{f^2(t)} \tag{7.14}$$

The useful signal power at the output is S_o, which can be found easily by observing the fact that the synchronous detector (for DSB-SC) output is $\tfrac{1}{2}f(t)$ (see Eq. 3.2). Hence

$$S_o = \overline{[\tfrac{1}{2}f(t)]^2} = \tfrac{1}{4}\overline{f^2(t)}$$
$$= \tfrac{1}{2}S_i \tag{7.15}$$

To calculate N_i and N_o, the noise powers at the input and the output of the demodulator, we use the bandpass noise model in Eq. 7.10, for $n_i(t)$, the noise signal at the input of the demodulator:

$$n_i(t) = n_c(t) \cos \omega_c t + n_s(t) \sin \omega_c t \tag{7.16}$$

where $n_i(t)$ is the bandpass noise at the input of the demodulator and N_i is given by

$$N_i = \overline{n_i{}^2(t)} \tag{7.17}$$

If $n_i(t)$ is applied at the input of a synchronous detector (which multiplies the incoming signal by $\cos \omega_c t$), then $n_d(t)$, noise output of the demodulator is given by $n_i(t) \cos \omega_c t$.

$$n_d(t) = n_i(t) \cos \omega_c t$$
$$= n_c(t) \cos^2 \omega_c t + n_s(t) \sin \omega_c t \cos \omega_c t$$
$$= \tfrac{1}{2}[n_c(t) + n_c(t) \cos 2\omega_c t + n_s(t) \sin 2\omega_c t]$$

The terms $n_c(t) \cos 2\omega_c t$ and $n_s(t) 2\omega_c t$ represent the spectra of $n_c(t)$ and $n_s(t)$ shifted at $\pm 2\omega_c$ and are filtered out by the low-pass filter at the demodulator output (see Fig. 7.7). Hence $n_o(t)$, the final noise output, is $\frac{1}{2} n_c(t)$.

$$n_o(t) = \tfrac{1}{2} n_c(t)$$

and

$$N_o = \overline{n_o^2(t)} = \tfrac{1}{4} \overline{n_c^2(t)}$$

Use of Eqs. 7.11 and 7.17 now yields

$$N_o = \tfrac{1}{4} \overline{n_i^2(t)} = \tfrac{1}{4} N_i \tag{7.18}$$

From Eqs. 7.15 and 7.18, we obtain

$$\frac{S_o/N_o}{S_i/N_i} = 2 \tag{7.19}$$

Hence for DSB-SC, the signal-to-noise ratio at the output of the demodulator is twice that at the input of the demodulator. This represents an improvement in the S/N ratio by a factor of 2 for DSB-SC systems.

The improvement factor 2 for DSB-SC can be explained qualitatively. Since the noise signal is random it has both sine and cosine components (Eq. 7.10). The demodulator multiplies these components by $\cos \omega_c t$. This causes the sine component to shift to twice the frequency ($\sin \omega_c t \cos \omega_c t = \frac{1}{2} \sin 2\omega_c t$) and is completely eliminated by a low-pass filter. Thus half the noise power (in sine component) is eliminated. This results in the signal-to-noise power ratio improvement of 2.

2. SSB-SC

The receiver for SSB-SC is identical to that for DSB-SC in all respects except the input bandpass filter (see Fig. 3.20). For DSB-SC, the filter must accept all frequencies in the range $\omega_c \pm \omega_m$. For SSB-SC, on the other hand, the incoming signal has only one sideband and hence the bandwidth of the filter will be half that of the filter in DSB-SC.

The spectrum of the input signal and the output signal at the demodulator of SSB-SC is given in Fig. 3.20 and is reproduced in Fig. 7.8. The spectrum of $f(t)$ is $F(\omega)$ shown in Fig. 7.8a. When $f(t)$ is applied at the input of the system in Fig. 3.18, the output signal is SSB-SC with spectrum as shown in Fig. 7.8b (see Eq. 3.27). This is $f_i(t)$, the signal

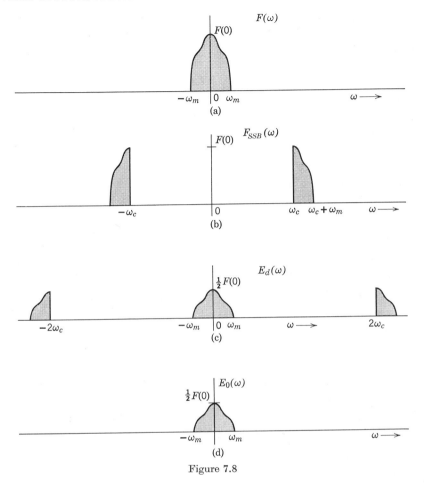

Figure 7.8

at the demodulator input. The spectrum of the demodulator output is shown in Fig. 7.8c (see Eq. 3.31). This output is passed through a low-pass filter to filter out the spectrum at $\pm 2\omega_c$. The final output spectrum is shown in Fig. 7.8d.

The power of a signal (the mean square value) is $1/2\pi$ times the area under the power density spectrum. From Fig. 7.8a and b it is obvious that the area under the power density of $f(t)$ and $f_{\text{SSB}}(t)$ are identical. Hence the power of the SSB modulated signal (Fig. 7.8b) and the original signal are identical. Thus S_i, the signal power at the demodulator, is given by

$$S_i = \overline{f^2(t)} \qquad (7.20)$$

The demodulator output is $\frac{1}{2}f(t)$, as seen from Eq. 3.31. Hence $S_o(t)$, the signal power at the demodulator output, is given by

$$S_o = [\tfrac{1}{2}f(t)]^2$$
$$= \tfrac{1}{4}\overline{f^2(t)}$$
$$= \tfrac{1}{4}S_i$$

Hence

$$\frac{S_o}{S_i} = \frac{1}{4} \tag{7.21}$$

To calculate N_i and N_o, we observe that the synchronous demodulators for DSB and SSB are identical. The demodulation in each case is accomplished by multiplying the incoming signal by $\cos \omega_c t$ and filtering the product through a low-pass filter to eliminate the spectrum at $\pm 2\omega_c$. The noise input in both DSB and SSB are narrowband noise signals. Obviously, the ratio N_o/N_i must be identical for both DSB and SSB. Hence from Eq. 7.18, we have

$$\frac{N_o}{N_i} = \frac{1}{4} \tag{7.22}$$

From Eq. 7.21 and 7.22, we have for SSB-SC systems

$$\frac{S_o/N_o}{S_i/N_i} = 1 \tag{7.23}$$

We therefore conclude that the signal-to-noise power ratios at the input and the output of the demodulator are identical. Hence there is no improvement in the signal-to-noise ratio. Superficially, this will give the impression that DSB is superior to SSB because DSB has twice the signal-to-noise improvement ratio of SSB. Close examination, however, shows that this conclusion is unwarranted. The SSB signal requires only half the bandwidth of that required for DSB. Hence N_i, the input noise power, is in the case of DSB twice* that of SSB. Although the S/N ratio in DSB improves by a factor 2, it has twice the noise power at the input to start with. The improvement in the demodulation is

---

* If the noise power density is symmetrical about $\omega = \omega_c$ (as in the case of white noise) the input noise power in DSB is exactly twice that in SSB. For nonsymmetrical noise power density, this factor may be greater than or less than 2 depending upon which sideband is suppressed (see Problem 5). However, the modulated signal bandwidth is usually so small that a uniform noise power density over the band may be assumed without significant error.

obviously nullified by the larger input noise. It can easily be seen that for a given signal power at the input, the signal-to-noise ratio at the output is identical for DSB and SSB. Hence the performance of SSB is identical to that of DSB from the point of view of noise improvement.

3. AM with Carrier (Envelope Detector)

For the case of AM with large carrier, the input to the demodulator is $f_i(t)$, given by

$$f_i(t) = [A + f(t)] \cos \omega_c t + n_i(t) \tag{7.24}$$

The signal in this case is $[A + f(t)] \cos \omega_c t$. It is obvious that S_i and N_i, the signal power and the noise power at the input of the detector, are given by

$$S_i = \frac{A^2}{2} + \frac{\overline{f^2(t)}}{2} \tag{7.25a}$$

$$N_i = \overline{n_i{}^2(t)} \tag{7.25b}$$

To compute S_o and N_o, we find the envelope of $f_i(t)$. Using the representation in Eq. 7.10 for the narrowband noise in Eq. 7.24, we get

$$f_i(t) = [A + f(t)] \cos \omega_c t + n_c(t) \cos \omega_c t + n_s(t) \sin \omega_c t$$

$$= [A + f(t) + n_c(t)] \cos \omega_c t + n_s(t) \sin \omega_c t \tag{7.26}$$

$$= E(t) \cos [\omega_c t + \psi(t)] \tag{7.27}$$

where

$$E(t) = \sqrt{[A + f(t) + n_c(t)]^2 + n_s{}^2(t)} \tag{7.28a}$$

and

$$\psi(t) = -\tan^{-1}\left[\frac{n_s(t)}{A + f(t) + n_c(t)}\right] \tag{7.28b}$$

It is evident that $E(t)$ is the envelope of $f_i(t)$ and $\psi(t)$ is the phase angle. The output of the envelope detector is obviously $E(t)$. We shall now consider the two cases where (a) small noise; $A + f(t) \gg n_i(t)$, and (b) large noise; $n_i(t) \gg A + f(t)$.

a. Small Noise Case. We can use phasors to represent $f_i(t)$ in Eq. 7.26 (Fig. 7.9). If $A + f(t) \gg n_i(t)$, then $A + f(t) \gg n_c(t)$ and $n_s(t)$. The resultant $E(t)$ in this case can be approximated by $A + f(t) + n_c(t)$

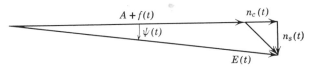

Figure 7.9

as shown in Fig. 7.9.

$$E(t) \simeq A + f(t) + n_c(t)$$

and

$$\psi(t) \simeq 0$$

We come to the same conclusion analytically. If $A + f(t) \gg n_c(t)$ and $n_s(t)$ both,* then Eq. 7.28a can be approximated as

$$E(t) \simeq \sqrt{[A + f(t)]^2 + 2[A + f(t)]n_c(t)}$$

$$= [A + f(t)]\left[1 + \frac{2n_c(t)}{A + f(t)}\right]^{1/2}$$

$$\simeq [A + f(t)]\left[1 + \frac{n_c(t)}{A + f(t)}\right]$$

$$= A + f(t) + n_c(t)$$

It is evident from this equation that the useful signal in the output is $f(t)$ and the noise is $n_c(t)$. Hence

$$S_o = \overline{f^2(t)} \tag{7.29a}$$

and

$$N_o = \overline{n_c^2(t)} = N_i \tag{7.29b}$$

Using Eqs. 7.25a, 7.25b, 7.29a, and 7.29b, we get

$$\frac{S_o/N_o}{S_i/N_i} = \frac{2\overline{f^2(t)}}{A^2 + \overline{f^2(t)}} \tag{7.30}$$

The improvement ratio increases as A is reduced. But for the envelope detector, A can not be reduced below $|f(t)|_{\max}$

$$A \geqslant |f(t)|_{\max}$$

---

* This statement should be interpreted with some caution. Since the signals $n_c(t)$ and $n_s(t)$ are random signals with some amplitude distribution, there will be instances when $n_c(t)$ and $n_s(t)$ will be greater than $A + f(t)$. However, if $A + f(t)$ is much larger, such instances will be rare. A correct statement would be $A + f(t) \gg n_c(t)$ and $n_s(t)$ most of the time.

It can be easily seen that the output signal-to-noise power ratio in AM is maximum for highest possible degree of modulation (100% modulation).

For a special case when $f(t)$ is a sinusoidal signal, the amplitude of $f(t)$ is A for 100% modulation. Hence

$$f^2(t) = \frac{A^2}{2}$$

and

$$\frac{S_o/N_o}{S_i/N_i} = \frac{2}{3}$$

Thus the maximum improvement in the signal-to-noise power ratio that can be achieved in this case is $\frac{2}{3}$.

If synchronous detection is used for the demodulation of AM with large carrier, the results are identical to those obtained for envelope detector. This can be easily seen from the fact that S_i and N_i, the input signal and noise powers, are identical in both cases:

$$\frac{S_i}{N_i} = \frac{A^2 + \overline{f^2(t)}}{2\overline{n_i^2(t)}}$$

The synchronous detector multiplies the incoming signal $f_i(t)$ by $\cos \omega_c t$. Hence $e_d(t)$, the output, is given by

$$e_d(t) = f_i(t) \cos \omega_c t$$

Substituting Eq. 7.26 for $f_i(t)$ and eliminating the terms with spectra at $2\omega_c$, we get the final output $e_o(t)$,

$$e_o(t) = \tfrac{1}{2}[A + f(t) + n_c(t)] \qquad (7.31)$$

The output contains the useful signal $\tfrac{1}{2}f(t)$ and the noise $\tfrac{1}{2}n_c(t)$. Hence

$$S_o = \tfrac{1}{4}\overline{f^2(t)}$$

$$N_o = \tfrac{1}{4}\overline{n_c^2(t)} = \tfrac{1}{4}\overline{n_i^2(t)}$$

Thus

$$\frac{S_o}{N_o} = \frac{\overline{f^2(t)}}{\overline{n_i^2(t)}} \qquad (7.32)$$

and

$$\frac{S_o/N_o}{S_i/N_i} = \frac{2\overline{f^2(t)}}{A^2 + \overline{f^2(t)}} \qquad (7.33)$$

It is therefore obvious that for AM, when the noise is small compared to the signal, the performance of the envelope detector is identical to that of the synchronous detector. Note that in deriving Eq. 7.33 we made no assumption regarding relative magnitudes of the signal and noise. Hence Eq. 7.33 is valid for all noise conditions for synchronous demodulation.

b. Large Noise Case. Next we consider the performance of the envelope detector in AM with large noise, $n_i(t) \gg [A + f(t)]$. This implies that $n_c(t)$ and $n_s(t) \gg [A + f(t)]$. Under these conditions, Eq. 7.28a becomes

$$E(t) \simeq \sqrt{n_c^2(t) + n_s^2 + 2n_c(t)[A + f(t)]}$$

$$= R(t) \sqrt{1 + \frac{2[A + f(t)]}{R(t)} \cos \theta(t)} \qquad (7.34)$$

where $R(t)$ and $\theta(t)$ are the envelope and the phase of $n_i(t)$ as given in Eqs. 7.13a and 7.13b.

$$R(t) = \sqrt{n_c^2(t) + n_s^2(t)}$$

$$\theta(t) = -\tan^{-1}\left[\frac{n_s(t)}{n_c(t)}\right]$$

Since $R(t) \gg [A + f(t)]$, Eq. 7.34 may be further approximated as

$$E(t) \simeq R(t)\left[1 + \frac{A + f(t)}{R(t)} \cos \theta(t)\right]$$

$$= R(t) + [A + f(t)] \cos \theta(t) \qquad (7.35)$$

A glance at Eq. 7.35 shows that the output contains no term proportional to $f(t)$. The signal $f(t) \cos \theta(t)$ represents $f(t)$ multiplied by a time-varying function (actually a noise) $\cos \theta(t)$ and is of no use in recovering $f(t)$. Thus the output contains no useful signal.

It is evident from this discussion that for a large noise, the signal is completely mutilated by the envelope detector. This behavior accounts for the so-called threshold effect in envelope detectors. By threshold we mean the value of an input signal-to-noise ratio below which the output signal-to-noise ratio deteriorates much more rapidly than the input signal-to-noise ratio. The threshold effect starts appearing in the region where the carrier power to noise power ratio approaches unity.

It should be stressed that the threshold effect is a property of envelope detectors. We observed no such effect for synchronous detectors. The output signal of the synchronous detector is given by Eq. 7.31:

$$e_o(t) = \tfrac{1}{2}[A + f(t) + n_c(t)]$$

In deriving this equation, we placed no restrictions on the signal or noise magnitudes. Hence it is true under all noise conditions. The output $e_o(t)$ always contains a term $\tfrac{1}{2}f(t)$ and hence the threshold effect does not appear. The S/N improvement ratio in Eq. 7.33 holds under all noise conditions. We have also seen that for DSB-SC and SSB-SC (which use synchronous detectors) there were no threshold effects.

We conclude that for AM with small noise, the performance of the envelope detector is almost equal to that of the synchronous detector. But for large noise, the envelope detector shows the threshold effect and proves inferior to the synchronous detector.

7.4 NOISE IN ANGLE-MODULATED SYSTEMS

I. Frequency Modulation

A schematic diagram of the modulator demodulator for FM is shown in Fig. 7.10. The first filter at the receiver filters out the noise that lies outside the band $(\omega_c \pm \Delta\omega)$ over which the useful signal exists. If $\Delta\omega$ is the carrier frequency deviation, then obviously the passband of this filter is, according to Eq. 4.27, $(\omega_c - \Delta\omega, \omega_c + \Delta\omega)$. The output of the demodulator $e_d(t)$ contains the message signal and noise of bandwidth $\Delta\omega$. Since the message signal has a bandwidth ω_m, we can remove the noise outside the signal band by a low-pass filter with cutoff frequency ω_m (Fig. 7.10).

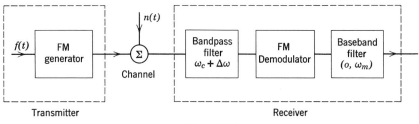

Figure 7.10

To calculate the output signal power and the output noise power, we shall assume that each can be calculated independently of the other. Thus to calculate the output signal power, the noise over the channel will be assumed to be zero, and to calculate the output noise power, the message signal $f(t)$ will be assumed to be zero. The justification for this procedure is given in Appendix A of this chapter.

Consider first the signal without noise. The FM carrier is given by

$$f_c(t) = A \cos \left[\omega_c t + k_f \int f(t)\, dt \right]$$

We observed in Section 4.7 that for FM, the carrier power with or without modulation is the same and is given by $A^2/2$. Thus

$$S_i = \frac{A^2}{2} \tag{7.36}$$

The output of the demodulator is proportional to the instantaneous frequency ω_i. If the constant of proportionality is α, then the output signal is

$$s_o(t) = \alpha \omega_i = \alpha \frac{d}{dt} \left[\omega_c t + k_f \int f(t)\, dt \right]$$

$$= \alpha \omega_c + \alpha k_f f(t)$$

The useful signal is $\alpha k_f f(t)$ and

$$S_o = \alpha^2 k_f{}^2 \overline{f^2(t)} \tag{7.37}$$

To compute N_i and N_o, we observe that the bandwidth of the signal at the demodulator input is $2\,\Delta\omega$ where $\Delta\omega$ is the maximum deviation of the carrier frequency (see Eq. 4.27). Thus

$$N_i = \frac{1}{\pi} \int_{\omega_c + \Delta\omega}^{\omega_c + \Delta\omega} S_n(\omega)\, d\omega \tag{7.38}$$

where $S_n(\omega)$ is the power density spectrum of $n_i(t)$. If the noise is white with power density spectrum of magnitude $\mathcal{N}/2$, then

$$N_i = \frac{1}{\pi} \int_{\omega_c - \Delta\omega}^{\omega_c + \Delta\omega} \frac{\mathcal{N}}{2}\, d\omega$$

$$= 2\mathcal{N}\,\Delta f \qquad (\Delta\omega = 2\pi\,\Delta f) \tag{7.39}$$

To compute N_o, we assume the message signal $f(t)$ to be zero. The demodulator input $f_i(t)$ is the sum of the carrier and the noise $n_i(t)$:

$$f_i(t) = A \cos \omega_c t + n_i(t)$$

$$= A \cos \omega_c t + n_c(t) \cos \omega_c t + n_s(t) \sin \omega_c t$$

$$= [A + n_c(t)] \cos \omega_c t + n_s(t) \sin \omega_c t$$

$$= E(t) \cos [\omega_c t + \psi(t)] \qquad (7.40)$$

where

$$E(t) = \sqrt{[A + n_c(t)]^2 + n_s^2(t)} \qquad (7.40a)$$

and

$$\psi(t) = -\tan^{-1} \left[\frac{n_s(t)}{A + n_c(t)} \right] \qquad (7.40b)$$

We have assumed a small noise case where $A \gg n_c(t)$ and $n_s(t)$. Hence

$$\psi(t) \simeq -\tan^{-1} \left[\frac{n_s(t)}{A} \right]$$

$$\simeq \frac{-n_s(t)}{A} \qquad (7.41)$$

The demodulator output is $\alpha \omega_i$ where ω_i is the instantaneous frequency of the input signal and α is the constant of proportionality.

From Eq. 7.40, we have

$$\omega_i = \frac{d}{dt} [\theta(t)] = \frac{d}{dt} [\omega_c t + \psi(t)]$$

$$= \omega_c + \dot{\psi}(t)$$

From Eq. 7.41, it now follows that

$$\omega_i = \omega_c - \frac{\dot{n}_s(t)}{A} \qquad (7.42)$$

The output of the demodulator is $\alpha \omega_i$ given by

$$f_d(t) = \alpha \left[\omega_c - \frac{\dot{n}_s(t)}{A} \right]$$

The noise component $n_o(t)$ is obviously

$$n_o(t) = - \frac{\alpha \dot{n}_s(t)}{A} \qquad (7.43)$$

If $n_s(t)$ has a power density spectrum $S_{n_s}(\omega)$, then its derivative $\dot{n}(s)$ has a spectrum $S_{\dot{n}_s}(\omega)$ given by (see Eq. 2.39)

$$S_{\dot{n}_s}(\omega) = \omega^2 S_{n_s}(\omega)$$

and $S_{n_o}(\omega)$, the power density spectrum of the output noise $[-\alpha\dot{n}(t)]/A$, is given by

$$S_{n_o}(\omega) = \frac{\alpha^2}{A^2} S_{\dot{n}_s}(\omega)$$

$$= \frac{\alpha^2\omega^2}{A^2} S_{n_s}(\omega) \tag{7.44}$$

The demodulator output is further passed through a low-pass filter of cutoff frequency ω_m to remove the excess noise lying outside the useful signal band.

After substituting Eq. 7.7 (and Eq. 7.8) in Eq. 7.44 and remembering that the output signal is filtered by a low-pass filter (ω_m), we have

$$S_{n_o}(\omega) = \begin{cases} \dfrac{\alpha^2\omega^2}{A^2}\,[S_n(\omega + \omega_c) + S_n(\omega - \omega_c)] & |\omega| < \omega_m \\[4mm] 0 & |\omega| > \omega_m \end{cases} \tag{7.45}$$

If the channel noise is white,

$$S_n(\omega) = \frac{\mathcal{N}}{2}$$

and

$$S_{n_o}(\omega) = \begin{cases} \dfrac{\alpha^2\omega^2\mathcal{N}}{A^2} & |\omega| < \omega_m \\[4mm] 0 & |\omega| > \omega_m \end{cases} \tag{7.46}$$

This is a parabolic spectrum as shown in Fig. 7.11.

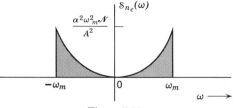

Figure 7.11

The output noise power N_o is given by

$$N_o = \frac{\alpha^2 \mathcal{N}}{\pi A^2} \int_0^{\omega_m} \omega^2 \, d\omega$$

$$= \frac{2}{3} \left(\frac{\alpha}{A}\right)^2 \frac{\mathcal{N}\omega_m^3}{2\pi} \qquad (7.47)$$

Using Eqs. 7.36, 7.37, 7.39, and 7.47, we get

$$\frac{S_o/N_o}{S_i/N_i} = \frac{6k_f^2 \overline{f^2(t)} \, \Delta\omega}{\omega_m^3} \qquad (7.48)$$

Note that the signal-to-noise power ratio at the demodulator output is given by (Eq. 7.37 and 7.47)

$$\frac{S_o}{N_o} = \frac{3}{2} \left(\frac{2\pi A^2}{\mathcal{N}}\right) \frac{k_f^2 \overline{f^2(t)}}{\omega_m^2} \qquad (7.49)$$

The carrier frequency deviation is directly proportional to k_f. Hence the bandwidth of the FM signal is proportional to k_f. It is therefore evident from Eq. 7.49 that the signal-to-noise power ratio at the FM demodulator output is proportional to the square of the bandwidth of transmission.

It is interesting to compare the S/N ratio at the demodulator output for FM and AM. If the signal $f(t)$ were transmitted by AM, the signal output of the envelope detector would be given by (Eq. 7.29a)

$$S_o = \overline{f^2(t)}$$

and the noise output N_o of the envelope detector would be given by (Eq. 7.29b)

$$N_o = \overline{n_c^2(t)} = N_i$$

where N_i is the noise power at the detector input. For AM, the bandwidth of transmission is $2f_m$. Hence for a white noise of density $\mathcal{N}/2$,

$$N_o = N_i = 2\mathcal{N}f_m$$

and

$$\left(\frac{S_o}{N_o}\right)_{\text{AM}} = \frac{\overline{f^2(t)}}{2\mathcal{N}f_m}$$

Substituting this result in Eq. 7.49, we obtain

$$\frac{(S_o/N_o)_{\text{FM}}}{(S_o/N_o)_{\text{AM}}} = 3\left(\frac{Ak_f}{\omega_m}\right)^2 \qquad (7.50)$$

In Eq. 7.50, A is the FM carrier amplitude. For a meaningful comparison between AM and FM, we shall assume the amplitude of the AM carrier also to be A. To obtain some quantitative results, let us consider a specific case when the modulating signal $f(t)$ is a sinusoidal signal. For comparison, we must consider AM under the most favorable condition, 100% modulation. In this case the amplitude of $f(t)$ is A (the same as the carrier). Also the FM carrier frequency deviation $\Delta\omega$ when $f(t)$ is a sinusoidal signal of amplitude A is given by (Eq. 4.20)

$$\Delta\omega = Ak_f$$

and the modulation index m_f is given by

$$m_f = \frac{\Delta\omega}{\omega_m} = \frac{Ak_f}{\omega_m}$$

where ω_m is the frequency of the modulating signal $f(t)$. Substituting this result in Eq. 7.50 yields

$$\frac{(S_o/N_o)_{\text{FM}}}{(S_o/N_o)_{\text{AM}}} = 3m_f{}^2 \tag{7.51a}$$

If we consider signal-to-noise root mean square voltage ratios rather than power ratios, we get

$$\frac{[(S_o/N_o)_{\text{FM}}]_{\text{vr}}}{[(S_o/N_o)_{\text{AM}}]_{\text{vr}}} = \sqrt{3}\, m_f \tag{7.51b}$$

It is obvious from Eq. 7.51 that the signal-to-noise ratio can be made much higher in FM than in AM by increasing the modulation index m_f. Note, however, that increasing the m_f increases the required bandwidth. Thus when $m_f = 5$, the S_o/N_o ratio in FM is 75 times that of AM, but the bandwidth required is increased roughly 8 times. This can be illustrated as follows. For AM the bandwidth is $2\omega_m$. For FM the bandwidth corresponding to $m_f = 5$ is 3.3 $\Delta\omega$ (see Fig. 4.5). But $\Delta\omega = m_f\omega_m$. Hence the bandwidth required for FM is 3.3 $\times$ $5\omega_m \simeq 16\omega_m$.

For a large value of m_f $(m_f > 10)$ the bandwidth B of the FM signal is about $2\,\Delta f$

$$B \simeq 2\,\Delta f$$

and

$$m_f = \frac{\Delta f}{f_m} = \frac{1}{2}\frac{B}{f_m}$$

thus

$$\frac{(S_o/N_o)_{\text{FM}}}{(S_o/N_o)_{\text{AM}}} = \frac{3}{4}\left(\frac{B}{f_m}\right)^2 \tag{7.52}$$

From this equation we have

$$\left(\frac{S_o}{N_o}\right)_{\text{FM}} = \frac{3}{4}\left(\frac{B}{f_m}\right)^2\left(\frac{S_o}{N_o}\right)_{\text{AM}}$$

If we wish to improve S_o/N_o in the FM system by a factor of 4 (6 db), we must increase the bandwidth B by a factor of 2. Thus the signal-to-noise power ratio improves 6 db for each two-to-one increase in the bandwidth occupancy. It must be remembered, however, that these results are derived under the small noise assumption.

We can write Eq. 7.52 in terms of output signal to noise root mean square voltage ratios.

$$\frac{[(S_o/N_o)_{\text{FM}}]_{\text{vr}}}{[(S_o/N_o)_{\text{AM}}]_{\text{vr}}} = \frac{\sqrt{3}}{2}\frac{B}{f_m} \tag{7.53}$$

Thus the output S/N root mean square voltage ratio in FM is directly proportional to the bandwidth.

The property of exchange of bandwidth for signal-to-noise ratio is generally true for all communication systems. Transforming a signal to occupy a larger bandwidth makes it more immune to noise interference. The theoretical basis for this property will be developed in Chapter 8.

It should be observed that an FM system allows one to exchange the signal-to-noise ratio for the bandwidth of transmission. Such exchange is not possible in AM systems where the bandwidth of transmission is fixed.

From Eq. 7.51b, it is obvious that in order to realize S/N ratio improvement in FM over AM, we must have

$$m_f > \frac{1}{\sqrt{3}} \simeq 0.6$$

It is interesting to note that $m_f = 0.6$ is about the transition point between the narrowband FM and the wideband FM. The narrowband FM therefore provides no improvement over AM. This is an expected result since the narrowband FM and AM have the same bandwidth.

The signal-to-noise ratio in FM cannot be improved indefinitely by increasing the bandwidth. As the bandwidth increases the input noise also increases and eventually a point is reached where the carrier power is of the order of the input noise power. Our results derived earlier under the assumption of small input signal no longer hold. Under sufficiently large input noise condition the phenomenon of threshold, mentioned earlier, appears.

Threshold in FM. The threshold effect in FM is much more pronounced and significant than that in AM. To study this effect, we express the signal at the demodulator input as

$$f_i(t) = A \cos [\omega_c t + \psi(t)] + n_i(t)]$$

where

$$\psi(t) = k_f \int f(t) \, dt$$

Using Eq. 7.12 to express $n_i(t)$, we obtain

$$f_i(t) = A \cos [\omega_c t + \psi(t)] + R(t) \cos [\omega_c t + \theta(t)] \qquad (7.54)$$

The phasor diagram representing Eq. 7.54 is shown in Fig. 7.12. It is obvious from this figure that

$$f_i(t) = E(t) \cos [\omega_c t + \theta(t) + \beta(t)] \qquad (7.55)$$

where

$$\beta(t) = \tan^{-1} \left\{ \frac{A \sin [\psi(t) - \theta(t)]}{R(t) + A \cos [\psi(t) - \theta(t)]} \right\} \qquad (7.56)$$

For large noise case, $R(t) \gg A$ and Eq. 7.56 can be expressed as

$$\beta(t) \simeq \tan^{-1} \left\{ \frac{A \sin [\psi(t) - \theta(t)]}{R(t)} \right\}$$

$$\simeq \frac{A}{R(t)} \sin [\psi(t) - \theta(t)] \qquad (7.57)$$

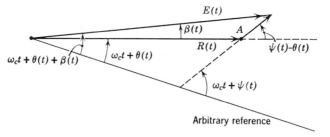

Figure 7.12

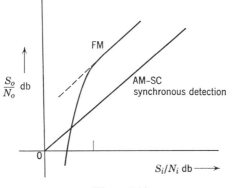

Figure 7.13

Note that $e_d(t)$, the output of the demodulator, is given by

$$e_d(t) = \alpha \frac{d}{dt} [\omega_c t + \theta(t) + \beta(t)]$$

$$= \alpha[\omega_c + \dot{\theta}(t) + \dot{\beta}(t)]$$

The signal $\alpha\dot{\theta}(t)$ is the noise signal. The information about $f(t)$ is contained in $\dot{\beta}(t)$. It can be seen from Eq. 7.57 that $\beta(t)$ contains a multiplying factor $1/R(t)$ which is a noise signal. Hence $\dot{\beta}(t)$ does not contain any extractable signal that is proportional to $f(t)$, hence the output signal will be distorted. This gives rise to the threshold effect as shown in Fig. 7.13.

Threshold Improvement through De-Emphasis   We have observed that the noise power density at the demodulator output rises parabolically with frequency (Fig. 7.11). This is rather unfortunate because for all practical message signals, the power density decreases with frequency. Thus the noise is strongest in the frequency range where the signal is weakest. The high frequency components of the message signal therefore suffer the most because of the noise interference on the channel. This difficulty can be alleviated by the so called pre-emphasis and de-emphasis technique.

At the transmitter, the high frequency components of the message signals $f(t)$ are boosted (pre-emphasis). This transformed signal $f'(t)$ is now used to frequency modulate the carrier. At the receiver, the demodulator yields the transformed signal $f'(t)$ and the parabolic noise. The desired signal is obtained by transmitting the demodulator output through a filter which restores the high frequency components to the original level (de-emphasis). This yields the original signal $f(t)$ and the noise whose power density spectrum is also reduced at higher frequencies correspondingly. The technique of pre-emphasis, de-emphasis not only reduces the undesirable high noise in the low signal level region but also cuts down the overall noise at the output and improves the output signal-to-noise ratio. This, in effect, improves the threshold of FM. We shall now calculate the improvement in the signal-to-noise ratio due to pre-emphasis and de-emphasis.

We shall consider a rather simple pre-emphasis filter as shown in Fig. 7.14a. This is a single pole filter whose frequency response is shown in Fig. 7.14b. The break points ω_1 and ω_2 are given by

$$\omega_1 = \frac{1}{R_1 C} \quad \text{and} \quad \omega_2 \simeq \frac{1}{R_2 C} \quad (R_1 \gg R_2)$$

A reasonable choice of ω_1 is where the spectrum of $f(t)$ drops to about 3 db from its low frequency value. For broadcast systems f_1, $(\omega_1/2\pi)$

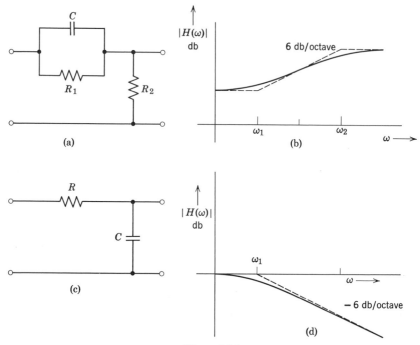

Figure 7.14

is taken as 2.1 kHz. One should choose ω_2 well beyond the highest frequency in $f(t)$. For broadcast systems f_2, $(\omega_2/2\pi)$ is typically $\geqslant 30$ kHz. The corresponding de-emphasis network is shown in Eq. 7.14c, and its frequency response is shown in Fig. 7.14d. Since the desired signal vanishes for $\omega > \omega_2$, there is no need for correction at frequencies beyond ω_2. For the de-emphasis filter

$$H(\omega) = \frac{1}{1 + j\omega/\omega_1}$$

$$|H(\omega)|^2 = \frac{\omega_1^2}{\omega^2 + \omega_1^2} \tag{7.58}$$

If $n_0'(t)$ represents the final noise output (output of the de-emphasis network), then the output noise density spectrum $S_{n_0'}(\omega)$ for white channel noise is given by (see Eq. 7.46)

$$S_{n_0'}(\omega) = \frac{\alpha^2\omega^2\mathcal{N}}{A^2}\,|H(\omega)|^2$$

$$= \frac{\alpha^2\omega_1^2\omega^2\mathcal{N}}{A^2(\omega^2 + \omega_1^2)}$$

The output noise power N_o' is given by

$$N_o' = \frac{1}{\pi}\int_0^{\omega_m} S_{n_0'}(\omega)\,d\omega$$

$$= \frac{\alpha^2\omega_1^2\mathcal{N}}{\pi A^2}\int_0^{\omega_m}\frac{\omega^2}{\omega^2 + \omega_1^2}\,d\omega$$

$$= \frac{\alpha^2\omega_1^3\mathcal{N}}{\pi A^2}\left[\frac{\omega_m}{\omega_1} - \tan^{-1}\left(\frac{\omega_m}{\omega_1}\right)\right] \tag{7.59}$$

If we define the noise reduction factor ρ as

$$\rho = \frac{N_o}{N_o'}$$

then from Eqs. 7.47 and 7.59 we get

$$\rho = \frac{1}{3}\frac{(\omega_m/\omega_1)^3}{(\omega_m/\omega_1) - \tan^{-1}(\omega_m/\omega_1)} \tag{7.60}$$

The improvement ratio ρ is plotted in Fig. 7.15 as a function of ω_m/ω_1.

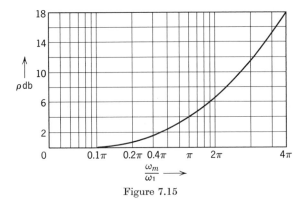

Figure 7.15

It is important to realize that pre-emphasis does not cause any increase in the transmitted signal power. This is because for FM the transmitted signal power is identical to the unmodulated carrier power $(A^2/2)$. The noise power N_i at the demodulator and the output signal power S_o are also unchanged. Hence ρ in Eq. 7.60 actually represents the additional improvement in S/N in FM.

The improvement ratio ρ increases as ω_m/ω_1 is increased. This means that decreasing the first break frequency ω_1 improves the performance. But decreasing the break frequency ω_1 causes the increase in the bandwidth of transmission. This is because as ω_1 is made smaller and smaller, the pre-emphasis network approximates more and more to an ideal differentiator. Indeed, for $\omega_1 = 0$ (and $\omega_2 \gg \omega_m$), the pre-emphasis network acts as an ideal differentiator. The original signal $f(t)$ is therefore differentiated before it is used to frequency modulate the carrier. The process of differentiation tends to give rise to sharp peaks of large amplitudes (in contrast to the smoothing effect of an integrator). Since the modulation constant k_f is constant, the increased maximum signal amplitude causes an increase in $\Delta\omega$, the deviation of the carrier frequency. Clearly, this increases the bandwidth of transmission. It is therefore obvious that the signal-to-noise ratio improvement effected by pre-emphasis, de-emphasis is not without its price. It should be mentioned in passing that if $\omega_1 = 0$ in the pre-emphasis network, then the FM with pre-emphasis and de-emphasis networks is equivalent to phase modulation (PM). This is because for $\omega_1 = 0$, the pre-emphasis network acts as a differentiator. We know that if a signal is differentiated and then allowed to frequency modulate a carrier, the resulting signal is a phase-modulated carrier.

2. Phase Modulation

Phase modulation may be treated as a special case of FM. As explained earlier, PM can be generated by differentiating $f(t)$ and allowing it to frequency modulate the carrier. If we use an FM demodulator at the receiver, the signal output will be df/dt. The desired signal can be recovered by adding an ideal integrator to the regular FM receiver. The PM carrier may be expressed as

$$\varphi_{PM}(t) = A \cos (\omega_c t + k_p f(t))$$

$$= A \cos \left[\omega_c t + k_p \int \frac{df}{dt} dt \right]$$

$$S_i = \frac{A^2}{2}$$

$$N_i = 2\mathcal{N}\, \Delta f \qquad \text{(for white noise)}$$

The output signal of the FM demodulator is $\alpha k_p (df/df)$ where α is the demodulator constant. The output signal is now integrated to obtain the output $\alpha k_p f(t)$. Hence

$$S_o = \alpha^2 k_p{}^2 \overline{f^2(t)}$$

The noise at the output of the FM demodulator is given by $[-\alpha \dot{n}_s(t)/A]$ (see Eq. 7.43). The ideal integrator following the FM demodulator yields $n_o(t)$, the final noise signal at the output:

$$n_o(t) = \frac{-\alpha n_s(t)}{A}$$

and

$$S_{n_o}(\omega) = \frac{\alpha^2}{A^2}\, S_{n_s}(\omega)$$

$$= \begin{cases} \dfrac{\alpha^2}{A^2} [S_n(\omega - \omega_c) + S_n(\omega + \omega_c)] & |\omega| < \omega_m \\[2mm] 0 & |\omega| > \omega_m \end{cases}$$

For white channel noise

$$S_n(\omega) = \frac{\mathcal{N}}{2}$$

and

$$S_{n_o}(\omega) = \begin{cases} \dfrac{\alpha^2 \mathcal{N}}{A^2} & |\omega| < \omega_m \\[2mm] 0 & |\omega| > \omega_m \end{cases}$$

Note that the output noise for PM has a uniform power density unlike FM. Hence there is no need for additional pre-emphasis, de-emphasis. In fact, as mentioned earlier, PM may be considered an FM system with perfect pre-emphasis, de-emphasis.

The noise power output N_o is given by

$$N_o = \frac{1}{\pi} \int_0^{\omega_m} S_{n_o}(\omega)\, d\omega$$

$$= \frac{2\alpha^2 \mathcal{N} f_m}{A^2}$$

Hence

$$\frac{S_o}{N_o} = \frac{(Ak_p)^2}{2\mathcal{N} f_m} \overline{f^2(t)}$$

and

$$\frac{S_o/N_o}{S_i/N_i} = \frac{2k_p{}^2 \overline{f^2(t)}\, \Delta f}{f_m}$$

and

$$\frac{(S_o/N_o)_{\text{PM}}}{(S_o/N_o)_{\text{AM}}} = (Ak_p)^2$$

Here A is the amplitude of the PM carrier. For a meaningful comparison of AM and PM, we must let the AM carrier be A. For a most favorable condition in AM, the modulation is 100%. Let us consider a special case when $f(t)$ is a sinusoidal signal. For 100% modulation,

$$f(t) = A \cos \omega_m t$$

and a PM carrier is given by

$$\varphi_{\text{PM}}(t) = A \cos (\omega_c t + Ak_p \cos \omega_m t)$$

$$\omega_i = \omega_c + Ak_p \omega_m \cos \omega_m t$$

and

$$\Delta \omega = Ak_p \omega_m$$

Hence

$$\frac{(S_o/N_o)_{\text{PM}}}{(S_o/N_o)_{\text{AM}}} = \left(\frac{\Delta \omega}{\omega_m}\right)^2$$

The S/N root mean square voltage ratio improvement is given by

$$\frac{[(S_o/N_o)_{\text{PM}}]_{\text{vr}}}{[(S_o/N_o)_{\text{AM}}]_{\text{vr}}} = \left(\frac{\Delta \omega}{\omega_m}\right)$$

Note that the nature of the improvement of the S/N ratio in PM over AM is similar to that of FM over AM. In both cases the S/N power ratio improvement is proportional to the square of the bandwidth of transmission.

7.5 NOISE IN PULSE-MODULATED SYSTEMS

In pulse-modulated systems, we shall study pulse amplitude modulation (PAM), pulse position modulation (PPM), and pulse code modulation (PCM).

I. PAM

The signal-to-noise behavior of PAM systems is identical to that of AM/SC systems. The S/N improvement ratio in PAM is unity, the same as that of SSB-SC. Similarly, the S/N improvement ratio in PAM/AM is 2, the same as that of DSB-SC. The result is not surprising since the bandwidth requirements of PAM and PAM/AM are identical to SSB and DSB, respectively (see Section 5.4).

In PAM, the signal $f(t)$ is sampled. The sampled signal $f_s(t)$ is then transmitted through a low-pass filter* which yields the original continuous signal. So, in effect, we are transmitting $f(t)$ directly. In the process of transmission it is corrupted by the noise. At the receiver, a low-pass filter is used. It is evident that the signal-to-noise ratio in this case remains unchanged. Thus

$$\frac{S_o/N_o}{S_i/N_i} = 1 \qquad (7.61)$$

It can be seen that

$$S_o = S_i = \overline{f^2(t)}$$

and

$$N_o = N_i = \mathcal{N} f_m \qquad \text{(for white noise)}$$

The result in Eq. 7.61 is proved in Appendix B for M time division multiplexed signals. If PAM is transmitted by PAM-AM, it can be readily shown that the signal-to-noise improvement ratio is 2 (same as DSB-SC).

---

* Actually, there are M messages whose samples are interleaved for the purpose of time division multiplexing. These combined samples are then transmitted through a low-pass filter of cutoff frequency Mf_m. Here, for convenience, we shall consider the case of a single signal only.

2. PPM (Pulse Position Modulation)

In PPM, a message signal $f(t)$ is sampled and the sample values are transmitted in the form of pulse positions. The kth pulse is shifted from its quiescent position by an amount proportional to $f(kT)$, the kth sample of $f(t)$ (see Fig. 5.9). The pulses are transmitted over a channel of bandwidth B. This causes the dispersion of the pulse as seen in Section 2.6 (Fig. 2.9). The received pulse can be approximated with fair accuracy by a trapezoidal pulse. This is shown in Fig. 7.16 (see Fig. 2.9). The rise time t_r of a pulse is given by (see Eq. 2.11)

$$t_r \simeq \frac{1}{B} \qquad (7.62)$$

where B is the bandwidth (in Hz) of the channel. The position of the trapezoidal pulse is sensitive to the additive channel noise. This can be seen from Fig. 7.16b. An addition of a constant signal of magnitude x will shift the pulse position by ε, where

$$\frac{\varepsilon}{x} = \frac{t_r}{A} \qquad (7.63)$$

If x varies randomly, then ε will also vary randomly. Since ε is proportional to x, the root mean square value of ε will be proportional to the root mean square value of x. From Eq. 7.63 it follows that

$$\frac{\overline{\varepsilon^2}}{\overline{x^2}} = \left(\frac{t_r}{A}\right)^2$$

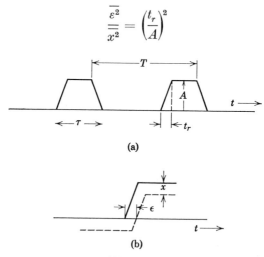

(a)

(b)

Figure 7.16

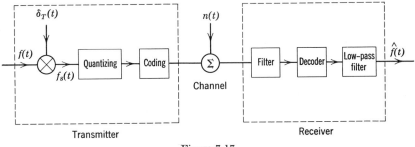

Figure 7.17

Thus if a noise signal $n(t)$ with mean square value $\overline{n^2(t)}$ is added to PPM signal, the mean square value of ε is given by

$$\overline{\varepsilon^2} = \left(\frac{t_r}{A}\right)^2 \overline{n^2(t)} \tag{7.64}$$

In the absence of noise, the pulse position of the kth pulse varies in proportion to $f(kT)$, the kth sample of $f(t)$. If we denote the change in the position of the kth pulse by y_k, then

$$y_k = k_1 f(kT)$$

where k_1 is a constant of proportionality of the modulator. However, because of the channel noise, the change in the pulse position is not y_k but $y_k + \varepsilon_k$, where ε_k is random and has a mean square value given by Eq. 7.64. Let

$$\hat{y}_k = y_k + \varepsilon_k = k_1 f(kT) + \varepsilon_k \tag{7.65}$$

At the receiver, the pulse positions are converted back into samples. The kth sample at the receiver is $\hat{y}_k$ (Eq. 7.65). To obtain the original message $f(t)$, we transmit the samples through a low-pass filter of cut-off frequency ω_m (see Fig. 1.47). The output of the low-pass filter is $\hat{y}(t)$, given by (see Eq. 1.40b)

$$\hat{y}(t) = \sum_k \hat{y}_k Sa(\omega_m t - n\pi)$$

$$= \sum_k [k_1 f(kT) + \varepsilon_k] Sa(\omega_m t - n\pi)$$

$$= \sum_k k_1 f(kT) Sa(\omega_m t - n\pi) + \sum_k \varepsilon_k Sa(\omega_m t - n\pi)$$

Each of the summations represent a signal bandlimited to ω_m radians per second. Since $f(kT)$ represents the sample of $f(t)$, it is evident from

Eq. 1.140b that the first summation is equal to $k_1 f(t)$. The second summation represents a noise signal $\varepsilon(t)$. Thus the output $\hat{y}(t)$ is given by

$$\hat{y}(t) = k_1 f(t) + \varepsilon(t)$$

Thus

$$s_o(t) = k_1 f(t)$$

$$S_o = k_1^2 \overline{f^2(t)}$$

$$n_o(t) = \varepsilon(t)$$

and

$$N_o = \overline{n_o^2(t)} = \overline{\varepsilon^2(t)}$$

We now use the result of Example 2.2 (Eq. 2.31), which states that the mean square value of a signal is equal to the mean square value of its samples. Thus $\overline{\varepsilon^2(t)}$, the mean square value of $\varepsilon(t)$, is equal to $\overline{\varepsilon_k^2}$. But the mean square value of ε_k is given by Eq. 7.64. Hence

$$\overline{\varepsilon^2(t)} = \overline{\varepsilon_k^2} = \left(\frac{t_r}{A}\right)^2 \overline{n^2(t)}$$

Thus

$$N_o = \left(\frac{t_r}{A}\right)^2 \overline{n^2(t)}$$

and

$$\frac{S_o}{N_o} = \frac{k_1^2}{\overline{n^2(t)}} \left(\frac{A}{t_r}\right)^2 \overline{f^2(t)}$$

$$= k_1^2 \frac{A^2}{\overline{n^2(t)}} \overline{f^2(t)} \, B^2$$

where B is the channel bandwidth (in Hz).

If the ratio of the pulse width τ to the period T is α, then the input signal power S_i is given by (assuming $t_r \ll T$)

$$S_i = \alpha A^2$$

and the input noise power N_i is given by

$$N_i = \overline{n^2(t)}$$

Hence

$$\frac{S_o/N_o}{S_i/N_i} = \frac{k_1^2}{\alpha} \overline{f^2(t)} \, B^2 \qquad\qquad (7.66a)$$

If we consider signal-to-noise root mean square voltage ratios rather than power ratios, we obtain

$$\frac{(S_o/N_o)_{\mathrm{vr}}}{(S_i/N_i)_{\mathrm{vr}}} = k_1 \left(\frac{\overline{f^2(t)}}{\alpha}\right)^{1/2} B \qquad (7.66b)$$

Thus the S/N power ratio improvement is proportional to the square of the bandwidth. This behavior is identical to that of wideband FM as seen in Eq. 7.49. In both these systems it is possible to trade bandwidth for the signal-to-noise ratio. The S/N voltage ratio is linearly proportional to the bandwidth B. This is the characteristic of uncoded systems. We shall now show that in the coded systems such as PCM, the S/N voltage ratio can be increased exponentially with bandwidth. Thus the coded systems are inherently capable of better transmission efficiency than the uncoded systems.

3. Coded Systems

The modulation systems can be broadly classified into two classes: (a) uncoded systems, and (b) coded systems. In uncoded systems one symbol in message space is transformed into one symbol in the modulated signal space. Thus for AM each possible amplitude of the original message is transformed into a particular amplitude of the modulated signal. Similarly, for PPM each amplitude of the message results in a particular displacement of the pulse position. In coded systems (such as PCM), however, each message symbol or amplitude is transformed into a number of signal symbols. Tuller* has shown that only coded systems can achieve the most efficient exchange of bandwidth for signal to noise ratio that is theoretically possible.† The uncoded systems (such as FM or PPM) are inherently incapable of efficient exchange of the bandwidth and signal-to-noise ratio. We shall now consider PCM from the point of view of exchange of bandwidth and signal-to-noise ratio.

---

* W. G. Tuller, "Theoretical Limits on the Rate of Transmission of Information," *Proc. IRE*, **37**, 468, 1949.

† It can, however, be shown that by using optimum phase demodulator (phase-locked loop), one can attain performance close to the upper theoretical bound in the limit as the bandwidth of transmission goes to infinity and the transmitted signal power density is uniform over the passband. See, for instance, A. Viterbi, *Principles of Coherent Communication*, McGraw-Hill, New York, 1966.

Binary PCM

In PCM, the modulating signal is sampled. Each of the samples is transmitted by a code formed by a group of pulses as discussed in Section 5.2. In general, the samples can assume any value in a certain continuous range of the amplitude. However, we allow the samples to take any one of the finite number of levels in this range by approximating each sample to the nearest allowable level as shown in Fig. 5.10a. This approximation, known as quantization, obviously is a source of error. The signal that is transmitted is an approximation and there is an uncertainty in the signal amplitude of the amount equal to the difference between the allowed levels. The error or the uncertainty can be made as small as possible by increasing the number of levels in a given amplitude range. In Fig. 5.10a there are 16 allowable levels, separated by 0.1 volts each.

We shall first consider the case of a binary pulse code where the pulses can assume amplitudes of 0 and A volts only. As observed in Fig. 5.10b, a group of 4 binary pulses are required to represent a sample that can assume any one of the 16 possible values. This is because 4 binary pulses can form $2 \times 2 \times 2 \times 2 = 16$ distinct patterns. Similarly, we need a group of 5 binary pulses to represent a sample which can assume one of the 32 possible values. It is easy to see that to represent a sample that can assume any one of the M possible values, we need a group of $\log_2 M$ number of binary pulses per sample. Note that it will take an infinite number of pulses per sample if the samples are transmitted exactly without quantization. This is because the samples can assume any value in a continuous range making $M = \infty$.

At the receiver, the binary pulses are converted back into the samples (the quantized samples) and are now passed through a low-pass filter to recover $f(t)$ (see Fig. 7.17).

Let us assume that the samples of the signal $f(t)$ under consideration are quantized to M levels, each separated by Δv volts. Each sample is now represented by a group of $\log_2 M$ binary pulses. In Fig. 5.10a, $M = 16$, $\Delta v = 0.1$ volt, and $\log_2 M = 4$. Let the mean square value of the noise be σ_n^2. We shall now calculate the signal power at the input of the demodulator. This input consists of binary pulses. In order to calculate the power of the coded signal, we must know the amplitude distribution of the modulating signal $f(t)$. For convenience,

let us assume that the amplitudes of $f(t)$ are uniformly distributed in*
the given range (0 to 1.5 volts in Fig. 5.10a). This implies that the likelihood of observing the signal amplitude at any one value in the given
range is the same as that at any other value in the same range. This leads
to the conclusion that the chances of occurrence of any one quantized
level is the same as any other (equal to $1/M$). It is therefore obvious that
the likelihood of observing any one code pattern is the same as that of
any other pattern (this is 1/16 for patterns in Fig. 5.10b). Thus each of
the code patterns occurs with the same frequency. Hence the likelihood
of observing 0 volt pulse is equal to that of observing A volt pulse (equal
to $\frac{1}{2}$). On the average, half the time the pulse of A volts will be present,
and the remaining half of the time there will be no pulse. It is obvious
that the average signal power S_i is given by

$$S_i = \frac{A^2}{2}$$

What should be the magnitude of A? It is obvious that the choice of
A is governed by noise considerations. If A is large compared to the
root mean square noise voltage, it will be possible to recognize the
pulse in the presence of noise with very small chance of error.

For PCM the signal detection is merely a question of recognizing the
presence or absence of a pulse. If we let $A = K\sigma_n$, then making K
sufficiently large, the probability of error can be reduced as much as
desired. This topic is considered in Chapter 9. It is shown that for
$K = 10$, the error probability is reduced to the order of 10^{-6}. In this
chapter however, we shall leave K as an unknown constant. If the
pulse is equally likely to be present and absent, then

$$S_i = \frac{A^2}{2} = \frac{K^2\sigma_n^2}{2}$$

and

$$N_i = \sigma_n^2$$

Therefore

$$\frac{S_i}{N_i} = \frac{K^2}{2} \qquad\qquad (7.67)$$

By making K large enough, we can make the error probability in this
detection as low as possible. Hence, in effect, the channel noise is

---

* This assumption is not necessary. The results derived here can be shown to be
valid for any amplitude distribution provided the quantizing interval is small enough.

eliminated. However, we do have an error in the received signal due to quantization. There is an uncertainty introduced at the transmitter itself, and the error between the actual and approximated signals acts as a noise. We call this the quantization noise.

To calculate the quantization noise, we use the results of Example 2.2, Eq. 2.31. This states that the mean square value of a bandlimited signal is equal to the mean square value of its samples (taken at a rate greater than or equal to the Nyquist rate):

$$\overline{f^2(t)} = \overline{f_k{}^2}$$

We shall now calculate the mean square value of the error (quantization noise) introduced by the process of quantization. Each sample of $f(t)$ is approximated to the nearest allowed level. Let the kth sample of $f(t)$ be f_k and the approximated kth sample be $\hat{f}_k$. Obviously $\hat{f}_k$ is one of the M allowed levels closest to f_k. We can express the signal $f(t)$ in terms of its samples f_k (Eq. 1.140b),

$$f(t) = \sum_k f_k Sa(\omega_m t - k\pi) \tag{7.68a}$$

At the receiver, when the pulse train is decoded, it yields samples $\hat{f}_k$. These samples, when passed through a low-pass filter, yield a continuous signal $\hat{f}(t)$ (Eq. 1.140b),

$$\hat{f}(t) = \sum_k \hat{f}_k Sa(\omega_m t - k\pi) \tag{7.68b}$$

The error signal $e(t)$ (which acts as a noise) is given by

$$e(t) = f(t) - \hat{f}(t) = \sum_k (f_k - \hat{f}_k)Sa(\omega_m t - k\pi) \tag{7.68c}$$

If we denote

$$e_k = f_k - \hat{f}_k$$

then

$$e(t) = \sum_k e_k Sa(\omega_m t - k\pi)$$

It is obvious from the results of Example 2.2 that the mean square value of $e(t)$ is equal to the mean square value of its samples e_k.

$$\overline{e^2(t)} = \overline{e_k{}^2}$$

We note that e_k is the difference between the actual sample value $f(t)$ and the quantized sample value $\hat{f}_k$. Since the quantized levels are spaced Δv volts apart, e_k must lie in the range $(-\Delta v/2)$, $(\Delta v/2)$. Since the amplitude distribution of $f(t)$ is assumed to be uniform in the range $[0, (M-1)\Delta v]$, the distribution of e_k will also be uniform in the range $(-\Delta v/2)$, $(\Delta v/2)$; that is, the likelihood of observing e_k, a typical quantization error at one value in the range $(-\Delta v/2)$, $(\Delta v/2)$, is the same as that of observing at any other value in the same range. To calculate the mean square value of e_k, we divide the range $(-\Delta v/2)$, $(\Delta v/2)$ into N small increments each of width $\Delta v/N$. The quantity e_k will assume these values spaced $\Delta v/N$ volts with equal frequency. We must now find the mean square of all these values* in the limit as $N \to \infty$ (or the increment goes to zero). Thus on the average, e_k will take on values 0, $\pm\Delta v/N$, $\pm 2\,\Delta v/N, \ldots, \pm r\,\Delta v/N, \ldots, \pm\Delta v/2$ with equal frequency. The mean square value of e_k is obviously given by

$$\overline{e_k^2} = \lim_{N\to\infty} \frac{1}{N} \sum_{n=-N/2}^{N/2} \left(\frac{n\,\Delta v}{N}\right)^2$$

$$= \lim_{N\to\infty} \frac{2}{N} \sum_{N=0}^{N/2} \left(\frac{n\,\Delta v}{N}\right)^2$$

$$= \lim_{N\to\infty} \frac{2(\Delta v)^2}{N^3} \sum_{N=0}^{N/2} n^2$$

$$= \lim_{N\to\infty} \frac{2(\Delta v)^2}{N^3} \left[\frac{\dfrac{N}{2}\left(\dfrac{N}{2}+1\right)(N+1)}{6}\right]$$

$$= \frac{(\Delta v)^2}{12}$$

---

* Those who are familiar with the probability theory will recognize that

$$\overline{e_k^2} = \int_{-\infty}^{\infty} e_k^2 p(e_k)\,de_k$$

where $p(e_k)$ is the probability density function of e_k. In this case

$$p(e_k) = \begin{cases} \dfrac{1}{\Delta v} & \dfrac{-\Delta v}{2} < e_k < \dfrac{\Delta v}{2} \\ 0 & \text{otherwise} \end{cases}$$

Hence

$$\overline{e_k^2} = \frac{1}{\Delta v}\int_{-\Delta v/2}^{\Delta v/2} e_k^2\,de_k = \frac{(\Delta v)^2}{12}$$

Thus

$$\overline{e^2(t)} = \frac{(\Delta v)^2}{12}$$

But $\overline{e^2(t)}$ is the mean square of the quantization error or quantization noise at the output. We shall denote the quantization noise power by N_q. Hence

$$N_o = N_q = \frac{(\Delta v)^2}{12} \tag{7.69}$$

Next we calculate the signal power at the output. The output signal $f(t)$ is given by (Eq. 7.68b)

$$\hat{f}(t) = \sum_k \hat{f}_k Sa(\omega_m t - k\pi)$$

Using the results in Example 2.2 (Eq. 2.31), we have

$$\overline{\hat{f}^2(t)} = \overline{(\hat{f}_k)^2}$$

Thus the mean square value of the output signal is equal to the mean square value of the quantized samples. The quantized samples can assume M discrete values, and each is equally probable. This is because we have assumed the amplitude of signal $f(t)$ equally likely to occur in the range 0 to $(M - 1)\,\Delta v$. The sample $\hat{f}_k$ can assume any of the M values $0, \Delta v, 2\,\Delta v, \ldots, (M - 1)\,\Delta v$, with the same probability. Clearly the mean square value of f_k is

$$\overline{(\hat{f}_k)^2} = \frac{1}{M}\{0^2 + (\Delta v)^2 + (2\,\Delta v)^2 + \cdots + [(M - 1)\,\Delta v]^2\}$$

$$= \frac{(\Delta v)^2}{M}\sum_{k=0}^{M-1} k^2$$

$$= \frac{\Delta v^2}{M}\frac{M(M - 1)(2M + 1)}{6}$$

$$= \frac{\Delta v^2(M - 1)(2M + 1)}{6}$$

In practice, $M \gg 1$ and

$$\overline{(\hat{f}_k)^2} \simeq \frac{M^2}{3}(\Delta v)^2$$

Hence

$$S_o = \overline{(\hat{f}_k{}^2)} = \frac{M^2}{3}(\Delta v)^2 \tag{7.70}$$

From Eqs. 7.67, 7.69, and 7.70, we obtain*

$$\frac{(S_o/N_o)}{(S_i/N_i)} = \frac{4M^2}{K^2/2} = \frac{8M^2}{K^2} \tag{7.71}$$

We must now relate the signal-to-noise improvement ratio to the bandwidth of the system. We have seen that for M quantization levels, we need $\log_2 M$ pulses per sample. Since for a signal bandlimited to f_m Hz, we need $2f_m$ samples per second, the system must be capable of transmitting $2f_m \log_2 M$ pulses per second. From discussion in Section 5.4, we observe that a system of bandwidth B is capable of transmitting $2B$ independent pulses per second. Hence to transmit $2f_m \log_2 M$ pulses per second, the bandwidth B required is $f_m \log_2 M$.

$$B = f_m \log_2 M$$

and

$$M^2 = (2)^{2B/f_m}$$

Thus

$$\frac{(S_o/N_o)}{(S_i/N_i)} = \frac{8}{K^2}(2)^{2B/f_m} \tag{7.72}$$

If we consider the S/N root mean square voltage ratio rather than the power ratio, we get

$$\frac{(S_o/N_o)_{\mathrm{vr}}}{(S_i/N_i)_{\mathrm{vr}}} = \frac{\sqrt{8}}{K}(2)^{B/f_m} \tag{7.73}$$

s'ary PCM

We have discussed binary PCM, where the pulses can assume two values only. One can use s'ary pulses instead. In s'ary pulses, the pulses can assume s levels. Thus for a 3'ary PCM, we use pulses with three amplitudes; for example, 0 volt, A volts, and $2A$ volts. For the s'ary case we need a group of $\log_s M$ pulses to represent a sample that can assume any one of M possible value. This is easily seen when we

---

* In this derivation it is implicitly assumed that $f(t) > 0$ (see Fig. 5.10a). Actually, $f(t)$ may be both positive and negative and may have a zero mean value. In such a case $S_o = (M^2/12)(\Delta v)^2$.

consider the fact that a group of k, s'ary pulses can form

$$s \times s \times s \times \cdots \times s = s^k$$
$$\underset{k \text{ times}}{}$$

distinct patterns. Hence to form M distinct pattern we need

$$M = s^k \quad \text{and} \quad k = \log_s M \quad \text{pulses per sample}$$

We assume each level of the s'ary pulse is separated by $K\sigma_n$. The rth-level pulse has an amplitude $rK\sigma_n$. If we assume that the amplitudes of $f(t)$ are equally likely to be anywhere in a given range, then in the transmitted signal (using s'ary pulses), the likelihood of observing a pulse of one level is the same as that of observing a pulse of any other level. Hence the mean square value of the input signal is given by

$$S_i = \frac{1}{s}[(A_0)^2 + (A_1)^2 + \cdots + (A_{s-1})^2] \quad \text{where} \quad A_r = rK\sigma_n$$

$$= \frac{1}{s}\sum_{r=0}^{s-1} A_r{}^2$$

$$= \frac{1}{s}\sum_{r=0}^{s-1} r^2 K^2 \sigma_n{}^2$$

$$= K^2 \sigma_n{}^2 \frac{(s-1)(2s-1)}{6} \tag{7.74}$$

At the receiver, the pulse signal is decoded to obtain quantized samples. These are of course identical to those obtained in binary PCM. It is obvious that the output signal power S_o and the output noise power (quantizing noise power) are the same as in binary PCM. Thus

$$S_o = \frac{M^2}{3}(\Delta v)^2$$

$$N_o = \frac{(\Delta v)^2}{12}$$

Also

$$N_i = \sigma_n{}^2$$

Hence

$$\frac{S_o/N_o}{S_i/N_i} = \frac{24M^2}{K^2(s-1)(2s-1)} \tag{7.75}$$

For an s'ary pulse we need only $\log_s M$ pulses per sample and a total of $2f_m \log_s M$ pulses per second. This calls for a bandwidth B given by

$$B = f_m \log_s M$$

Hence

$$M^2 = s^{2B/f_m}$$

and

$$\frac{S_o/N_o}{S_i/N_i} = \frac{24}{K^2(s-1)(2s-1)} s^{2B/f_m} \tag{7.76}$$

Note that substitution of $s = 2$ in Eq. 7.76 yields Eq. 7.72 as expected. The S/N voltage ratio relationship is given by

$$\frac{(S_o/N_o)_{\mathrm{vr}}}{(S_i/N_i)_{\mathrm{vr}}} = \frac{2\sqrt{6}}{K\sqrt{(s-1)(2s-1)}} s^{B/f_m} \tag{7.77}$$

In this discussion we have used s'ary pulses whose amplitudes ranged from 0 to $(s-1)K\sigma_n$ in steps of $K\sigma_n$. All the pulses here have positive values. The coded signal thus has a d-c value (in this case $K\sigma_n(s-1)/2$). Transmission of this d-c component serves no useful purpose whatsoever. Hence we can subtract this component and reduce the signal power by an amount $[K\sigma_n(s-1)/2]^2$. The new signal power S_i' is given by

$$S_i' = S_i - \left[\frac{K\sigma_n(s-1)}{2}\right]^2$$

$$= K^2\sigma_n^2\left[\frac{(s-1)(2s-1)}{6} - \frac{(s-1)^2}{4}\right]$$

$$= \frac{K^2\sigma_n^2(s^2-1)}{12} \tag{7.78}$$

The subtraction of d-c level $K\sigma_n(s-1)/2$ from the coded signal is equivalent to using s pulses in steps of $K\sigma_n$ in the range $-K\sigma_n(s-1)/2$ to $K\sigma_n(s-1)/2$. Thus merely by using binary pulses (allowing the pulses to assume positive and negative values with a given separation between levels), we have reduced the power by the amount $[K\sigma_n(s-1)/2]^2$. For a binary case, this means using two pulses of height $A/2$ and $-A/2$ instead of 0 and A. The average power of the bipolar scheme is obviously $A^2/4$ and that of the unipolar scheme is $A^2/2$.

The improvement ratio with the reduced input signal power now becomes

$$\frac{S_o/N_o}{S_i{'}/N_i{'}} = \frac{48}{K^2(s^2 - 1)} s^{2B/f_m} \tag{7.79}$$

The S/N voltage ratio relationship is given by

$$\frac{(S_0/N_0)_{\mathrm{vr}}}{(S_i{'}/N_0{'})_{\mathrm{vr}}} = \frac{4\sqrt{3}}{K\sqrt{s^2 - 1}} s^{B/f_m} \tag{7.80}$$

Note that the input signal-to-noise ratio S'/N' is given by (Eq. 7.78)

$$\frac{S_i{'}}{N_i{'}} = \frac{S_i{'}}{\sigma_n{}^2} = \frac{K^2(s^2 - 1)}{12} \tag{7.81}$$

7.6 COMPARISON OF CODED AND UNCODED SYSTEMS

We have now discussed a number of wideband systems which allow exchange of S/N ratio and bandwidth. Examples of uncoded wideband systems are FM (also PM) and PPM. The pulse code modulation is an example of a coded wideband system. In uncoded systems (FM and PPM) we have observed that the improvement in S/N voltage ratio is linear with bandwidth (Eq. 7.53, 7.66b). For the coded system (PCM), the improvement of S/N voltage ratio was observed to be exponential with bandwidth (Eq. 7.80). Thus trading bandwidth and S/N ratio is much more efficient in coded systems than in uncoded systems. Coded systems are inherently capable of better transmission efficiency than uncoded systems.

It will be shown in Chapter 8 that the maximum theoretical attainable improvement in S/N ratio is exponential with bandwidth. Thus coded systems come close to realization of the theoretical maximum efficiency that can be attained. The qualitative reason for the exponential relationship in PCM is not difficult to understand. For uncoded systems, doubling the bandwidth doubles the S/N voltage ratio. But in PCM doubling the bandwidth allows twice the number of pulses to be transmitted. If we double the number of pulses, the possible patterns that can be formed increases as a square. For example, a group of 2 binary pulses can form 4 patterns. But a group of 4 binary pulses can form 16 patterns. Hence doubling the bandwidth squares the quantized level. Similarly, tripling the bandwidth will allow three times the number of pulses. But tripling the number of pulses increases the number of patterns to the power 3. In general, increasing the bandwidth n times increases the number of patterns exponentially (to the nth power). This allows quantized levels to increase exponentially in a given range. Obviously, Δv, the quantum gap, decreases exponentially with the bandwidth. The quantizing noise power (the output noise power) is $(\Delta v/12)^2$ and reduces exponentially. This causes the S/N ratio to improve exponentially.

The most important advantage of a coded system is its convenience for long distance communication by using several repeater stations. At each repeater station, the signal mixed with noise is cleaned up by regenerating the pulses. Thus at each repeater station a noise-free signal is transmitted. In effect, this implies that for PCM one needs to be concerned only about the noise on the link between the repeater stations. The noise does not accumulate throughout the transmission path. Thus the transmission requirements for PCM are almost independent of the total length of the system. Indeed, PCM can be transmitted over any distance by placing enough repeater stations. This is the basic reason for using coded systems. In contrast, for uncoded systems or analog modulation systems the signal is continuously affected by the noise and cannot be cleaned up or regenerated periodically along the path. The noise accumulates throughout the transmission path. The larger the transmission path, the larger will be the noise. This restricts the distance over which such signals can be transmitted.

The reader will recognize the similarity between coded and uncoded systems with the digital and analog systems, respectively. In analog systems, say, analog computers, increasing the voltage representing the physical variable increases the precision linearly. In the digital systems, however, increasing the number of digits increases the precision exponentially. The errors in analog systems accumulate, whereas in digital systems (the digital computer, for example) accumulation does not occur and the noise below a certain level has no influence.

APPENDIX A. JUSTIFICATION FOR CALCULATING OUTPUT SIGNAL AND NOISE POWER INDIVIDUALLY IN FM

At the input of the demodulator, the noise is a bandpass noise with a bandwidth of approximately $2\,\Delta\omega$. The carrier frequency, on the other hand, varies in proportion to the modulating signal $f(t)$ which is bandlimited to ω_m radians per second. For wideband FM, $\Delta\omega \gg \omega_m$. Hence the variation of noise signal is much more rapid than the variations in carrier frequency. It is therefore reasonable to assume the carrier frequency to be constant over a Nyquist interval $1/2f_m$ seconds. This amounts to approximately $f(t)$ by a staircase function as shown in Fig. 4.3a. Thus between any two successive sampling instants, the carrier has a constant frequency. Let us concentrate on one such interval. Over this interval, the carrier is constant. This may be interpreted as if the modulating signal $f(t) = 0$ and the carrier frequency $\omega_k = \omega_c + k_f f(t_k)$. The bandpass noise signal $n_i(t)$ can be expressed with ω_k as the center frequency as follows.

$$n_i(t) = n_c(t) \cos \omega_k t + n_s(t) \sin \omega_k t$$

By following the development of Eq. 7.43 in the text, we obtain

$$S_{n_0}(\omega) = \frac{\alpha^2 \omega^2}{A^2} S_{n_s}(\omega)$$

where

$$S_{n_s}(\omega) = \begin{cases} S_n(\omega - \omega_k) + S_n(\omega + \omega_k) & |\omega| < \Delta\omega + 2\omega_m + k_f f(t_k) \\ 0 & \text{otherwise} \end{cases}$$

Note that the bandwidth of transmission is $2(\Delta\omega + 2\omega_m)$. For white noise

$$S_n(\omega) = \frac{\mathcal{N}}{2}$$

The final baseband filter suppresses all noise in the range $|\omega| > \omega_m$. If $S_n(\omega) = \mathcal{N}/2$, then obviously $S_{n_0}(\omega)$, the power density of the output noise, is

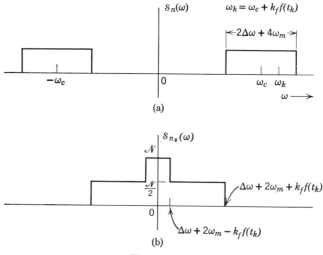

Figure A7.1

given by (Fig. A7.1)

$$S_{n_s}(\omega) = \begin{cases} \dfrac{\alpha^2\omega^2\,\mathcal{N}}{A^2} & |\omega| < \omega_m \\[2mm] 0 & \text{otherwise} \end{cases} \qquad\qquad (A7.1)$$

This result is identical to that in Eq. 7.46. The output noise power N_o is given by

$$N_o = \frac{\alpha^2\,\mathcal{N}}{\pi A^2}\int_0^{\omega_m}\omega^2\,d\omega$$

$$= \frac{2}{3}\left(\frac{\alpha}{A}\right)^2\frac{\mathcal{N}\omega_m{}^3}{2\pi} \qquad\qquad (A7.2)$$

This result is the same as that in Eq. 7.47. Thus the noise power over each interval is identical and is given by Eq. 7.47. Obviously, the noise power over the entire interval will be the same.

It should be noted that the justification given here holds only if $\Delta\omega \gg \omega_m$ (wideband FM) and the noise is white.

APPENDIX B. SIGNAL-TO-NOISE RATIO IN TIME DIVISION MULTIPLEXED PAM SYSTEMS

Let us consider the time division multiplexed PAM system shown in Fig. A7.2. This is the same arrangement as the one discussed in Section 5.4. There are M message signals $f_1(t), f_2(t), \ldots, f_M(t)$, each bandlimited to B Hz. These signals are sampled individually and all the samples are interleaved as in Section 5.3. The multiplexed samples are now transmitted through a low-pass filter of cutoff frequency MB Hz. The output of this filter is a continuous signal whose amplitudes at the sampling instants are identical to the sample values.* Let us denote the output of this filter by $\varphi(t)$. To calculate the power of $\varphi(t)$, we use the result in Eq. 2.31, which states that for a bandlimited signal $f(t)$, the mean square value of the signal is equal to the mean square value of its samples

$$\overline{f^2(t)} = \overline{f_k{}^2}$$

where $\overline{f_k{}^2}$ is the mean square value of the samples. This is obtained by adding the square of the samples (taken over a large time interval) and dividing by the

---

* Actually, the amplitudes of the continuous signal are $1/T$ times the sample values where T is the sampling interval (in this case $T = 1/2MB$). This can be seen from Eq. 5.1. For convenience, we shall assume an ideal filter with a gain of T, so that the output of the filter is a continuous signal whose amplitudes at the sampling instant are identical to the corresponding samples. This assumption does not affect our calculations in any way; it merely adds a gain factor in the system.

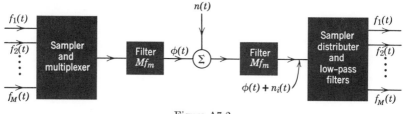

Figure A7.2

total number of samples. If $f(t)$ is bandlimited to B Hz, then

$$\overline{f_k{}^2} = \lim_{\tau \to \infty} \frac{1}{2B\tau} \sum_k f_k{}^2$$

Here f_k is the kth sample of $f(t)$, and $2B\tau$ is the total number of samples over τ seconds (at a rate $2B$ samples per second).

It is now obvious that the mean square value of $\varphi(t)$ is the mean square value of its samples

$$\overline{\varphi^2(t)} = \overline{\varphi_k{}^2}$$

But the samples of $\varphi(t)$ are the samples of M signals $f_1(t), f_2(t), \ldots, f_M(t)$ interleaved. Let us, for convenience, assume that all these signals have identical powers (identical mean square values):

$$\overline{f_1{}^2(t)} = \overline{f_2{}^2(t)} = \cdots = \overline{f_M{}^2(t)} = \overline{f^2(t)}$$

Obviously, the mean of sample squares of each signal will be $\overline{f^2(t)}$. If all these samples are interleaved, the mean (of the square) will remain changed since all samples have the same mean (of the square). Thus

$$\overline{\varphi^2(t)} = \overline{f^2(t)}$$

Thus the power (mean square value) of $\varphi(t)$ is the same as that of individual message signals. The mean square of $\varphi(t)$ is the input signal power S_i

$$S_i = \overline{f^2(t)}$$

The bandwidth of $\varphi(t)$ is MB and hence the input filter at the receiver has a bandwidth MB. If we assume the channel noise to be white, with a power density $\mathcal{N}/2$, then the power (the mean square value) of the input noise $n_i(t)$ is given by

$$\overline{n_i{}^2(t)} = 2(MB)\frac{\mathcal{N}}{2} = M\mathcal{N}B$$

This is the input noise power N_i. Hence

$$N_i = M\mathcal{N}B$$

Note that the mean square value of the noise samples will also be $M\mathcal{N}B$. Thus

$$\overline{n_k{}^2} = M\mathcal{N}B$$

Next, at the detector, the continuous input signal $\varphi(t) + n_i(t)$ is sampled at a rate $2MB$ samples per second. The kth sample is $(\varphi_k + n_k)$, the sum of the kth sample of $\varphi(t)$ and the kth sample of $n_i(t)$. The mean square value of φ_k is $\overline{f^2(t)}$ and the mean square value of n_k is $M\mathcal{N}B$. These samples are distributed to M channels, successively, by the detector. The mean square values of the samples in each of the M channels, however, remain unchanged. These samples are now transmitted through a low-pass filter to yield M individual signals. The mean square value of each signal is $\overline{\varphi_k{}^2} = \overline{f^2(t)}$ and the mean square value of the noise output in each channel will be $\overline{n_k{}^2} = M\mathcal{N}f_m$. Hence

$$S_o = \overline{f^2(t)}$$

$$N_o = M\mathcal{N}B$$

Therefore

$$\frac{S_o/N_o}{S_i/N_i} = 1$$

PROBLEMS

1. A white noise of power density $\mathcal{N}/2$ is transmitted through a bandpass filter whose transfer function is shown in Fig. P-7.1. The output of the

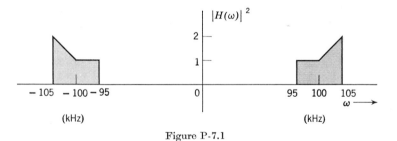

Figure P-7.1

filter is a bandpass noise. Express this signal in terms of quadrature components as in Eq. 7.10. Determine the power density spectra and the mean square values of $n_c(t)$ and $n_s(t)$.

2. A noise signal with a power density spectrum $S_n(\omega)$ as shown in Fig. P-7.2a is transmitted through an ideal bandpass filter in Fig. P-7.2b. Express the output signal in terms of quadrature components as in Eq. 7.10.

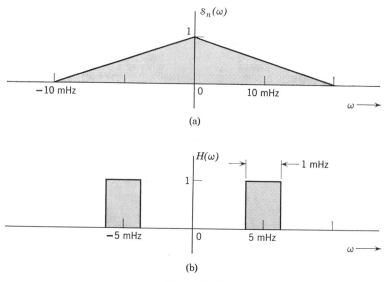

Figure P-7.2

3. A certain channel has a uniform noise power density spectrum $S_n(\omega) = 0.5 \times 10^{-3}$. Over this channel a DSB-SC signal is transmitted. The modulating signal $f(t)$ is bandlimited to 5 kHz and the carrier frequency is 100 kHz. The power of the modulated signal (sidebands) is given to be 10 kW. The incoming signal at the receiver is filtered through an ideal bandpass filter before it is fed to the demodulator.

(a) What must be the transfer function of this filter?
(b) What is the signal-to-noise power ratio at the demodulator input?
(c) What is the signal-to-noise power ratio at the demodulator output?
(d) Find and sketch the noise power density spectrum at the output of the demodulator.

4. Consider the channel in Problem 3 with the uniform noise power density $S_n(\omega) = 0.5 \times 10^{-3}$. Over this channel a SSB-SC signal is transmitted. The modulating signal $f(t)$ is bandlimited to 5 kHz and the carrier frequency is 100 kHz. The power of the modulated signal is 10 kW. The incoming signal at the receiver is filtered through an ideal bandpass filter before it fed to the demodulator. Consider the upper sideband (USB) transmission.

(a) What must be the transfer function of the input bandpass filter?
(b) What is the signal-to-noise power ratio at the demodulator input?
(c) What is the signal-to-noise power ratio at the demodulator output?

(d) How does this ratio compare with that for DSB-SC in Problem 3?
Comment.

Repeat this problem for the lower sideband (LSB) case. Comment.

5. The power density spectrum $S_n(\omega)$ of a certain channel noise is shown
in Fig. P-7.5. Over this channel a DSB-SC signal is transmitted. The
modulating signal $f(t)$ is bandlimited to 5 kHz and the carrier frequency is
100 kHz. The power of the modulated signal is given to be 10 kW. The

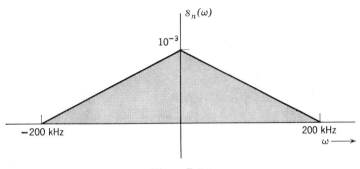

Figure P-7.5

incoming signal at the receiver is filtered through an ideal bandpass filter
before it is fed to the demodulator.

(a) What must be the transfer function of this filter?

(b) What is the signal-to-noise power ratio at the demodulator input?·

(c) What is the signal-to-noise power ratio at the demodulator output?

(d) Find and sketch the noise power density at the demodulator output.

6. Consider the channel in Problem 5 with the noise power density
spectrum shown in Fig. P-7.5. Over this channel an SSB-SC signal is transmitted. The modulating signal $f(t)$ (same as in Problem 5) is bandlimited to
5 kHz and the carrier frequency is 100 kHz. The modulated signal power
is given to be 10 kW. The incoming signal at the receiver is filtered through
an ideal bandpass filter before it is fed to the demodulator.

Consider the upper sideband (USB) transmission.

(a) What must be the transfer function of the bandpass filter at the
receiver input?

(b) What is the signal-to-noise power ratio at the demodulator input?

(c) What is the signal-to-noise power ratio at the demodulator output?

(d) How does the ratio in (c) compare with the output signal-to-noise
ratio for DSB-SC in Problem 5? Comment.

(e) Find and sketch the noise power density at the demodulator output.

(f) Repeat the problem for lower sideband (LSB) case. Comment.

7. A channel with a white noise (power density $S_n(\omega) = 0.5 \times 10^{-3}$) is used to transmit an AM signal. The modulating signal $f(t)$ is bandlimited to 5 kHz and the carrier frequency is 100 kHz. The sideband power is given to be 10 kW and the carrier amplitude is 200 volts (carrier power $= 40$ kW). At the receiver, the signal is demodulated by an envelope detector after being passed through an appropriate ideal bandpass filter.

(a) Find the signal to-noise ratio at the demodulator input.

(b) Find the signal-to-noise ratio at the demodulator output.

8. Repeat Problem 7 if the power density spectrum of the noise is as shown in Fig. P-7.5.

9. A signal $f(t)$ has a mean square value $\frac{4}{3}[\overline{f^2(t)} = \frac{4}{3}]$. The positive peak value of $f(t)$ is 2 volts and the negative peak value is -2 volts. It is also given that $f(t)$ is bandlimited to 5 kHz. The signal is transmitted by FM on a channel with a white noise. Find the bandwidth of transmission if the improvement in the signal-to-noise ratio at the demodulator output is required to be 60. (*Hint:* Use Eq. 7.48 and recall that $\Delta\omega = k_f|f(t)|_{\max} = 2k_f$.)

10. In a wideband FM system, the carrier amplitude is 100 volts and the frequency is 100 mHz. The modulating signal is bandlimited to 5 kHz and its mean square value is 5000. The modulation constant k_f is 500π and the maximum carrier frequency deviation $\Delta f = 75$ kHz. The noise power density on the channel is uniform and is given by $S_n(\omega) = 10^{-3}$.

(a) Find the transfer function of the ideal bandpass filter at the receiver input.

(b) Find the signal-to-noise power ratio at the input of the demodulator.

(c) Find the signal-to-noise power ratio at the output of the demodulator.

(d) Find and sketch the noise power density at the demodulator output.

(e) If the signal $f(t)$ is transmitted by AM, find the signal-to-noise ratio at the output and compare it with that obtained in FM.

11. Consider the FM system in Problem 10. It is decided to use pre-emphasis, de-emphasis to reduce the noise at the output. In the pre-emphasis network, the break frequencies are 1.5 kHz and 30 kHz, respectively. Find the signal-to-noise ratio at the demodulator output. Compare this with the one obtained in Problem 10.

12. The signal-to-noise improvement ratio in PPM is given by Eq. 7.66a. The improvement ratio is proportional to $k_1^2 \overline{f^2(t)} B$. Thus the ratio may be improved by increasing B, $\overline{f^2(t)}$, or k_1. By increasing B we increase the bandwidth of the transmission, and by increasing $\overline{f^2(t)}$ we increase the signal power. Thus we must sacrifice something to improve the S/N ratio. What is being sacrificed if the improvement is achieved by increasing k_1? Can

one increase k_1 indefinitely and improve the S/N ratio? From Eq. 7.66a, it is seen that the S/N ratio can also be improved by reducing α. What is being sacrificed in reducing α? Can α be reduced indefinitely? Discuss. Also discuss what will happen if the channel bandwidth is made too small.

13. A modulating signal $f(t)$ is transmitted by PPM. It is given that $|f(t)|_{max} = 2$ volt, $\overline{f^2(t)} = \frac{4}{3}$. The signal $f(t)$ is bandlimited to 5 kHz. It is sampled at a rate 10,000 samples per second and these sample values are transmitted by PPM. The channel bandwidth is 300 kHz and the pulse width $\tau = 20$ μs. What is the maximum allowable value of k_1? For this value of k_1, determine the improvement in the signal-to-noise ratio.

To obtain this improvement by using FM, how much bandwidth will be required?

14. The signal-to-noise power improvement ratio in Eq. 7.72 is derived for unipolar binary pulse, that is, the two pulses assume values 0 and $K\sigma_n$. Derive this ratio when bipolar pulses of amplitude

$$\frac{-K\sigma_n}{2} \quad \text{and} \quad \frac{K\sigma_n}{2}$$

are used instead.

15. Consider again the signal $f(t)$ in Problem 13 where $|f(t)|_{max} = 2$ volts and $f^2(t) = \frac{4}{3}$. The signal $f(t)$ is bandlimited to 5 kHz and is sampled at a rate 10,000 samples per second. These samples are quantized and coded into binary pulses. The quantizing levels are $\frac{1}{32}$ volts apart. The root mean square value of the channel noise is 0.1 volt. (Assume the amplitudes of signal $f(t)$ to be uniformly distributed in the range -2 to 2.) Assume unipolar binary pulses of amplitude 0 and $K\sigma_n$ where $K = 10$. Find the following:

(a) The bandwidth of transmission.

(b) The signal-to-noise ratio at the demodulator input.

(c) The signal-to-(quantizing) noise ratio at the demodulator output.

(d) The S/N improvement ratio.

(e) To achieve this S/N improvement by FM, what bandwidth will be required?

Repeat the problem for bipolar binary pulses of amplitude

$$\frac{-K\sigma_n}{2} \quad \text{and} \quad \frac{K\sigma_n}{2} \quad (K = 10)$$

Compare this system with the PPM in Problem 13.

16. Repeat Problem 15 if 4'ary pulses are used to transmit the signal instead of binary pulses and the quantizing level is set at $\frac{1}{64}$ volt. Calculate for unipolar and bipolar 4'ary pulses. Assume $K = 10$ for error-free detection.

chapter 8

Introduction to Information Transmission

8.1 MEASURE OF INFORMATION

The purpose of communication is to convey information. We shall discuss the nature of information content in a message from an intuitive angle and also from an engineering point of view. Surprisingly, both viewpoints lead to the same quantitative definition of the unit of information.

1. Intuitive Point of View

Every message conveys some information, but some messages convey more information than others. A close scrutiny shows that the probability of occurrence of an event is closely related to the amount of information. If someone tells us the occurrence of a highly probable event, he conveys little information compared to that which will be conveyed if the event were less probable. The element of surprise or uncertainty in the occurrence of an event appears to be proportional to the amount of information. If someone says that the sun rises in the east, it conveys absolutely no information because everyone knows that it is true. There is no uncertainty in the event that the sun rises in the east every day. In other words, the probability of the event that the sun rises in the east every day is unity. However, if on a day in

January, a weatherman from a national broadcasting service says that the temperature in Minneapolis today reached 150°F, the statement would convey a tremendous amount of information. This is because this event is totally unexpected and its probability of occurrence is very small ($P \rightarrow 0$). In other words, this event is very uncertain and so its information tends to be very large.

Consider now the statement from a news service, "United States invades Cuba." The statement definitely conveys a large amount of information because the event has a very small probability, and hence the statement comes as a surprise. But the surprise is not nearly as great as if the statement were "Cuba invades United States." This is because the probability of the latter statement is extremely small compared to that of the former. The surprise, of course, comes as a result of uncertainty or unexpectedness. The more unexpected the event is, more the surprise and hence more the information. The probability of an event is the measure of expectedness, hence it is related to the information content of the event.

On an intuitive basis, the amount of the information received from the knowledge of occurrence of a certain event is inversely related to the probability of occurrence of that event. What must be the nature of this relationship? It is obvious that if the event is certain (probability 1), it conveys zero amount of information. On the other hand, if the event is impossible (probability zero), then its occurrence conveys an infinite amount of information. This suggests that the amount of information should be a logarithmic function of the reciprocal of the event probability.

$$\text{Information } I \sim \log \frac{1}{P} \tag{8.1}$$

where P is the probability of the occurrence of the event and I is the amount of information received from the knowledge of the occurrence of the event.

2. Engineering Point of View

We shall now show from the engineering point of view that the information of an event is identical to that obtained on the intuitive basis (Eq. 8.1). What do we mean by an engineering point of view? An engineer is concerned with communicating information bearing messages efficiently. From his point of view an amount of information in a message is proportional to the time required to transmit the message. We shall now see that this concept of information also leads to Eq. 8.1, which implies that it takes shorter time to transmit a message with higher certainty (or higher probability) compared to that required to transmit messages of lower probability. This fact may be easily verified from the example of transmission of alphabetic symbols in the English language by using Morse code.

Morse code is made up of various combinations of two symbols (such as mark and space or pulses of height a and $-a$ volts). Each alphabet is represented by a certain combination of these symbols and has a certain length. Obviously, shorter code words are assigned to letters e, t, a, and o, which occur more frequently. The longer code words are assigned to letters x, k, q, and z, which occur with less frequency. Each of the alphabets may be considered as a message. It is obvious that the letters with higher certainty (higher probability) need shorter time to transmit (shorter code words) than do those with smaller probability. We shall now show that the minimum time required to transmit a symbol (or a message) with probability P is indeed proportional to log $(1/P)$.

We begin by assuming that we are required to transmit either of the two equiprobable messages a and b. These may, for example, be weather reports such as 'sunny' and 'rainy.' We can transmit each of these messages by a suitable waveform. Assuming that we use binary pulses to transmit these messages, we may assign no pulse (pulse of zero volt) to message a (sunny) and a pulse (of 1 volt) to the message b (rainy).

It is evident that a minimum of 1 binary pulse is required to transmit any one of the two equiprobable messages. The information in either of them is accordingly defined as 1 bit (**BI**nary uni**T**). Note that the length of the message has nothing to do with its information content. We will always need one binary pulse to transmit any one of two possible equiprobable messages, regardless of their length or any other characteristic. It is also evident that one binary pulse is capable of transmitting one bit of information.

Next consider the case of 4 equiprobable messages. If these messages are transmitted by binary pulses, we need a group of two binary pulses per message. Each binary pulse can assume two states, hence a combination of two pulses form four distinct patterns which are assigned to each of the four messages (Fig. 8.1). We therefore need two binary pulses to transmit any of the four equiprobable messages. Each of these

Symbol	Binary digit equivalent	Binary pulse waveform	Quaternary digit equivalent	Quaternary pulse waveform
A	00	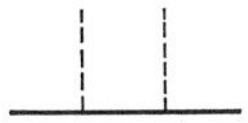	0	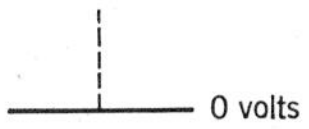 0 volts
B	01		1	 1 volt
C	10		2	2 volts
D	11		3	3 volts

Figure 8.1

messages takes twice as much time as that required to transmit one of the two equiprobable messages and hence it contains twice as much information, that is, 2 bits.

Similarly, we can transmit any one of eight equiprobable messages by a group of 3 binary pulses. This is because 3 binary pulses form 8 distinct patterns which can be assigned to each of the eight messages. Hence each one of the eight equiprobable messages contains 3 bits of information. It can be easily seen that, in general, any one of the n equiprobable messages contains $\log_2 n$ bits of information according to the engineering basis of information. To reiterate, the amount of information contained in any one of the n equiprobable messages is equal to $\log_2 n$ bits. This implies that a minimum of $\log_2 n$ binary pulses are required to transmit such a message. Note that P, the probability of occurrence of any one of these events, is $1/n$. Hence

$$\text{Information } I = \log_2 n$$
$$= \log_2 \frac{1}{P} \tag{8.2}$$

We have shown this result for a highly special case, that of equiprobable messages. It can also be shown that even if the messages are not equiprobable, an average of $\log_2 (1/P)$ binary pulses are required to transmit a message with probability P. The proof is given in the appendix (at the end of this chapter).

From this discussion it is evident that the information measure (in bits) of a message is equal to the minimum number of binary pulses required to encode that message.

This definition of information may appear, superficially, to be too restrictive since it applies only to information of a discrete nature, such as the transmission of some discrete and finite number of symbols or messages. However, it is a major result of information theory that any form of information to be transmitted can always be represented in binary form without a loss of generality. We have already seen in the preceding chapter that the information in a continuous bandlimited signal can be represented by a discrete number of sample values per second. It is now possible to represent these sample values by a code that uses binary pulses.

We shall show in the next section that each communication system (or channel) is capable of transmitting a certain amount of information per second. This is called the channel capacity C. Thus a given channel can transmit a quantity of information which is, at most, C bits per second. The channel capacity will be shown to be limited by the bandwidth and the signal-to-noise power ratio of the system.

Instead of binary pulses, we may use M'ary pulses (pulses which can assume M distinct values) for the purpose of encoding. We shall now show that each M'ary pulse can carry an information of amount $\log_2 M$ bits.

To show this, let us assume that we are required to transmit any one of four equiprobable messages instead of two. Obviously, it is not possible to transmit this information by a single binary pulse since it can assume only two states. But we can transmit any of the four messages by a group of two binary pulses. Each binary pulse can assume two states, and hence a combination of two pulses will form four distinct patterns as shown in Fig. 8.1. The zero state of a pulse (absence of a pulse) is shown by a dotted line. We therefore need two binary pulses to transmit any of the four equiprobable messages. Hence, the information transmitted per message is 2 bits.

Alternatively, we may transmit this information by a quaternary pulse which can assume four states or four levels, for example, 0, 1, 2, and 3 volts. Each state corresponds to one of the four possible symbols. It is obvious that any one of the four possible symbols can be transmitted by a single quaternary pulse (Fig. 8.1). It follows that a single quaternary pulse can transmit the information carried by two binary pulses and hence carries 2 bits of information. Similarly, if it is desired to transmit any one of eight possible messages, we need a group of three binary pulses for each symbol. Since each binary pulse has two states, a combination of three pulses will yield eight distinct patterns. Each of the possible eight messages may also be transmitted by a single 8'ary pulse (a pulse which can assume eight states or eight values). Hence a single 8'ary pulse carries 3 bits of information. It is easy to see that a pulse which can assume M distinct states or M distinct levels carries an information of $\log_2 M$ bits.

It therefore follows that the larger the number of distinct levels that a pulse can assume, the larger the information carried by each pulse. A pulse that can assume an infinite number of distinct levels carries an infinite amount of information. This means that it is possible to transmit any amount of information by a single pulse which can assume an infinite number of distinct levels. This result appears to be fantastic, but it is perfectly logical and sound. If a pulse can assume an infinite number of distinct levels, then it is possible to assign one of the levels to any conceivable message or signal no matter how large. For example, we may assign one of the infinite levels to represent the complete contents of this book. Now, if it is desired to transmit the complete contents of this book, all that is required is to transmit one pulse of the corresponding level. Since there are an infinite number of levels available, it is possible to assign one level to any conceivable message or signal of any length in this universe. Cataloging of the code in such a case may prove to be next to impossible, nevertheless, it illustrates the possibility of transmission of an infinite amount of information by a single pulse.

At this point one may wonder why we do not use pulses that can assume an infinite number of distinct levels. There are limitations imposed upon the system by practical considerations. It should be remembered that in all of our discussion when we talk about the transmission of information by transmitting pulses we refer to the composite system that transmits the information at the transmitter and receives it at the destination. Hence, to transmit certain information, we must be able to transmit as well as receive these pulses. Moreover, we must be able to recognize the distinct levels of the pulses. Now, what can prevent us from transmitting pulses that assume an infinite number of distinct states? It is obvious that since practical considerations require that the pulses must have finite amplitudes, the infinite number of distinct states implies that each state is separated from the neighboring state by an infinitesimal amount. Since, in any practical channel, there is always a certain amount of noise signal present, it will be impossible to distinguish at the receiver levels lying within the noise signal amplitude. Hence the noise consideration requires that the levels must be separated at least by the noise signal amplitude.

8.2 CHANNEL CAPACITY

It was stated earlier that the bandwidth and the noise power place a restriction upon the rate of information that can be transmitted by a channel. It can be shown rigorously that in a channel disturbed by a white Gaussian noise, one can transmit information at a rate of, at most, C bits per second, where C is the channel capacity given by

$$C = B \log_2 \left(1 + \frac{S}{N} \right) \tag{8.3}$$

B is the channel bandwidth in Hz, S is the signal power, and N is the noise power. The expression for channel capacity in Eq. 8.3 is valid for white Gaussian noise. For other types of noise, the expression is modified.

The rigorous proof of the channel capacity formula is somewhat involved and is beyond the scope of this book.* Here we shall present a very crude proof based upon the plausible assumption that if a signal is mixed with noise, the signal amplitude can be recognized only within the root mean square noise voltage. In other words, the uncertainty in recognizing the exact signal amplitude is equal to the root mean square noise voltage.

Let us assume that the average signal power and the noise power are S watts and N watts, respectively. If we assume a load of 1 ohm, then the root mean square value of the received signal is $\sqrt{S + N}$ volts and the root mean square value of the noise voltage is $\sqrt{N}$ volts. We now want to distinguish the received signal of the amplitude $\sqrt{S + N}$ volts in the presence of the noise amplitude $\sqrt{N}$ volts. It follows from our

* See Bibliography.

assumption that the input signal variation of less than $\sqrt{N}$ volts will be indistinguishable at the receiver. Consequently, the number of the distinct levels that can be distinguished without error will be given by

$$M = \frac{\sqrt{S+N}}{\sqrt{N}} = \sqrt{1 + \frac{S}{N}} \qquad (8.4)$$

Hence the maximum value of M is given by Eq. 8.4. The maximum amount of information carried by each pulse having $\sqrt{1 + S/N}$ distinct levels is given by

$$I = \log_2 \sqrt{1 + \frac{S}{N}} \qquad (8.5)$$

$$= \tfrac{1}{2} \log_2 \left(1 + \frac{S}{N}\right) \qquad \text{bits} \qquad (8.6)$$

We are now in a position to determine the channel capacity. The channel capacity is the maximum amount of information that can be transmitted per second by a channel. If a channel can transmit a maximum of K pulses per second, then, obviously, the channel capacity C is given by

$$C = \frac{K}{2} \log_2 \left(1 + \frac{S}{N}\right) \qquad \text{bits per second} \qquad (8.7)$$

In Chapter 5, in connection with the bandwidth requirements of PAM signals, we showed that a system of bandwidth nf_m Hz can transmit $2nf_m$ independent pulses per second. It was shown that under these conditions the received signal will yield the correct values of the amplitudes of the pulses but will not reproduce the details of the pulse shapes. Since we are interested only in the pulse amplitudes and not their shapes, it follows that a system with bandwidth B Hz can transmit a maximum of $2B$ pulses per second. Since each pulse can carry a maximum information of $\tfrac{1}{2} \log_2 (1 + S/N)$ bits, it follows that a system of bandwidth B can transmit the information at a maximum rate of

$$C = B \log_2 \left(1 + \frac{S}{N}\right) \qquad \text{bit per second} \qquad (8.8)$$

The channel capacity C is thus limited by the bandwidth of the channel (or system) and the noise signal. For a noiseless channel, $N = 0$ and the channel capacity is infinite. In practice, however, N is always finite and hence the channel capacity is finite.*

Equation 8.8 is known as the Hartley-Shannon law and is considered the central theorem of information theory.† It is evident from this theorem that the bandwidth and the signal power can be exchanged for one another. To transmit the information at a given rate, we may reduce the signal power transmitted, provided that the bandwidth is increased correspondingly. Similarly, the bandwidth may be reduced if we are willing to increase the signal power. As stated before, the process of modulation is really a means of effecting this exchange between the bandwidth and the signal-to-noise ratio. The improvement in the signal-to-noise ratio in wideband FM and PCM can be properly understood in the light of this theorem.

It must be remembered, however, that the channel capacity represents the maximum amount of information that can be transmitted by a channel per second. To achieve this rate of transmission, the information has to be processed properly or coded in the most efficient manner. The fact that such a coding is possible is one of the significant results

---

* This is true even if the bandwidth B is infinite. The noise signal is a white noise with a uniform power density spectrum over the entire frequency range. Hence as the bandwidth B is increased, N also increases and thus the channel capacity remains finite even if $B = \infty$. If $\mathcal{N}/2$ is the power density, then $N = \mathcal{N}B$ and

$$C = B \log_2 \left(1 + \frac{S}{\mathcal{N}B} \right)$$

and

$$\lim_{B \to \infty} C = \frac{S}{\mathcal{N}} \frac{\mathcal{N}B}{S} \log_2 \left(1 + \frac{S}{\mathcal{N}B} \right)$$

The last limit can be found by noting that

$$\lim_{x \to 0} \frac{1}{x} \log_2 (1 + x) = \log_2 e = 1.44$$

Hence

$$\lim_{B \to \infty} C = \frac{S}{\mathcal{N}} \log_2 e = 1.44 \frac{S}{\mathcal{N}}$$

† Information theory is a body of results based on a particular quantitative definition of an amount of information and is a subdivision of a broader field—the statistical theory of communication—which includes all of the probabilistic analysis of communications problems. See, for instance, P. Elias, "Information Theory" in Grabbe, Ramo, and Wooldridge, *Handbook of Automation, Computation and Control*, Vol. 1, Ch. 16, John Wiley and Sons, New York, 1961.

attributed to Shannon in information theory. Actually, however, not all of the communication systems used (uncoded systems such as AM, FM, etc.) achieve this maximum rate.

8.3 TRANSMISSION OF CONTINUOUS SIGNALS

We shall illustrate the implications of the Hartley-Shannon law regarding the exchange of bandwidth and signal-to-noise ratio by a continuous signal bandlimited to f_m Hz. From the sampling theorem, the information of such a signal is completely specified by $2f_m$ samples per second. Thus to transmit the information of such a signal, it is necessary to transmit only these discrete samples.

The next important question is: How much information does each sample contain? It depends upon how many discrete levels or values the samples may assume. Actually, these samples can assume any value, and hence to transmit such samples we need pulses capable of assuming infinite levels. Obviously, the information carried by each sample is infinite bits. Hence the information contained in a continuous bandlimited signal is infinite. In the presence of noise (finite value of N), the channel capacity is finite. It is therefore impossible to transmit complete information in a bandlimited signal by a physical channel in the presence of noise (existing over the same band). In the absence of noise, $N = 0$, the channel capacity is infinite, and any desired signal can be transmitted. Clearly, it is impossible to transmit the complete information contained in a continuous signal unless the transmitted signal power is made infinite. Due to the presence of noise, there is always a certain amount of uncertainty in the received signal. The transmission of complete information in a signal would mean a zero amount of uncertainty. Actually, the amount of uncertainty can be made arbitrarily small by increasing the channel capacity (by increasing the bandwidth and/or increasing the signal power), but it can never be made zero.

It is important to realize that the uncertainty is introduced in the process of transmission. Hence, although it is possible to transmit the complete information in a continuous signal at the transmitter end, it is impossible to recover this infinite amount of information at the receiver. The amount of information that can be recovered per second at the receiver is, at most, C bits per second where C is the channel capacity. This is precisely what happens when we transmit a continuous signal by direct transmission, such as by AM and FM. In these cases, the complete information in a signal is transmitted at the transmitter. But since a channel has a finite capacity C bits per second, at most, information of only C bits per second can be recovered at the receiver.

Alternatively, instead of transmitting all of the information at the transmitter, we can approximate the signal so that its information contents are reduced to C bits per second and transmit this approximated signal which has a finite information content. It will now be possible to recover all of the information that has been transmitted. This is precisely what is done in pulse code modulation. How can we approximate a signal so that the approximated signal has a finite information content per second? This can be done by a process of quantization discussed in Chapter 7. Consider the continuous signal bandlimited to f_m Hz as shown in Fig. 8.2. To transmit the information in this signal, we need to transmit only $2f_m$ samples per second. The samples are also shown in the figure. As stated before, the samples can take any value, and to transmit them directly we need pulses which can assume an infinite number of levels. (We have seen that although it is possible to transmit such pulses at the transmitter end, it is impossible to recover the exact heights of these pulses at the receiver due to noise.) Therefore, instead of transmitting the exact values of these pulses, we round off the amplitudes to the nearest one of the finite number of permitted

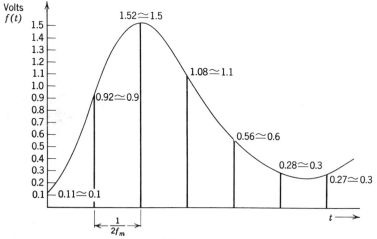

Figure 8.2

values. In this example, all of the pulses are approximated to the nearest tenth of a volt. It is evident from the figure that each of the pulses transmitted assumes any one of the 16 levels, and hence carries an amount of information of $\log_2 16 = 4$ bits. Since there are $2f_m$ samples per second, the total information content of the approximated signal is $8f_m$ bits per second. If the channel capacity is greater than or equal to $8f_m$ bits per second, all of the information that has been transmitted will be recovered completely without any uncertainty. This means that the received signal will be an exact replica of the approximated signal that was transmitted.

It is now problematic whether the noise introduced in the process of transmission may not further cause an additional degree of uncertainty and thus increase the total uncertainty more than 0.1 volt in the received signal amplitude. We can easily show that, if the channel capacity is $8f_m$ bits per second, the process of transmission does not introduce an additional degree of uncertainty. Suppose we are using a channel of bandwidth f_m Hz to transmit these samples, then, since the channel capacity required will be $8f_m$ bits per second, the required signal-to-noise power ratio will be given by

$$8f_m = f_m \log_2 \left(\frac{S + N}{N} \right)$$

Hence

$$\frac{S + N}{N} = 256$$

We have seen before that the number of levels that can be distinguished at the receiver is $\sqrt{(S + N)/N}$. It follows that in this case the receiver can distinguish the 16 states without error. Hence, although the process of transmission introduces some noise in the desired signal, the levels are far enough apart to be distinguishable at the receiver. This is another way of saying that a channel of the capacity of $8f_m$ bits per second can transmit information of $8f_m$ bits per second virtually without error.

8.4 EXCHANGE OF BANDWIDTH FOR SIGNAL-TO-NOISE RATIO

A given signal can be transmitted with a given amount of uncertainty by a channel of finite capacity. We have seen that a given channel

capacity may be obtained by any number of combinations of bandwidth and signal power. In fact, it is possible to exchange one for the
other. We shall now demonstrate how such an exchange can be effected.

Consider the transmission of signal $f(t)$ shown in Fig. 8.2. We have
seen that if an uncertainty of 0.1 volt is tolerated, the information content of the signal is given by $8f_m$ bits per second. It will now be shown
that this information can be transmitted by various combinations of
bandwidths and signal power.

One possible way of transmission is to send $2f_m$ samples per second
directly. Each of the samples can assume any of the 16 states
(sixteenary pulse). In this case we must have a signal-to-noise ratio
which allows us to distinguish 16 states. Obviously, $\sqrt{S+N}/\sqrt{N} = 16$.
Moreover, in order to transmit $2f_m$ pulses per second, we need a channel
of bandwidth f_m Hz. Hence the required channel capacity C is given by
(Eq. 8.8)

$$C = f_m \log_2 \frac{S+N}{N}$$

$$= f_m \log_2 (16)^2$$

$$= 8f_m \qquad \text{bits per second}$$

Thus the channel capacity is exactly equal to the amount of information
per second in the signal $f(t)$.

Alternatively, we may transmit the samples in Fig. 8.2 by quaternary
pulse (pulses that can assume four states). It is evident that we need a
group of two quaternary pulses to transmit each sample that can assume
16 states. Now the signal-to-noise ratio required at the receiver to
distinguish pulses that assume four distinct states is $\sqrt{S+N}/\sqrt{N} = 4$.
It is evident that in this mode of transmission the required signal power
is reduced. However, we now have to transmit twice as many pulses
per second, that is, $4f_m$ pulses per second. Hence the bandwidth
required is $2f_m$ Hz. The channel capacity C in this case is

$$C = 2f_m \log_2 \frac{S+N}{N}$$

$$= 2f_m \log_2 (4)^2$$

$$= 8f_m \qquad \text{bits per second}$$

It is evident from this example that a given amount of information can
be transmitted by various combinations of signal power and bandwidth

and one can be exchanged for the other. The signal $f(t)$ can also be transmitted by binary pulses ($8f_m$ pulses per second), which requires $\sqrt{S + N}/\sqrt{N} = 2$ and a channel bandwidth of $4f_m$ Hz. It is interesting to note that $f(t)$ may also be transmitted by a channel of bandwidth lower than f_m Hz if enough signal power is transmitted (see Problem 3 at the end of this chapter).

It should be observed that the process of exchange of bandwidth and signal power is not automatic. We have to modify or transform the signal information (coding) to occupy the desired bandwidth. In practice, this is accomplished by various types of modulation. However, it should be recognized that not every communication system achieves the full capabilities inherent in its bandwidth and the signal power used. Some modulation forms prove superior to others in utilizing the channel capacity. We have demonstrated in Chapter 7 that coded systems are superior to uncoded systems (such as FM, PPM) in effecting the exchange of bandwidth for S/N ratio.

From the Shannon-Hartley law we can derive the ideal law for the exchange between bandwidth and S/N ratio. Consider a message which has a bandwidth f_m Hz. Let us assume that the information contents of this signal are I bits per second. Assume that this signal is so encoded (or modulated) that the resulting bandwidth is B Hz. The modulation of the signal does not change the information contents of the signal in any way. The modulated signal is now applied to the demodulator input (Fig. 8.3). Let the signal and the noise power at the demodulator input be S_i and N_i, respectively. It is obvious that

$$I = B \log_2 \left(1 + \frac{S_i}{N_i}\right) \tag{8.9}$$

The demodulator output yields the original signal $f(t)$ of bandwidth f_m and some noise. Let the output signal and noise powers be S_o and N_o, respectively. For an ideal demodulator, the information I of the output signal must be identical to that of the input signal. Hence

$$I = f_m \log_2 \left(1 + \frac{S_o}{N_o}\right) \tag{8.10}$$

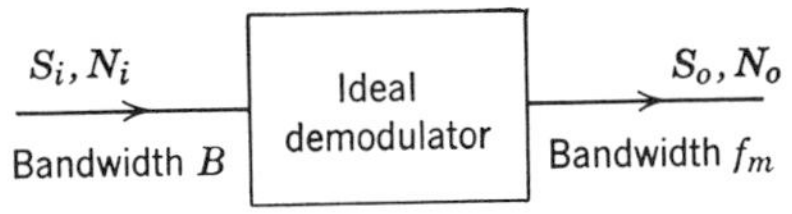

S_i, N_i — Bandwidth B → Ideal demodulator → S_o, N_o — Bandwidth f_m

Figure 8.3

From Eqs. 8.9 and 8.10, we obtain

$$B \log_2 \left(1 + \frac{S_i}{N_i}\right) = f_m \log_2 \left(1 + \frac{S_o}{N_o}\right)$$

Hence

$$\left(1 + \frac{S_o}{N_o}\right) = \left(1 + \frac{S_i}{N_i}\right)^{B/f_m} \tag{8.11}$$

In practice, S_o/N_o and $S_i/N_i \gg 1$ and we have

$$\frac{S_o}{N_o} \simeq \left(\frac{S_i}{N_i}\right)^{B/f_m} \tag{8.12}$$

If we consider signal-to-noise voltage ratios rather than power ratios, we obtain

$$\left(\frac{S_o}{N_o}\right)_{vr} \simeq \left(\frac{S_i}{N_i}\right)_{vr}^{B/2f_m} \tag{8.13}$$

Thus in an ideal system the output signal-to-noise power ratio (S_o/N_o) increases exponentially with the bandwidth B. The same result holds for signal-to-noise voltage ratio. This performance is obviously far superior to that of uncoded wideband systems such as FM and PPM discussed in Chapter 7. For these systems, it was shown that the signal-to-noise power ratio increases as the square of the bandwidth B. For the case of coded systems (PCM), however, we have observed that the improvement in the signal-to-noise ratio was indeed exponential with bandwidth (Eq. 7.79).

Example 8.1

In this example, using the concepts of the information theory, we shall calculate the bandwidth of the video (picture) signal in TV.

A television picture may be considered as composed of approximately 300,000 small picture elements. Each of these elements can assume 10 distinguishable brightness levels (such as black and shades of gray) for proper contrast. We assume that for any picture element, the 10 brightness levels are equally likely to occur. There are 30 picture frames being transmitted per second. It is also given that for a satisfactory reproduction of the picture a signal-to-noise ratio of 1000 (30 db) is required.

Based on this information, we shall calculate the bandwidth required to transmit the video signal of TV. First we calculate the information per picture element. Since each picture element can assume 10 levels with equal likelihood,

Information per picture element $= \log_2 10 = 3.32$ bits/element

Information per picture frame $= 300,000 \times 3.32$
$$= 996,000 \text{ bits/picture frame}$$

Since there are 30 picture frames per second, we have

Information per second $= 996,000 \times 30 = 29.9 \times 10^6$ bits/second

The video signal thus has an information of 29.9×10^6 bits per second. To transmit this information, the channel capacity C must be equal to 29.9×10^6 bits per second.
$$C = 29.9 \times 10^6 \text{ bits/second}$$

But for any channel of bandwidth B (in Hz) is given by

$$C = B \log_2 \left(1 + \frac{S}{N} \right)$$

However, we are given $S/N = 1000$. Hence

$$C = 29.9 \times 10^6 \simeq B \log_2 1000$$
$$= 9.95\,B$$

Therefore
$$B = 3.02 \times 10^6$$
$$= 3 \text{ mHz}$$

8.5 EFFICIENCY OF PCM SYSTEMS

Although the increase of the output signal-to-noise ratio in PCM is exponential with bandwidth, as required for an ideal system, the performance of PCM still falls short of that predicted by the Shannon-Hartley law. Let us consider an s'ary PCM system. The message signal $f(t)$ has a bandwidth f_m Hz. There are $2f_m$ samples per second. We shall assume M quantizing levels. Thus a sample of the signal $f(t)$ is approximated to any one of the M levels. Assuming all the levels to be equiprobable, the information per sample is $\log_2 M$ bits. Thus the total information content of the signal is $2f_m \log_2 M$ bits per second. To transmit this signal, ideally we need a channel of capacity given by

$$C = 2f_m \log_2 M \text{ bits/second} \qquad (8.14)$$

If we use s'ary pulses (pulses that can assume s states), we shall require a group of $\log_s M$ s'ary pulses to represent one sample that can assume one of the M values. Hence we need to transmit a total of $2f_m \log_s M$ s'ary pulses per second. The bandwidth B required to transmit these pulses is given by

$$B = f_m \log_s M \qquad \text{Hz} \tag{8.15}$$

From Eq. 8.14, we have

$$C = (f_m \log_s M)(2 \log_2 s)$$
$$= B \log_2 s^2 \tag{8.16}$$

But for s'ary PCM, the input signal-to-noise power ratio is given by (Eq. 7.81)

$$\frac{S_i}{N_i} = \frac{K^2(s^2 - 1)}{12}$$

Hence

$$s^2 = 1 + \frac{12}{K^2} \frac{S_i}{N_i}$$

and

$$C = B \log \left(1 + \frac{12}{K^2} \frac{S_i}{N_i} \right) \tag{8.17}$$

Equation 8.17 represents the theoretical channel capacity required to transmit the quantized message $f(t)$. Actually, however, we are transmitting this message over a channel of bandwidth B (Eq. 8.15) and the signal and noise powers S_i and N_i, respectively. Hence C', the capacity of the actual channel used, is given by (Eq. 8.8)

$$C' = B \log \left(1 + \frac{S_i}{N_i} \right) \text{ bits/second} \tag{8.18}$$

Thus C (Eq. 8.17) is the optimum or theoretical channel capacity required to transmit the quantized message $f(t)$, and C' (Eq. 8.18) is the actual capacity of the channel required to transmit the quantized $f(t)$ by PCM. Evidently the S_i/N_i of PCM is $K^2/12$ times that required of the ideal system. It is shown in Chapter 9 that $K = 10$ is a reasonable value for an acceptable error probability (10^{-6}). Hence the power requirements of PCM are roughly $\frac{100}{12}$ times (9.2 db) that required of the ideal system.

For the error probability of 10^{-5}, the discrepancy is about 8 db. The performance of an ideal system and PCM for various values of s are

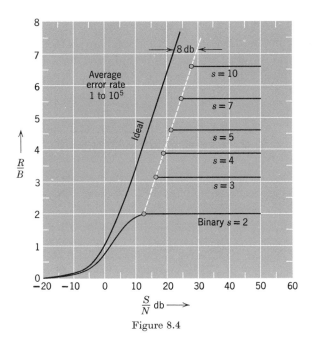

Figure 8.4

shown in Fig. 8.4. The average error rate of detection is 1 in 10^{-5}. The actual curve is displaced to the right of the ideal curve by 8 db.

In Fig. 8.4, we observe a saturation effect with respect to maximum transmission rate for a given s. Thus for $s = 2$ (binary PCM) the transmission rate per unit bandwidth cannot be increased beyond 2 bits. The reason is not difficult to understand. A binary pulse can transmit a maximum of 1 bit per second. If a system has a bandwidth of B Hz, then we can transmit $2B$ pulses per second. Thus R, the transmission rate, is given by

$$R = 2B \text{ bits/second}$$

and

$$\frac{R}{B} = 2 \text{ bits/second/bandwidth} \tag{8.19}$$

When the signal-to-noise ratio is small, the pulses cannot be distinguished properly at the receiver, and hence there is a large error probability. This causes the rate R/B to fall below 2 for low signal-to-noise ratio. As the signal-to-noise ratio improves, the error probability becomes less and R/B approaches its ideal value of 2. Once the signal-to-noise level reaches a certain value where the pulses can be clearly distinguished in the noise, any further increase in the signal power (or the signal-to-noise ratio) has negligible influence on the performance of the system. This is the reason for the saturation effect. The saturation level can be increased by increasing s as shown in Fig. 8.4. In general, the maximum information transmitted by one s'ary pulse is $\log_2 s$ bits per second. A system of bandwidth B can transmit $2B$ pulses per second. Hence R, the maximum rate of information transmission, is given by

$$R = 2B \log_2 s \text{ bits/second}$$

and

$$\frac{R}{B} = 2 \log_2 s \text{ bits/second/bandwidth} \qquad (8.20)$$

Thus for $s = 3$,

$$\left(\frac{R}{B}\right)_{\text{max}} = 2 \log_2 3 = 3.16 \text{ bits/second/bandwidth}$$

for $s = 4$

$$\left(\frac{R}{B}\right)_{\text{max}} = 2 \log_2 4 = 4 \text{ bits/second/bandwidth}$$

Hence the saturation levels for $s = 3$ and 4 are given by 3.16 and 4 bits/second/bandwidth, respectively.

We note here that the channel capacity predicted by Shannon-Hartley law cannot be attained by finite coding. It can be realized by block coding where a sequence of N symbols is considered as one symbol and in the limit $N \rightarrow \infty$. The complexity in such a coding is so prohibitive that in practice one gladly settles for something less than ideal.

APPENDIX. INFORMATION CONTENT OF NONEQUIPROBABLE MESSAGES

Consider a source which generates messages $a_1, a_2, \ldots, a_n$ with probability $P_1, P_2, \ldots, P_n$. Let the source generate a sequence of N messages. If N is made very large, then according to the law of large numbers, this sequence will contain message a_1, NP_1 number of times, message a_2, NP_2 number of times, and so on. Since the occurrence of each message is independent, the probability $P(S)$ of occurrence of any one sequence S of N messages is

$$P(S) = (P_1)^{NP_1}(P_2)^{NP_2} \cdots (P_n)^{NP_n} \qquad (A8.1)$$

Note that since N is very large, each of the possible sequences has the same number of messages a_1 (NP_1), a_2 (NP_2), and so on. Hence all these sequences are equiprobable with probability $P(S)$ given in Eq. A8.1. One such sequence has an information $I(S)$.

$$I(S) = \log_2 \frac{1}{P(S)}$$

$$= N \sum_{i=1}^{n} P_i \log_2 \frac{1}{P_i} \qquad (A8.2)$$

Since the sequence S is composed of N messages, the average information per message is $I(s)/N$:

$$\frac{I(s)}{N} = \sum_{i=1}^{n} P_i \log_2 \frac{1}{P_i} \text{ bits} \qquad (A8.3)$$

This is the average information per message. Since messages $a_1, a_2, \ldots, a_n$ occur with probability $P_1, P_2, \ldots, P_n$, it is obvious that a message a_k with probability P_k carries an information of $\log 1/P_k$ bits, which is the result we wanted.

PROBLEMS

1. A voltage waveform $Sa(2000\pi t)$ is to be transmitted with an uncertainty not exceeding $\frac{1}{80}$ volt. Determine the channel capacity required (See Fig. 1.12).

2. Repeat Problem 1 for a waveform $[Sa(2000\pi t)]^2$ if the uncertainty is not to exceed $\frac{1}{64}$ volt (see Fig. 1.12).

3. Devise a scheme to transmit a continuous signal $f(t)$, shown in Fig. 8.2, with an uncertainty not exceeding 0.1 volt using a channel of bandwidth $(f_m/2)$ Hz. Assume that the signal $f(t)$ is bandlimited to f_m Hz.

4. Repeat Problem 3 if the uncertainty is not to exceed 0.025 volt.

5. In Example 8.1 in the text, the amount of information per picture frame on television was found to be about $9.96 \times 10^5 \simeq 10^6$ bits. An announcer on radio tries to describe a television picture orally by 1000 words out of his vocabulary of 10,000 words. Assume that each of the 10,000 words in his vocabulary is equally likely to occur in describing this picture (a crude approximation, but good enough to give the idea). Determine the amount of information broadcast by the announcer in describing the picture. Would you say the broadcaster could do justice to the picture by using 1000 words? Is the old adage "one picture is worth 1000 words" exaggerating or underrating the reality?

6. In a facsimile transmission of picture, there are about 2.25×10^6 picture elements per frame. For a good reproduction, 12 brightness levels

are necessary. Assume all these levels equally likely to occur. Calculate the channel bandwidth required to transmit 1 picture every 3 minutes. Assume the signal to noise power ratio over the channel to be 30 db (1000).

7. Consider a weatherman transmitting the weather conditions over a wire. There are four possible messages; sunny, cloudy, rainy, and foggy. If each of the messages is equally likely, what is the minimum number of binary pulses required per message transmitted? Give a typical code pattern for the four messages using binary pulses.

Assume now that the probabilities of the four messages are 1/4, 1/8, 1/8, and 1/2, respectively. In this case the information per message on the average is less than 2 bits. This is seen from Eq. A8.3.

$$\text{Average information per message} = \sum_{i=1}^{4} P_i \log \frac{1}{P_i}$$

In this case, P_1, P_2, P_3, and P_4 are 1/4, 1/8, 1/8, and 1/2. It is therefore possible to use the code which will need on the average less than 2 bits. Verify that the use of the following code actually requires only 1.75 bits/ message.

Sunny	10
Cloudy	110
Rainy	111
Foggy	0

Note that this code is uniquely decodable, that is, any possible sequence formed by using this code is uniquely decodable. The reader can verify that no uniquely decodable code exists which can do better than this one. This is because the average information per message in this case (See Eq. A-8.3) is 1.75 bits.

8. There are 81 coins in the bag. Eighty of the coins are identical in all respects. The remaining one coin is identical to the rest, except that it is slightly heavier. What amount of information is required to locate this coin? (*Hint:* Consider the minimum number of measurements required to locate the heavier coin.)

9. If, in Problem 8, it is known that one of the coins differs in weight (but is not known whether it is heavier or lighter than the others), what amount of information is required to locate this coin and find out whether it is heavier or lighter than the others?

chapter 9

Elements of Digital Communication

Our concern thus far has been communication of continuous signals. There is an infinite number of possible waveforms that can be formed by continuous signals. In contrast, we have a case where we are interested in transmitting one of the finite number of waveforms or messages. A simple example of this is observed in transmission of an English text by using some code such as a Morse code. In this case there are in all 27 symbols or messages (26 alphabets and a space). These symbols are transmitted by various combinations of mark and space. Hence the problem of transmission is reduced to the problem of transmitting a sequence of waveforms, each of which is selected from a specified and finite set. This type of communication is called *digital communication*. This contrasts with the case of transmitting analog information, for example, radio or television broadcast where the resulting set has infinite possible waveforms.

The fundamental difference between the digital data and continuous data (or analog data) communication systems is now obvious. In digital, the communication involves transmission and detection of one of the finite number of known waveforms, whereas in continuous, there is an infinitely large number of messages and the corresponding waveforms are not at all known. It should be realized that the PCM discussed in an earlier chapter is a digital data communication used to transmit continuous data. This transformation was made possible through the process of quantization. This process in effect approximates the continuous signals so that they can assume only certain discrete amplitudes. In essence, this is digitizing the continuous signal. The messages now can be transmitted by a finite number of symbols (or levels).

In digital systems, the problem of detection is somewhat simpler than that in continuous systems. During transmission, the waveforms are corrupted by the channel noise. When such a signal is received at the receiver, we need to make decisions as to which of the n possible known waveforms has been transmitted. Once this decision is made, the transmitted waveform is recovered exactly without any noise. The channel noise therefore has no influence in this sense. However, the channel noise will cause certain error in our decision. We may, for example, be misled by the noise and make a wrong decision. So we must accept the fact that some of the decisions will be in error and that the error will increase with increasing noise. The likelihood of making an error (probability of error) is obviously a more meaningful criterion for signal detection in digital systems. In this chapter we shall study the problem of optimum detection. We shall restrict ourselves to binary communication, that is, communication using two symbols only. This type of communication is the one most frequently observed in practice.

9.1 DETECTION OF BINARY SIGNALS: THE MATCHED FILTER

In the case of binary communication, messages are transmitted by two symbols only (such as binary PCM). One of the symbols is represented by a pulse $s(t)$ shown in Fig. 9.1a, and the other symbol is represented by the absence of the pulse (no signal). Let the duration of the pulse $s(t)$ be T seconds. The duration of space [no $s(t)$] is also T seconds. It should once again be stressed that in this problem, the detection of the waveform is not important. The waveform is already known. We wish to determine whether the pulse is present or absent. The detector at the receiver must therefore be a decision making device. It must examine the contents of the incoming signal over each T seconds and then decide whether the pulse is present or absent. The optimum detector will be the one that has the least probability of error in making a decision.

The decision making can be facilitated if we pass the signal through a filter which will accentuate the useful signal $s(t)$ and suppress the noise

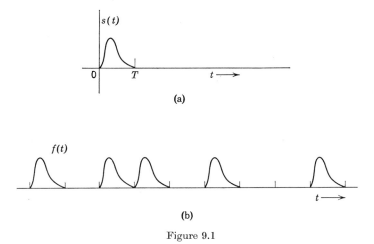

(a)

(b)

Figure 9.1

$n(t)$ at the same time. We therefore seek a filter which will peak out the signal component at some instant and suppress the noise amplitude at the same time. This will give a sharp contrast between the signal and the noise, and if the pulse $s(t)$ is present, the output will appear to have a large peak at this instant. If the pulse is absent, no such peak will appear. This arrangement will make it possible to decide whether the pulse is present or absent with a reduced error probability. The filter which accomplishes this is known as the *matched filter*. The purpose of this filter is to increase the signal component and decrease the noise component at the same instant. This is obviously equivalent to maximizing the ratio of the signal amplitude to the noise amplitude at some instant at the output. It proves

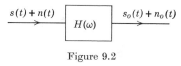

Figure 9.2

more convenient to use the square of amplitudes, so we shall design a filter which will maximize the ratio of the square of signal amplitude to the square of the noise amplitude.

Let the input signal be $s(t) + n(t)$, where $s(t)$ is the useful signal pulse, $n(t)$ is the channel noise, and $s_o(t) + n_o(t)$ is the output of the filter (Fig. 9.2). The signal component at the output is $s_o(t)$ and the noise component is $n_o(t)$. Thus we wish to maximize the ratio $s_o^2(t)/n_o^2(t)$ at some instant $t = t_m$ (decision making instant). Note that $n_o(t)$ is a random signal and hence cannot be determined exactly. Therefore we must be content with taking its mean square value $\overline{n_o^2(t)}$. Thus we wish to maximize the ratio ρ given by

$$\rho = \frac{s_o{}^2(t_m)}{n_o{}^2(t_m)} \tag{9.1}$$

Let $S(\omega)$ be the Fourier transform of $s(t)$ and $H(\omega)$ be the transfer function of the desired optimum filter. Then

$$s_o(t) = \mathscr{F}^{-1}[S(\omega)H(\omega)]$$

$$= \frac{1}{2\pi} \int_{-\infty}^{\infty} H(\omega)S(\omega)e^{j\omega t}\,d\omega$$

and

$$s_o(t_m) = \frac{1}{2\pi} \int_{-\infty}^{\infty} H(\omega)S(\omega)e^{j\omega t_m}\,d\omega \tag{9.2}$$

The mean square value of the noise signal can be expressed in terms of the noise power density spectrum at the output (Eq. 2.37). If $S_n(\omega)$ is the power density spectrum of the input noise signal $n(t)$, then $|H(\omega)|^2\,S_n(\omega)$ is the power density spectrum of $n_o(t)$. Hence

$$\overline{n_o{}^2(t)} = \frac{1}{2\pi} \int_{-\infty}^{\infty} S_n(\omega)\,|H(\omega)|^2\,d\omega \tag{9.3}$$

Note that the mean square value of $n_o(t)$ is independent of t. Hence

$$\overline{n_o{}^2(t_m)} = \frac{1}{2\pi} \int_{-\infty}^{\infty} S_n(\omega)\,|H(\omega)|^2\,d\omega \tag{9.4}$$

Let us assume the channel noise $n(t)$ to be a white noise with power density $\mathscr{N}/2$.

$$S_n(\omega) = \frac{\mathscr{N}}{2}$$

and

$$n_o{}^2(t_m) = \frac{\mathscr{N}}{4\pi} \int_{-\infty}^{\infty} |H(\omega)|^2\,d\omega \tag{9.5}$$

Substituting Eqs. 9.2 and 9.4 in Eq. 9.1, we get

$$\rho = \frac{s_o{}^2(t_m)}{n_o{}^2(t_m)} = \frac{\left| \int_{-\infty}^{\infty} H(\omega)S(\omega)e^{j\omega t_m}\,d\omega \right|^2}{\pi\mathscr{N} \int_{-\infty}^{\infty} |H(\omega)|^2\,d\omega} \tag{9.6}$$

Note that since $s_o(t)$ is a real number, $s_o{}^2(t) = |s_o(t)|^2$.

At this point we use the Schwarz inequality. One form of this inequality states that if $F_1(\omega)$ and $F_2(\omega)$ are complex functions, then

$$\left| \int_{-\infty}^{\infty} F_1(\omega) F_2(\omega) \, d\omega \right|^2 \leqslant \int_{-\infty}^{\infty} |F_1(\omega)|^2 \, d\omega \int_{-\infty}^{\infty} |F_2(\omega)|^2 \, d\omega \qquad (9.7a)$$

The equality holds only if

$$F_1(\omega) = kF_2^*(\omega) \qquad (9.7b)$$

where k is an arbitrary constant. If we let

$$F_1(\omega) = H(\omega) \qquad \text{and} \qquad F_2(\omega) = S(\omega)e^{j\omega t_m}$$

then

$$\left| \int_{-\infty}^{\infty} H(\omega) S(\omega) e^{j\omega t_m} \, d\omega \right|^2 \leqslant \int_{-\infty}^{\infty} |H(\omega)|^2 \, d\omega \int_{-\infty}^{\infty} |S(\omega)|^2 \, d\omega \qquad (9.8)$$

Substitution of inequality 9.8 in Eq. 9.6 yields

$$\frac{s_o^2(t_m)}{n_o^2(t_m)} \leqslant \frac{1}{\pi \mathcal{N}} \int_{-\infty}^{\infty} |S(\omega)|^2 \, d\omega$$

Hence

$$\rho_{\max} = \left. \frac{s_o^2(t_m)}{n_o^2(t_m)} \right|_{\max} = \frac{1}{\pi \mathcal{N}} \int_{-\infty}^{\infty} |S(\omega)|^2 \, d\omega \qquad (9.9)$$

and occurs when inequality in Eq. 9.8 becomes equality. This is possible only if (Eq. 9.7)

$$H(\omega) = kS^*(\omega)e^{-j\omega t_m}$$
$$= kS(-\omega)e^{-j\omega t_m} \qquad (9.10)$$

where k is an arbitrary constant.

The impulse response $h(t)$ of the optimum system is given by

$$h(t) = \mathcal{F}^{-1}[H(\omega)]$$
$$= \mathcal{F}^{-1}[kS(-\omega)e^{-j\omega t_m}]$$

Note that the inverse Fourier transform of $S(-\omega)$ is $s(-t)$, and the term $e^{-j\omega t_m}$ represents a time shift of t_m seconds. Hence

$$h(t) = ks(t_m - t) \qquad (9.11)$$

For the sake of convenience, we shall assume $k = 1$.

As mentioned earlier, the message signal $s(t)$ is of finite duration. Let $s(t)$ be zero outside the interval $(0, T)$ as shown in Fig. 9.3a. The signal $s(t_m - t)$ can be obtained by folding $s(t)$ about the vertical axis

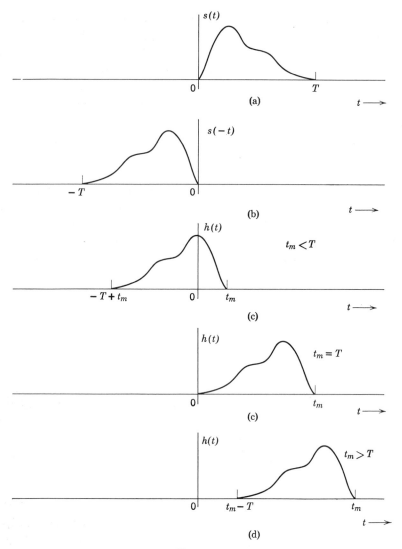

Figure 9.3

and shifting to the right by t_m seconds. Three cases where $t_m < T$, $t_m = T$, and $t_m > T$ are shown in Fig. 9.3. Observe that the impulse response $h(t)$ is noncausal for $t_m < T$ (Fig. 9.3c). This represents a physically unrealizable system. For physical realizability $t_m \geqslant T$, as shown in Figs. 9.3c and d. Both of these systems will yield the desired result. However, it is desirable to have the observation time t_m as small as possible in order to make the decision quickly. For larger values of t_m, one must wait a correspondingly longer time for the desired observation. Hence $t_m = T$ is preferable to $t_m > T$. We therefore conclude that the impulse response of the optimum system is the mirror image of the desired message signal $s(t)$ about the vertical axis and shifted to the right by T seconds. Such a receiver is called the *matched filter* or the *matched receiver*.

In a matched filter the signal-to-noise ratio (Eq. 9.1) becomes maximum at the instant t_m which is also the instant when all of the signal $s(t)$ has entered the receiver (Fig. 9.4). It should be realized that the matched filter is the optimum of all linear filters. In general, a better signal-to-noise ratio can be obtained if the restriction on linearity on the filter is removed. This will, of course, yield a nonlinear filter.

The maximum value of the signal-to-noise ratio attained by the matched filter is given by Eq. 9.9. Note that E, the energy of the signal $s(t)$, is given by

$$E = \int_{-\infty}^{\infty} s^2(t)\, dt = \frac{1}{2\pi} \int_{-\infty}^{\infty} |S(\omega)|^2\, d\omega$$

Hence

$$\rho = \frac{s_o^2(t_m)}{n_o^2(t_m)} = \frac{E}{\mathcal{N}/2} = \frac{2E}{\mathcal{N}} \tag{9.12}$$

$$= \frac{\text{Energy of the signal } s(t)}{\text{Power density spectrum of the input noise signal}}$$

The signal amplitude $s_o(t_m)$ is obtained by substituting Eq. 9.10 in Eq. 9.2

$$s_o(t_m) = \frac{1}{2\pi} \int_{-\infty}^{\infty} |S(\omega)|^2\, d\omega$$

$$= E \tag{9.13}$$

Hence the maximum amplitude of the signal component at the output occurs at $t = t_m$ and has magnitude E, the energy of the signal $s(t)$ (see

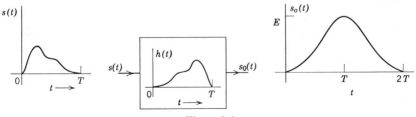

Figure 9.4

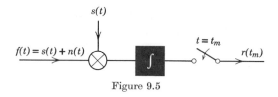

Figure 9.5

Fig. 9.4). This is a remarkable result. The maximum amplitude is independent of the waveform $s(t)$ and depends only upon its energy.

The mean square value of the noise signal at the output can be obtained by substituting Eq. 9.13 in Eq. 9.12.

$$\overline{n_o{}^2(t_m)} = \frac{\mathcal{N}E}{2} \tag{9.14}$$

The matched filter may be realized by an alternative arrangement shown in Fig. 9.5. If the input to the matched filter is $f(t)$, then $r(t)$, its output, is given by

$$r(t) = \int_{-\infty}^{\infty} f(x)h(t - x)\, dx$$

where $h(t)$ is the impulse response given by $h(t) = s(t_m - t)$ and $h(t - x) = s(x + t_m - t)$. Hence

$$r(t) = \int_{-\infty}^{\infty} f(x)s(x + t_m - t)\, dx$$

The decision at the output is made from $r(t_m)$, the output at $t = t_m$.

$$r(t_m) = \int_{-\infty}^{\infty} f(x)s(x)\, dx \tag{9.15}$$

We can obtain $r(t_m)$ by an arrangement (known as time correlator) shown in Fig. 9.5. In this arrangement, the incoming signal $s(t) + n(t)$ multiplies by $s(t)$. This is obviously synchronous detection (or coherent detection). Thus the matched filter detection is essentially a synchronous detection.

9.2 DECISION THRESHOLD IN A MATCHED FILTER

The matched filter is designed to maximize the signal-to-noise ratio at the instant t_m ($t_m = T$). Whether the signal $s(t)$ is present is therefore decided by the observation of the output at $t = T$. If $r(t)$ represents the output of the matched filter at $t = T$, then

$$r(T) = s_o(T) + n_o(T)$$

Substituting Eq. 9.13 in the last equation, we obtain

$$r(T) = E + n_o(T) \tag{9.16}$$

Since the input noise signal is random, $n_o(T)$ is also random. Hence if the signal $s(t)$ is present at the input, the output $r(T)$ is given by a constant E plus a random variable $n_o(T)$. The output will therefore differ from E by the noise amplitude. If the signal $s(t)$ is absent at the input, the output of the filter will be given solely by the noise term

$$r(T) = n_o(T) \tag{9.17}$$

Thus if the signal is present, the output will be $E + n_o(T)$, and if the signal is absent, the output will be $n_o(T)$. The decision whether the signal is present or absent could be made easily from this information, except that $n_o(T)$ is random and its exact value is unpredictable. It may have a large or a small value and can be either negative or positive. It is possible that the signal $s(t)$ is present at the input, but $n_o(T)$ may have a large negative value. This will make the output $r(T)$ very small. On the other hand, even if the signal $s(t)$ is absent, $n_o(T)$ may be quite large. This will cause the output $r(T)$ to be very large. Thus there is no sure way of deciding whether $s(t)$ is present or absent. However, it is evident that when $r(T)$ is large, it is more likely that $s(t)$ is present. On the other hand, if $r(T)$ is very small, it is more likely that the signal is absent. No matter what decision is made, there is always some likelihood of error. We must therefore find a decision rule which will minimize the likelihood of the error.

Let the decision rule be "signal present" if $r(T) > a$ and "no signal" if $r(T) < a$. We shall now find the optimum decision threshold a which will minimize the error probability (likelihood) of the decision.

To find the optimum threshold, we must first consider the nature of the noise amplitude $n_o(T)$. The noise is a random signal and its amplitudes have certain distribution. The most commonly observed amplitude distribution is the Gaussian distribution.* This means the

---

* It can be shown by using the central limit theorem in the probability theory that a signal composed of a large number of relatively independent signals tends to be Gaussian. Most of the noise signals are the result of relatively numerous independent perturbations. The shot noise and the thermal noise obviously fall in this category. Hence the Gaussian assumption for noise is justified in most cases.

relative frequency of occurrence of noise amplitudes has a Gaussian form. This distribution $p(x)$ is called the probability density function of the amplitude x and is given by

$$p(x) = \frac{1}{\sigma_x \sqrt{2\pi}} e^{-x^2/2\sigma_x^2} \qquad (9.18)$$

where σ_x^2 is the mean square value of the signal. This distribution is shown in Fig. 9.6a. It can be seen that the amplitude distribution is symmetrical about $x = 0$. The signal is equally likely to be positive and negative; hence it has a zero mean value. The probability density represents the relative frequency of occurrence of amplitudes. The function is so normalized that the area $p(x)\,dx$ (shown in Fig. 9.6a) represents the likelihood (or probability) of observing the signal amplitude in the range $(x, x + dx)$. Thus, if we observe a signal in the interval 0, T ($T \to \infty$) as shown in Fig. 9.6b, the amplitude is in the range $x, x + dx$ over a certain time dT given by

$$dT = \sum_{j=1}^{8} dt_i$$

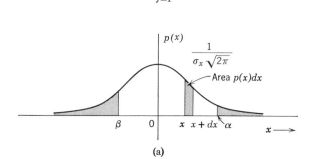

(a)

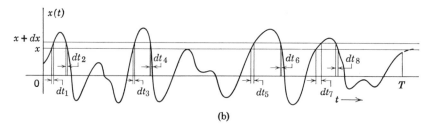

(b)

Figure 9.6

Hence the relative frequency of observing the signal in the range x, $x + dx$ is dT/T. Therefore

$$p(x)\, dx = \frac{\sum dt_i}{T}$$

It is now obvious that the probability of observing x in the range (x_1, x_2) is given by the area of $p(x)$ under (x_1, x_2):

$$\text{Probability } (x_1 < x < x_2) = \int_{x_1}^{x_2} p(x)\, dx \qquad (9.19)$$

Similarly, the probability of observing $x > \alpha$ is given by

$$\text{Probability } (x > \alpha) = \int_{x}^{\infty} p(x)\, dx \qquad (9.20)$$

and

$$\text{Probability } (x < \beta) = \int_{-\infty}^{\beta} p(x)\, dx \qquad (9.21)$$

For the output noise $n_o(t)$, the mean square value is given by $\mathcal{N}E/2$ (Eq. 9.14). In Eq. 9.18, σ_x^2 represents the mean square value. Hence

$$\sigma_x^2 = \frac{\mathcal{N}E}{2}$$

and

$$p(x) = \frac{1}{\sqrt{\pi \mathcal{N}E}}\, e^{-x^2/\mathcal{N}E} \qquad (9.22)$$

Let us now turn our attention to the matched filter output. When $s(t)$ is absent, the output is $n_o(T)$ and has amplitude distribution given by Eq. 9.22. This is shown in Fig. 9.7a. If we denote the output amplitude by r, then $r = n_o(T)$ and

$$p(r) = \frac{1}{\sqrt{\pi \mathcal{N}E}}\, e^{-r^2/\mathcal{N}E} \qquad (9.23)$$

If the signal $s(t)$ is present, then

$$r = E + n_o(T).$$

The output r consists of a constant E plus a random component $n_o(T)$. The amplitude distribution of r is obviously the same as in Eq. 9.23,

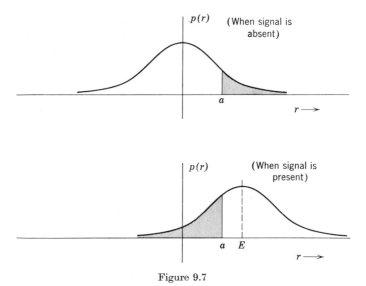

Figure 9.7

but shifted by a constant E. Thus in this case

$$p(r) = \frac{1}{\sqrt{\pi \mathcal{N} E}} e^{-(r-E)^2/\mathcal{N}E} \tag{9.24}$$

This distribution is shown in Fig. 9.7b. Both the distributions are shown together in Fig. 9.8.

Let a be the decision threshold. The decision is "signal present" if $r > a$ and is "signal absent" if $r < a$. From Fig. 9.7a it is obvious that there are instances when $r > a$ even if the signal is absent. The probability that $r > a$ when the signal is absent is given by the shaded area in Fig. 9.7a. It is evident that by using a as the threshold, we commit an error (called false alarm) with probability equal to the shaded area in Fig. 9.7a. On the other hand, even if the signal is present, the output

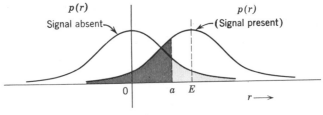

Figure 9.8

amplitude r can fall below a. In this instance our decision is "no signal present" even if the signal is actually present. This type of error is called false dismissal error and its probability is given by the shaded area in Fig. 9.7b. Thus for a given threshold a, we commit two different kinds of errors, the false alarm and the false dismissal.

If the signal $s(t)$ is equally likely to be present and absent, then, on the average, half the time $s(t)$ will be present and the remaining half time $s(t)$ will be absent. When $s(t)$ is present, we commit false dismissal type of error, and when $s(t)$ is absent, we commit false alarm type of error. Hence the error probability in the decision will be given by the mean of the two shaded areas in Figs. 9.7a and 9.7b. This is half the sum of two areas. From Fig. 9.8, it is obvious that the sum of areas is minimum if we choose

$$a = \frac{E}{2} \qquad (9.25)$$

Hence the optimum threshold is given by Eq. 9.25.

Error Probability

We have seen that when the signal $s(t)$ is equally likely to be present and absent, then the probability of the error in the decision is given by half the sum of the areas in Fig. 9.7a and 9.7b. Also, the optimum decision threshold $a = E/2$. Hence the two areas are identical. Therefore the error probability $P(\varepsilon)$ is given by either of the areas. We shall here use the area in Fig. 9.7a.

$$P(\varepsilon) = \int_a^\infty p(r)\, dr$$

$$= \frac{1}{\sqrt{\pi \mathcal{N} E}} \int_a^\infty e^{-r^2/\mathcal{N} E}\, dr \qquad (9.26)$$

The integral on the right-hand side of Eq. 9.26 cannot be evaluated in a closed form. It is, however, extensively tabulated in standard tables under probability integral or error function erf (x).

We define the error function erf (x) as*

$$\text{erf}\,(x) = \frac{1}{\sqrt{2\pi}} \int_{-\infty}^x e^{-y^2/2}\, dy \qquad (9.27)$$

---

* At present there exist in the literature several definitions of erf (x) and erfc (x) which are essentially equivalent with minor differences.

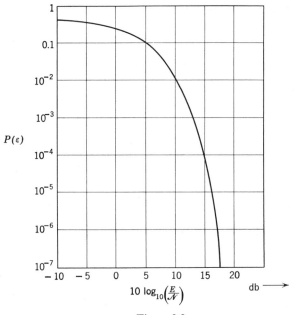

Figure 9.9

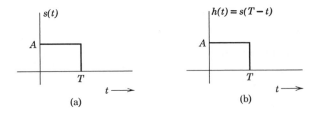

(a)

(b)

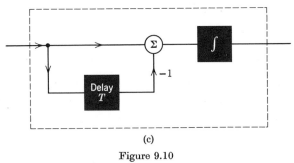

(c)

Figure 9.10

and the complementary error function erfc (x) as

$$\text{erfc } (x) = \frac{1}{\sqrt{2\pi}} \int_x^\infty e^{-y^2/2} \, dy \qquad (9.28)$$

It is obvious from these definitions that

$$\text{erf } (x) + \text{erfc } (x) = 1 \qquad (9.29)$$

A useful approximation for erfc (x) is given by

$$\text{erfc } (x) \simeq \frac{1}{x\sqrt{2\pi}} \left(1 - \frac{1}{x^2} \right) e^{-x^2/2} \qquad \text{for } x > 2 \qquad (9.30)$$

The error in this approximation is about 10 % for $x = 2$ and is less than 1 % for $x > 3$.

Using definition 9.27b, we can express Eq. 9.26 as

$$P(\varepsilon) = \text{erfc } \left(\frac{a}{\sqrt{\mathcal{N}E/2}} \right) \qquad (9.31)$$

But since $a = E/2$

$$P(\varepsilon) = \text{erfc } \left(\sqrt{\frac{E}{2\mathcal{N}}} \right) \qquad (9.32)$$

Fig.9.9 shows the error probability $P(\varepsilon)$ as a function of $\dfrac{E}{\mathcal{N}}$

How do we interpret the probability of error? The probability of an event implies the likelihood of the event or the relative frequency of the event. Thus if we have made N decisions ($N \to \infty$), then N_ε, the total number of wrong decisions, is given by

$$P(\varepsilon) = \frac{N_\varepsilon}{N}$$

and

$$N_\varepsilon = P(\varepsilon)N$$

Thus if $P(\varepsilon) = \frac{1}{100}$, on the average, one in 100 decisions will be in error.

Example 9.1 (Binary PCM)

For binary PCM (discussed in Chapter 7), $s(t)$ is a rectangular pulse of height A and width T. The impulse response of the matched filter is given by

$$h(t) = s(T - t)$$

Note that $s(T - t)$ is $s(t)$ folded about the vertical axis and shifted to the right by T seconds. This is identical to $s(t)$. Hence

$$h(t) = s(t)$$

This filter can be realized by an arrangement shown in Fig. 9.10c.

The energy E of $s(t)$ is given by

$$E = A^2 T$$

We are also given that

$$A = K\sigma_n$$

when σ_n is the root mean square value of noise signal.

$$\sigma_n{}^2 = N_i = \overline{n^2(t)}$$

Since the pulse duration is T, there are $1/T$ pulses per second. To transmit $1/T$ pulses per second, the bandwidth B required for transmission is $1/2T$

$$B = \frac{1}{2T}$$

If $\mathcal{N}/2$ is the power density spectrum of noise, then

$$N_i = \mathcal{N}B = \frac{\mathcal{N}}{2T}$$

or

$$\sigma_n{}^2 = \frac{\mathcal{N}}{2T}$$

and

$$\mathcal{N} = 2T\sigma_n{}^2 \qquad (9.33)$$

Obviously,

$$\frac{E}{\mathcal{N}} = \frac{A^2 T}{2T\sigma_n{}^2} = \frac{A^2}{2\sigma_n{}^2} = \frac{K^2 \sigma_n{}^2}{2\sigma_n{}^2} = \frac{K^2}{2}$$

For a value of $K = 10$,

$$\frac{E}{\mathcal{N}} = 50$$

and the probability of error $P(\varepsilon)$ is given by

$$P(\varepsilon) = \text{erfc } \sqrt{25}$$

$$= \text{erfc } (5) \qquad (9.34)$$

Use of Eq. 9.30 yields

$$P(\varepsilon) \simeq 0.284 \times 10^{-6} \qquad (9.35)$$

This result can also be read off directly from Fig. 9.9. For $E/\mathcal{N} = 50$, $10 \log_{10} E/\mathcal{N} = 16.9$ db. This yields $P(\varepsilon) \simeq 0.284 \times 10^{-6}$.

Thus if the pulse amplitude is made 10 times the root mean square value of noise ($K = 10$), the error probability is of the order of 10^{-6}, which is acceptable in most practical cases.

In this discussion, we have assumed an idealized rectangular pulse for $s(t)$. However, because of finite channel bandwidth, this pulse will become trapezoidal (see Section 2.6) in the process of transmission. Hence the matched filter impulse response should also be trapezoidal to match the signal waveform received. This point should be kept in mind in our future discussion where idealized rectangular pulses are used for $s(t)$.

9.3 AMPLITUDE SHIFT KEYING (ASK)

The binary PCM in Example 9.1 can be transmitted over wires easily. But when the transmission is through space via radiation, we must use amplitude modulated binary PCM in Example 9.1. The amplitude modulation shifts the low frequency spectrum of binary PCM to a high frequency (at carrier frequency). This scheme is known as amplitude shift keying (ASK). One of the binary symbols is transmitted by a sinusoidal pulse $s(t)$ given by

$$s(t) = \begin{cases} A \sin \omega_c t & 0 < t < T \\ 0 & \text{otherwise} \end{cases}$$

The remaining symbol is transmitted by a space (no signal). A typical ASK waveform is shown in Fig. 9.11.

We shall now find the optimum receiver and the error probability for ASK. It is assumed that the probability of pulse $s(t)$ being present is the same as being absent (0.5).

The pulse $s(t)$ is shown in Fig. 9.12a. The impulse response of the matched filter is

$$h(t) = s(T - t)$$

Observe that $s(T - t)$ is just $-s(t)$. Hence

$$h(t) = -s(t)$$

The output of the matched filter when $s(t)$ is present at the input is given as the convolution of $s(t)$ with $h(t)$. This is shown in Fig. 9.12c. The output is maximum at $t = T$ as expected and has a magnitude

Figure 9.11

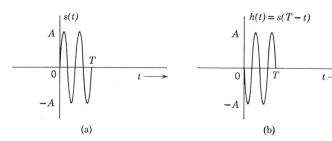

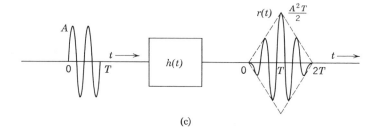

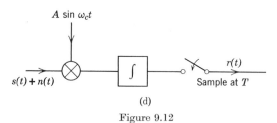

(d)

Figure 9.12

equal to the energy E of the signal $s(t)$. In this case

$$E = \frac{A^2}{2} T \tag{9.36}$$

The threshold of detection is $E/2$. In this case the threshold is

$$a = \frac{A^2 T}{4}$$

The probability of error is given by (Eq. 9.32)

$$P(\varepsilon) = \operatorname{erfc}\left(\sqrt{\frac{E}{2\mathcal{N}}}\right) \tag{9.37}$$

$$= \operatorname{erfc}\left(\frac{A}{2}\sqrt{\frac{T}{\mathcal{N}}}\right) \tag{9.38}$$

The error probability can also be read off directly from Fig. 9.9.

We can express the error probability in terms of average signal power. The signal $s(t)$ has energy E given by

$$E = \frac{A^2 T}{2}$$

The signal $s(t)$ is present half the time on the average and for the remaining half there is no signal. Hence the average signal power P_s is given by

$$P_s = \frac{1}{T} \frac{E}{2} = \frac{A^2}{4}$$

and

$$P(\varepsilon) = \text{erfc}\left(\sqrt{\frac{P_s T}{\mathcal{N}}}\right) \tag{9.39}$$

The matched filter has an impulse response (Fig. 9.12b)

$$h(t) = -s(t) = \begin{cases} -A \sin \omega_c t & 0 < t < T \\ 0 & \text{otherwise} \end{cases}$$

Alternatively, we may use the correlation arrangement for the matched filter (see Fig. 9.5). For this particular case, the appropriate arrangement is shown in Fig. 9.12d. Note that the matched filter detection is essentially a synchronous detection.

9.4 PHASE SHIFT KEYING (PSK)

In Chapter 7 we observed that for efficient binary PCM, one should use bipolar pulses (two pulses of height $A/2$ and $-A/2$) instead of two pulses of height 0 and A. Thus in bipolar PCM, the two symbols are represented by $s(t)$ and $-s(t)$. When we use the amplitude-modulated rectangular pulses (Fig. 9.13), the scheme is known as the phase shift keying (PSK). Phase shift keying may also be considered as a phased modulated binary PCM. A typical PSK waveform is shown in Fig. 9.13c. We shall find the optimum detector and the error probability for this case.

The two symbols are transmitted by waveforms $s_1(t)$ and $s_2(t)$ where

$$s_2(t) = -s_1(t)$$

Let

$$s_1(t) = -s_2(t) = s(t)$$

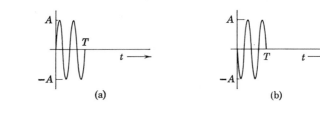

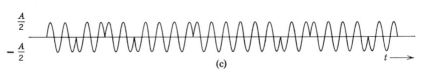

Figure 9.13

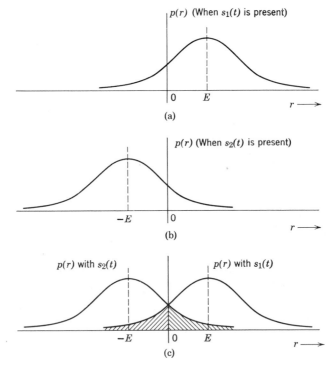

$p(r)$ (When $s_1(t)$ is present)

(a)

$p(r)$ (When $s_2(t)$ is present)

(b)

$p(r)$ with $s_2(t)$ $p(r)$ with $s_1(t)$

(c)

Figure 9.14

The waveform $s(t)$ will be assumed arbitrary for the sake of generality. The only restriction on $s(t)$ is that it has a duration of T seconds. Let the energy of $s(t)$ be E.

The matched filter for $s(t)$ will have an impulse response

$$h(t) = s(T - t)$$

If $s_1(t)$ is applied to the input of this filter, the output at $t = T$ is given by $r(T)$ in Eq. 9.16

$$r(T) = E + n_o(T) \tag{9.40}$$

The component $n_o(T)$ is random with a Gaussian distribution with mean square value $\mathcal{N}E/2$. Hence

$$p(r) = \frac{1}{\sqrt{\pi \mathcal{N}E}} e^{-(r-E)^2/\mathcal{N}E} \tag{9.41}$$

This distribution is shown in Fig. 9.14a. Similarly, when the signal $s_2(t)$ is applied at the input of the matched filter, the output $r(T)$ is given by

$$r(T) = -E + n_o(T) \tag{9.42}$$

This follows from the fact that $s_2(t) = -s_1(t)$. Hence the output due to $s_2(t)$ will be the negative of the output due to $s_1(t)$. The amplitude distribution $r(T)$ when $s_2(t)$ is present is shown in Fig. 9.14b. It is evident that in this case

$$r(T) = \frac{1}{\sqrt{\pi \mathcal{N}E}} e^{-(r+E)^2/\mathcal{N}E} \tag{9.43}$$

The two distributions are shown together in Fig. 9.14c. We must now determine the decision threshold which will minimize the error probability.

Using the argument parallel to those used earlier (see Eq. 9.25), it is obvious that the optimum detection threshold a is given by

$$a = 0 \tag{9.44}$$

Thus if $r(T) > 0$, the decision is "$s_1(t)$ present," and if $r(T) < 0$, the decision is "$s_2(t)$ present." The error probability is given by the area of $p(r)$ (in Fig. 9.14b) from 0 to ∞.

$$P(\varepsilon) = \frac{1}{\sqrt{\pi \mathcal{N}E}} \int_0^\infty e^{-(r+E)^2/\mathcal{N}E} \tag{9.45}$$

$$= \frac{1}{\sqrt{\pi \mathcal{N}E}} \int_E^\infty e^{-x^2/\mathcal{N}E} \, dx \tag{9.46}$$

This integral is exactly of the form in Eq. 9.26, except that a is replaced by E. Hence

$$P(\varepsilon) = \operatorname{erfc}\left(\sqrt{\frac{2E}{\mathcal{N}}}\right) \tag{9.47}$$

For PSK arrangement (Fig. 9.13)

$$E = \frac{A^2 T}{2}$$

and

$$P(\varepsilon) = \operatorname{erfc}\left(A\sqrt{\frac{T}{\mathcal{N}}}\right) \tag{9.48}$$

The average power of PSK signal is $A^2/2$,

$$P_s = \frac{A^2}{2} \tag{9.49}$$

Hence

$$P(\varepsilon) = \operatorname{erfc}\left(\sqrt{\frac{2P_s T}{\mathcal{N}}}\right) \tag{9.50}$$

Compare this with Eq. 9.39 for ASK. It is obvious from these equations that to attain a given error probability, the average power required for ASK is twice that required for PSK. Hence PSK is superior to ASK by 3 db in the average signal power requirement.

9.5 FREQUENCY SHIFT KEYING (FSK)

Frequency shift keying may be considered a frequency-modulated binary PCM. The two symbols are represented by two waveforms $s_1(t)$ and $s_2(t)$,

$$s_1(t) = \begin{cases} A \sin m\omega_0 t & 0 < t < T \\ 0 & \text{otherwise} \end{cases} \tag{9.51a}$$

$$s_2(t) = \begin{cases} A \sin n\omega_0 t & 0 < t < T \\ 0 & \text{otherwise} \end{cases} \tag{9.51b}$$

where

$$\omega_0 = \frac{2\pi}{T}$$

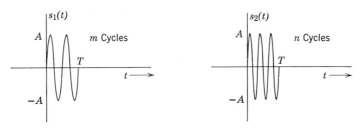

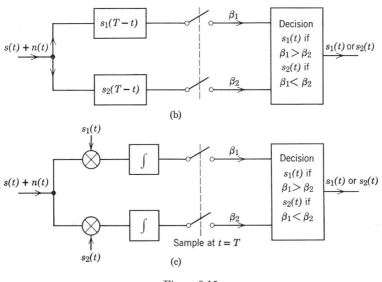

(b)

(c)

Figure 9.15

The two waveforms are shown in Fig. 9.15a. For this case the two waveforms are different, and hence we need two matched filters. We shall now show that the optimum receiver for FSK is as shown in Fig. 9.15b or c. The arrangement in Fig. 9.15b is the matched filter arrangement, whereas the one in Fig. 9.15c is the correlator arrangement (Fig. 9.5).

The incoming signal is $s(t) + n(t)$ where $s(t)$ is either $s_1(t)$ or $s_2(t)$. Let us denote the incoming signal by $f(t)$

$$f(t) = s(t) + n(t)$$

To the incoming signal we add $-s_2(t)$ as shown in Fig. 9.16a. Addition

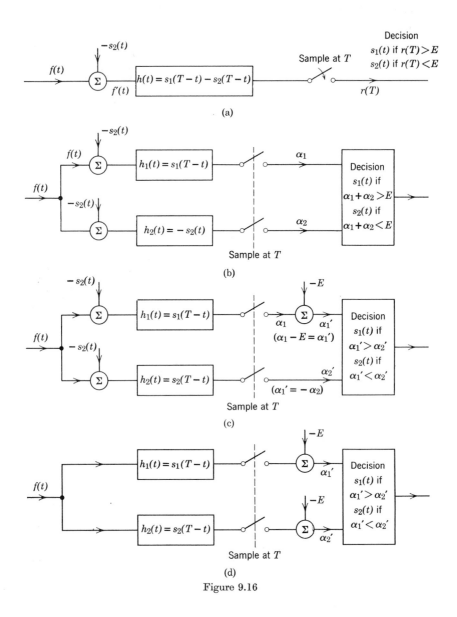

Figure 9.16

416

of a known signal to the incoming signal should not affect the optimum decision procedure.* The new signal $f'(t)$ is now given by

$$f'(t) = s(t) + n(t) - s_2(t)$$

But $s(t)$ is $s_1(t)$ or $s_2(t)$. If $s(t)$ is $s_1(t)$, then the useful signal in $f'(t)$ is $s_1(t) - s_2(t)$. If $s(t)$ is $s_2(t)$, then the useful signal in $f'(t)$ is zero. Thus $f'(t)$ is now reduced to the case where the two symbols are represented by "the pulse present" and "the pulse absent" (mark and space). This is identical to the case of binary signals, as discussed in Section 9.2. The new signal $s'(t)$ is now $s_1(t) - s_2(t)$. The matched filter for this signal has an impulse response

$$h(t) = s_1(T - t) - s_2(T - t) \tag{9.52}$$

The matched filter is shown in Fig. 9.16a. The energy E' of $s'(t)$ is given by

$$E' = \int_0^T [s_1(t) - s_2(t)]^2 \, dt$$

$$= \int_0^T s_1{}^2(t) \, dt + \int_0^T s_2{}^2(t) \, dt - 2 \int_0^T s_1(t)s_2(t) \, dt \tag{9.53}$$

where $s_1(t)$ and $s_2(t)$ are given by Eqs. 9.51a and 9.52b. Note that the last integral on the right-hand side of Eq. 9.53 is zero:

$$\int_0^T \sin m\omega_0 t \sin n\omega_0 t \, dt = 0 \qquad \omega_0 = \frac{2\pi}{T}$$

and

$$\int_0^T s_1{}^2(t) \, dt = \int_0^T s_2{}^2(t) \, dt = E = \frac{A^2 T}{2} \tag{9.54}$$

Hence

$$E' = 2E = A^2 T \tag{9.55}$$

Thus the energy E' of signal $s'(t)$ is $2E (= A^2 T)$ where E is the energy of $s_1(t)$ or $s_2(t)$.

The decision threshold of the matched filter is $E'/2 = E$. Hence the decision is "$s_1(t)$ present" if $r(t) > E$ and "$s_2(t)$ present" if $r(T) < E$.

---

* This result is a special case of the theorem on reversibility. If one performs any desired operations on a signal, the optimum operation may be obtained through an intermediate operation which is reversible. Subtracting $s_2(t)$ is a reversible operation because the original signal can be obtained by adding $s_2(t)$ to $f'(t)$ in Fig. 9.16a.

This matched filter is shown in Fig. 9.16a. The error probability of this filter is the same as that in Eq. 9.32 except that the energy E in this case is $E'(= 2E)$. Therefore

$$P(\varepsilon) = \operatorname{erfc} \sqrt{\frac{E}{\mathcal{N}}} \qquad (9.56)$$

The matched filter with impulse response $s_1(T - t) - s_2(T - t)$ can be broken into two matched filters in parallel with impulse response $s_1(T - t)$ and $-s_2(T - t)$ as shown in Fig. 9.16b. It is obvious that the two arrangements are equivalent. We make further transformation as shown in Fig. 9.16c. In Fig. 9.16b, the criterion is $\alpha_1 + \alpha_2 > E$ or $<E$. If we subtract E from the output of the upper filter as shown in Fig. 9.16c, the decision criterion reduces to $\alpha_1 + \alpha_2 > 0$ or <0. In addition, we change the sign on the impulse response of the lower filter. Therefore in Fig. 9.16c $\alpha_1' = \alpha_1 - E$ and $\alpha_2' = -\alpha_2$. This gives us the decision criterion $\alpha_1' - \alpha_2' > 0$ or $\alpha_1' - \alpha_2' < 0$.

Thus the new decision rule is "$s_1(t)$ present" if $\alpha_1' > \alpha_2'$ and "$s_2(t)$ present" if $\alpha_1' < \alpha_2'$. At this point we recognize the output of the upper filter to signal $s_2(t)$ at $t = T$ is zero. The response of the upper filter to $s_2(t)$ at $t = T$ is given by

$$s_2(t) * h_1(t) = \int_0^T A^2 \sin n\omega_0\tau \sin m\omega_0(t - \tau)\,d\tau = 0 \qquad \omega_0 = \frac{2\pi}{T}$$

Hence the signal $-s_2(t)$ at the input of the upper filter may be removed without affecting the arrangement. Further, we realize that the signal $-s_2(t)$ at the input of the lower filter [matched to $s_2(t)$] yields the output $-E$ at $t = T$ (see Eq. 9.13). Since our decision depends upon the output at $t = T$ only, we may remove $-s_2(t)$ at the input of the lower filter and subtract E from its output as shown in Fig. 9.16d. Thus the arrangement in Fig. 9.16d is equivalent to that in Fig. 9.16c. Next we recognize that the decision is based upon comparison of two outputs in Fig. 9.16d. Hence addition of $-E$ to both outputs may be removed. This yields the final arrangement shown in Fig. 9.15b. The corresponding correlator arrangement is shown in Fig. 9.15c.

Error Probability in FSK

Since all the arrangements in Fig. 9.16 and Fig. 9.15b are equivalent, all have the same error probability. The error probability for arrangement 9.16a was derived earlier (Eq. 9.56):

$$P(\varepsilon) = \text{erfc}\left(\sqrt{\frac{E}{\mathscr{N}}}\right) \tag{9.57}$$

The average power P_s for FSK is obviously given by

$$P_s = \frac{A^2}{2} = \frac{E}{T}$$

Hence

$$P(\varepsilon) = \text{erfc}\left(\sqrt{\frac{P_s T}{\mathscr{N}}}\right) \tag{9.58}$$

This is identical to the error probability of amplitude shift keying (ASK) as seen from Eq. 9.39. It is therefore obvious from this discussion that PSK (phase shift keying) is superior to both ASK (amplitude shift keying) and FSK (phase shift keying).

9.6 SOME COMMENTS ON MATCHED FILTER DETECTION

We have shown that a matched filter is equivalent to a time correlator arrangement (Fig. 9.5). In the correlator arrangement the incoming signal $[s(t) + n(t)]$ is multiplied by $s(t)$. This is obviously a synchronous detection (also known as coherent detection).

It was mentioned earlier that matched filter detection is optimum under the constraint of linear systems. In general, a better system may be found if we do not restrict ourselves to linear systems. It can, however, be shown that if the noise is Gaussian (as is the case for most noise signals), then the matched filter (or the correlator detector) receiver is the absolute optimum.

Throughout the discussion, we have implicitly assumed that the transmitter and the receiver are synchronized. For any given pulse, the decision is made at the instant the pulse is completely fed to the matched filter. Thus decisions at the receiver are made every T seconds and these instants must be properly synchronized. In addition, it is assumed that at the decision making instant, the output is entirely due to the pulse under consideration and the noise. We ignore the possibility of intersymbol interference which may arise because of residual response of the matched filter due to previous pulse. In an ideal case, the residual response due to the $(n-1)$th pulse is zero at $t = nT$. This can be seen

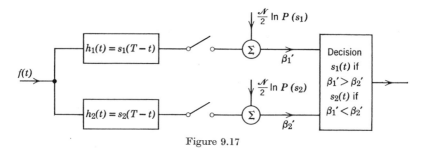

Figure 9.17

from Fig. 9.4. The response to a pulse of width T occupies an interval $2T$. Thus the response to the first pulse will occupy an interval $(0, 2T)$, whereas the response to the second pulse will occupy an interval $(T, 2T)$. The sampling instant for the second pulse is $t = 2T$, at which time the response due to the first pulse has vanished.

It should be realized that throughout our discussion, we assumed that the two binary symbols are equally likely to appear. In case the two signals are not equiprobable, the optimum receiver in Fig. 9.15b is modified as shown in Fig. 9.17.*

In this chapter we have discussed the method of coherent or synchronous detection. This detection is accomplished by a matched filter or a correlator. With ASK, PSK, and FSK, the phases of the incoming signals were assumed to be known. In many cases, however, the carrier phase of the received signal is not known. This may be caused by the instabilities in the transmitter and/or receiver oscillators or by an unknown propagation path length. Thus there is a certain amount of ignorance about the received signal carrier phase. Under such conditions coherent detection (matched filter or correlation detector) cannot be used since in these techniques the exact knowledge of the arrival of the incoming waveform is essential. In such cases incoherent detection is employed.

Incoherent detection is performed by feeding the incoming signal to an envelope detector. The outputs of the envelope detector is examined every T seconds for making the proper decision. It can be shown that this method has inferior performance compared to that of the coherent detection, particularly at lower signal-to-noise ratios.

---

* See, for instance, B. P. Lathi, *An Introduction to Random Signals and Communication Theory*, International Textbook Co., Scranton, Pa. 1968. Also see J. M. Wozencraft and I. M. Jacobs, *Principles of Communication Engineering*, John Wiley and Sons, New York, 1965.

APPENDIX A. SCHWARZ INEQUALITY

If $F_1(\omega)$ and $F_2(\omega)$ are complex functions of ω, then the Schwarz inequality states that

$$\left| \int_{-\infty}^{\infty} F_1(\omega) F_2(\omega)\, d\omega \right|^2 \leqslant \left(\int_{-\infty}^{\infty} |F_1(\omega)|^2\, d\omega \right)\left(\int_{-\infty}^{\infty} |F_2(\omega)|^2\, d\omega \right)$$

Proof: Let

$$\Phi(\omega) = \frac{F_2^*(\omega)}{\left[\int_{-\infty}^{\infty} |F_2(\omega)|^2\, d\omega \right]^{\frac{1}{2}}} \tag{A9.1a}$$

and

$$\alpha = \int_{-\infty}^{\infty} F_1(\omega)\Phi^*(\omega)\, d\omega \tag{A9.1b}$$

Then since

$$[F_1(\omega) - \alpha\Phi(\omega)][F_1^*(\omega) - \alpha^*\Phi^*(\omega)] = |F_1(\omega) - \alpha\Phi(\omega)|^2 \geqslant 0 \tag{A9.2}$$

We have

$$\int_{-\infty}^{\infty} |F_1|^2\, d\omega + |\alpha|^2 \int_{-\infty}^{\infty} |\Phi|^2\, d\omega - \alpha \int_{-\infty}^{\infty} \Phi F_1^*\, d\omega - \alpha^* \int_{-\infty}^{\infty} \Phi^* F_1\, d\omega \geqslant 0 \tag{A9.3}$$

But from A9.1a

$$\int_{-\infty}^{\infty} |\Phi|^2\, d\omega = 1$$

and from A9.1b, we have

$$\int_{-\infty}^{\infty} \Phi F_1^*\, d\omega = \alpha^*$$

Hence Eq. A9.3 becomes

$$\int_{-\infty}^{\infty} |F_1|^2\, d\omega + |\alpha|^2 - \alpha\alpha^* - \alpha^*\alpha \geqslant 0$$

or

$$\int_{-\infty}^{\infty} |F_1(\omega)|^2\, d\omega - |\alpha|^2 \geqslant 0 \tag{A9.4}$$

Substitution of Eq. A9.1a and A9.1b in Eq. A9.4 yields

$$\int_{-\infty}^{\infty} |F_1(\omega)|^2\, d\omega \geqslant \frac{\left| \int_{-\infty}^{\infty} F_1(\omega) F_2(\omega)\, d\omega \right|^2}{\int_{-\infty}^{\infty} |F_2(\omega)|^2\, d\omega}$$

Q.E.D.

Note that the inequality of A9.4 becomes equality if and only if

$$\int_{-\infty}^{\infty} |F_1(\omega)|^2 \, d\omega = |\alpha|^2$$

From Eq. A9.1 it can be seen that this is possible only if

$$F_1(\omega) = kF_2^*(\omega)$$

where k is an arbitrary constant.

PROBLEMS

1. In a binary transmission, one of the messages is represented by a rectangular pulse $s(t)$ shown in Fig. P-9.1a. The other message is transmitted by the absence of the pulse. The matched filter impulse response is $h(t) = s(T - t) = s(t)$. Calculate the signal-to-noise power ratio $s_o^2(t)/\overline{n_o^2(t)}$ at $t = T$. Assume white noise with a power density $\mathcal{N}/2$.

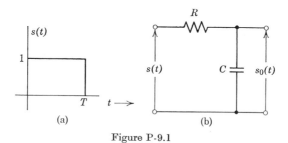

Figure P-9.1

It is decided to use a simple R-C filter (Fig. P-9.1b) instead of a matched filter at the receiver. Calculate the maximum signal-to-noise power ratio $[s_o^2(t)/\overline{n_o^2(t)}]$ that can be attained by this type of filter and compare it with that obtained by the corresponding matched filter. [*Hint:* Observe that $s_o(t)$ is maximum at $t = T$. The signal-to-noise ratio is a function of time constant RC. Find the value of RC which yields the maximum signal-to-noise ratio.]

2. Calculate the transfer function of the matched filter for a Gaussian signal pulse given by

$$s(t) = \frac{1}{\sigma\sqrt{2\pi}} e^{-t^2/2\sigma^2}$$

The noise on the channel is a white noise with power density spectrum $\mathcal{N}/2$. Calculate the maximum S/N ratio achieved by this filter.

3. Show that $s_o(t)$, the output of the matched filter to the input signal $s(t)$ is symmetrical about $t = T$.

4. Two messages are transmitted by mark and space using a single binary pulse shown in Fig. P-9.4.

(a) Design the optimum receiver if the channel noise is a white noise of power density $\mathcal{N}/2$ ($\mathcal{N} = 10^{-4}$).

(b) Find the error probability of the optimum receiver assuming that the probability of $s(t)$ being present is 0.5.

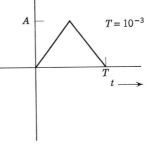

Figure P-9.4

5. If the messages in Problem 4 are transmitted by two binary pulses as shown in Fig. P-9.5, design the optimum receiver and find the error probability of the receiver. Compare this scheme with the one in Problem 4.

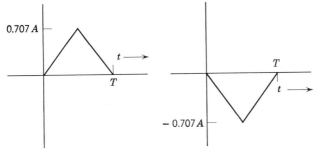

Figure P-9.5

6. A Gaussian signal has a zero mean and the mean square value is σ_n^2. Find the probability of observing the signal amplitude above $10\sigma_n$.

7. If two messages are transmitted by waveforms $s_1(t)$ and $s_2(t)$ shown in Fig. P-9.7, design the optimum receiver for a white channel noise.

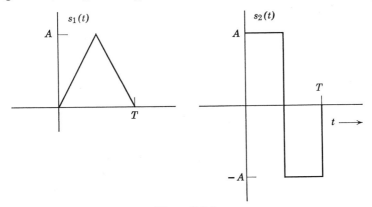

Figure P-9.7

Calculate the error probability of the optimum receiver. Compare this scheme with the one using only a single triangular pulse (as in Problem 4) or two triangular pulses (as in Problem 5). How does this scheme compare with FSK?

8. In the text, the matched filter was obtained for the case of white noise. Proceeding along the same lines, obtain the matched filter for a colored noise (nonuniform power density) with a given power density $S_n(\omega)$. [*Hint:* In Schwarz inequality Eq. 9.7a, let

$$F_1(\omega) = S(\omega)H(\omega) \qquad \text{and} \qquad F_2(\omega) = \frac{S(\omega)}{S(\omega)}$$

where $S(\omega)$ is obtained by factorizing $S_n(\omega) = S(\omega)S(-\omega)$, and $S(\omega)$ has all poles and zero in LHP of the complex frequency plane.]

Bibliography

Chapters 1, 2

Bracewell, R. M., *The Fourier Transform and Its Applications*, McGraw-Hill, New York, 1965.

Craig, E. J., *Laplace and Fourier Transforms for Electrical Engineers*, Holt, Rinehart, and Winston, New York, 1964.

Javid, M. and E. Brenner, *Analysis, Transmission and Filtering of Signals*, McGraw-Hill, New York, 1963.

Lathi, B. P., *Signals, Systems, and Communication*, John Wiley and Sons, New York, 1965.

Marshall, J. L., *Signal Theory*, International Textbook Co., Scranton, Pa.

Papoulis, A., *The Fourier Integral and its Applications*, McGraw-Hill, New York, 1962.

Chapters 3, 4, 5, 6, 7

Black, H. S., *Modulation Theory*, D. Van Nostrand Co., Princeton, N.J., 1953.

Bennett, W. R. and J. R. Davey, *Data Transmission*, McGraw-Hill, New York, 1965.

Downing, J. J., *Modulation Systems and Noise*, Prentice-Hall, Englewood Cliffs, N.J., 1964.

Freeman, J. J., *Principles of Noise*, John Wiley and Sons, New York, 1958.

Hancock, J., *Principles of Communication Theory*, McGraw-Hill, New York, 1961.

425

Panter, P. F., *Modulation, Noise and Spectral Analysis*, McGraw-Hill, New York, 1965.

Rowe, H. E., *Signals and Noise in Communication Systems*, D. Van Nostrand Co., Princeton, N.J., 1965.

Schwartz, M., *Information Transmission, Modulation and Noise*, McGraw-Hill, New York, 1959.

Chapters 8, 9

Abramson, N., *Information Theory and Coding*, McGraw-Hill, New York, 1963.

Harman, W. W., *Principles of the Statistical Theory of Communication*, McGraw-Hill, New York, 1963.

Lathi B. P., *An Introduction to Random Signals and Communication Theory*, International Textbook Co., 1968.

Reza, F. M., *An Introduction to Information Theory*, McGraw-Hill, New York, 1961.

Schwartz M., W. R. Bennett and S. Stein, *Communication Systems and Techniques*, McGraw-Hill, New York, 1966.

Wozencraft, J. M. and I. M. Jacobs, *Principles of Communication Engineering*, John Wiley and Sons, New York, 1965.

Index